ROUTLEDGE LIBRAI
20TH CENTURY

T0227524

Volume 7

NIELS BOHR: PHYSICS AND THE WORLD

NIELS BOHR: PHYSICS AND THE WORLD

Proceedings of the Niels Bohr Centennial Symposium

Edited by
HERMAN FESHBACH, TETSUO MATSUI AND
ALEXANDRA OLESON

Routledge
Taylor & Francis Group

LONDON AND NEW YORK

First published in 1988

This edition first published in 2014
by Routledge
2 Park Square, Milton Park, Abingdon, Oxfordshire OX14 4RN

and by Routledge
711 Third Avenue, New York, NY 10017, USA

First issued in paperback 2016

Routledge is an imprint of the Taylor & Francis Group, an informa business

British Library Cataloguing in Publication Data
A catalogue record for this book is available from the British Library

ISBN: 978-0-415-73519-3 (Set)
ISBN: 978-1-138-01354-4 (Volume 7)
ISBN 13: 978-1-138-97720-4 (pbk)
ISBN 13: 978-1-138-01354-4 (hbk)

Publisher's Note
The publisher has gone to great lengths to ensure the quality of this book but points out that some imperfections from the original may be apparent.

Disclaimer
The publisher has made every effort to trace copyright holders and would welcome correspondence from those they have been unable to trace.

Niels Bohr: Physics and the World

Proceedings of the Niels Bohr Centennial Symposium

Boston, MA, USA November 12–14, 1985

American Academy of Arts and Sciences Cambridge, Massachusetts

Edited by

Herman Feshbach

Tetsuo Matsui

Alexandra Oleson

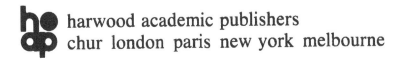
harwood academic publishers
chur london paris new york melbourne

© 1988 by Harwood Academic Publishers GmbH,
Poststrasse 22, 7000 Chur, Switzerland. All rights reserved.

Harwood Academic Publishers

Post Office Box 197
London WC2E 9PX
England

58, rue Lhomond
75005 Paris
France

Post Office Box 786
Cooper Station
New York, New York 10276
United States of America

Private Bag 8
Camberwell, Victoria 3124
Australia

Library of Congress Cataloging-in-Publication Data
Niels Bohr, physics and the world : proceedings of the Niels Bohr
   Centennial Symposium, American Academy of Arts and Sciences / edited
   by Herman Feshbach, Tetsuo Matsui, Alexandra Oleson.
      p. cm.
   Includes index.
   1. Bohr, Niels Henrick David, 1885–1962—Congresses. 2. Quantum
theory—Congresses. 3. Physics—Congresses. 4. Science—
Congresses. I. Feshbach, Herman. II. Matsui, Tetsuo.
III. Oleson, Alexandra   IV American Academy of Arts and
Sciences.
QC16.B63N54   1988   530—dc19   88-21213
ISBN 3-7186-0484-1

Disclaimer notice: The opinions, findings, conclusions and recommendations
expressed in this volume are those of the authors and do not necessarily reflect
the views of the members of the American Academy of Arts and Sciences or the
sponsoring foundations.

Contents

Reprinted with the permission of the American Institute of Physics.

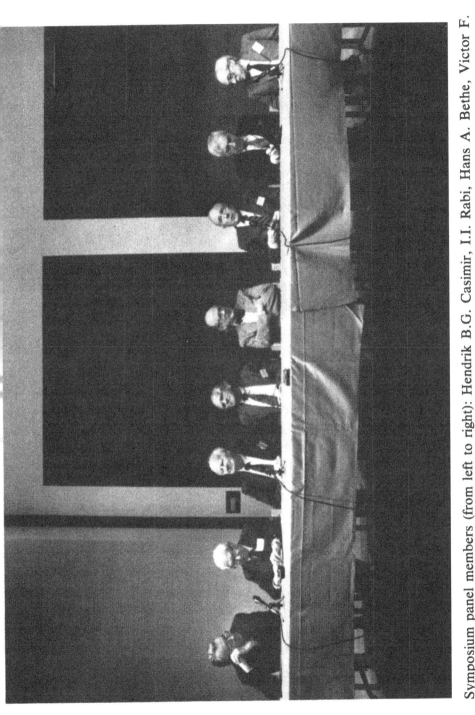

Symposium panel members (from left to right): Hendrik B.G. Casimir, I.I. Rabi, Hans A. Bethe, Victor F. Weisskopf, Herman Feshbach, John A. Wheeler, Erik Bohr, Abraham Pais.

Preface

This volume contains the proceedings of the symposium held by the American Academy of Arts and Sciences celebrating the hundredth anniversary of the birth of Niels Bohr. More than any other single individual, Bohr was responsible for the development of quantum mechanics and for many of its applications to the fundamental understanding of the physical world. Quantum mechanics is a remarkable theory. It provides a universal framework which permits the determination of the characteristic behavior of a physical system. Using quantum mechanics, one can understand the properties of atoms, of molecules, of chemical reactions, of conductors and insulators, and show that these are a consequence of the action of electromagnetic fields. Originally derived for the atom, quantum mechanics has been found to be valid for the atomic nucleus, a system roughly one hundred thousand times smaller than the atom, but in this case new non-electromagnetic forces operate. So far as one can tell, as we approach the frontiers, the subnuclear world of quark and gluon is describable in quantum mechanical terms.

In celebrating Bohr, his unique role in the discovery and elucidation of quantum theory forms a natural beginning to the symposium. This is followed by a description of the successes of quantum theory in various fields of modern science. The frontiers of each field are delineated and the important opportunities for new knowledge and new insights into the properties of matter and the history of the universe are described.

Certainly there are important problems. Are there any of such a fundamental nature that a generalization of present day quantum theory is required for their resolution? On the other hand, are the foundations of quantum mechanics sound? This latter question led to the famous Einstein–Bohr debate and to the formulation of the Einstein, Rosen, Podolsky paradox. The last scientific sessions of the symposium are devoted to these issues. Remarkably, our technical capability has progressed so far that the "thought" experiments proposed by Einstein, Rosen and Podolsky can be performed. They have been and are reported in this volume.

Bohr participated in World War II and made a strong personal effort to build mutual trust and reduce tensions between the East and West during and after the war. Bohr emphasized the importance of control of nuclear weapons, anticipating many of the concerns which have made it so very difficult to achieve. The history of that effort is an important story worth telling, together with a review of the contemporary situation.

The result is a unique volume providing a panoramic view of modern physics, some insights into the history of physics and the qualities of a unique scientist, Bohr, who had so much to do with its progress and finally an example of a scientist's concern with the human condition, one that every scientist should take to heart.

The first paper by Victor Weisskopf provides an overview of Bohr's contributions to atomic theory, to the discovery of quantum theory and its interpretation encapsulated in the concept of complementarity, to nuclear physics and especially nuclear fission. The last leads to a description of Bohr's efforts to achieve an international understanding on the problem of nuclear weapons. Weisskopf gives us an insight into Bohr's personality and the atmosphere at Bohr's institute at Copenhagen. The international nature of science was explicitly demonstrated as scientists from many countries met at Bohr's institute and engaged in remarkably fruitful collaborations. He later participated actively in the formation of CERN (European Center for Nuclear Research) in Geneva; perhaps the most successful international organization of this era.

A. Pais and Weisskopf were collaborators of Bohr, Weisskopf in the late 1930s, Pais after World War II. The latter's paper mixes personal reminiscences with some history and some physics. The overall impression is that of a truly great man, one who not only had a massive impact on physics, but also had compassion and concern for his fellow human beings. Arthur Miller examines the origins of the principle of complementarity relying on the investigation by Bohr and his colleagues of the propagation of light in a medium. Miller then goes to discuss Bohr's concern with intuition and the limitations of language. Bohr found a similar problem in painting and compared the quantum theory and complementarity with the cubism of Picasso, Braque, etc.

Following this paper, the next articles are concerned with modern physics.* The first of these by Daniel Kleppner describes the revolution in atomic physics through the use of lasers which permit an enormous increase in energy resolution. Substantial improvement in the accuracy with which fundamental constants is known is achieved. More precise values of the Lamb shift, measured now to an accuracy of 9 kHz, are in good agreement with the predictions of quantum electrodynamics. The hyperfine frequency in hydrogen has now been measured to one part in 10^{12}. These high precision determinations can be exploited for metrology, and for example is needed in the determination of the Earth's gravitational red shift. Laser radiation can be used to produce very large atoms known as Rydberg atoms by exciting one of the atomic electrons to a high state of excitation. As Kleppner points out, these atoms are used in the study of a wide range of phenomena.

A second feature of atomic physics discussed by Kleppner is the hydrogen-like systems which is profitable to study because of their intrinsic simplicity. These include, one-electron heavy atoms "high Z hydrogen" obtained by stripping all but one of the atomic electrons from the atom, positronium, the "atom" made up of the positron and electron, muonium, made up of a positive muon plus an electron, a single electron in a trap, and finally the effect of a cavity on the radiative properties of an atom.

Michael Fisher gives a "picture of the multifaceted character of modern condensed matter physics and the varying degrees to which quantum mechanics is of direct relevance *or* almost total irrelevance". Fisher provocatively points out that in many situations it is statistical mechanics only that counts, although quantum mechanics and the Fermi statistics are most important for the stability of many-body systems. Fisher illustrates his thesis by numerous beautiful examples of various types of matter consisting of ordered or disordered aggregates of atoms introducing the concept of scaling and emphasizing the importance of dimensionality. Spatially modulated matter, quasi-crystals, disordered matter are examples of the various types of order he describes. He concludes with the quantum Hall

---

* Papers were not presented for publication by Richard Zare, Arthur Kerman, Martin Klein and Bruno Zumino. Kerman's paper was replaced by a paper by this writer which was not delivered at the symposium.

effect which of course does involve quantum theory in an essential fashion.

The next paper on nuclear physics emphasizes principles and methods of general importance which have been learned from the study of nuclear systems. These include the role of symmetry and conservation rules, the treatment of strong interactions leading to the mean field approximation and to a theory of nuclear reactions. Nuclear physics has played an essential role in the understanding of astronomical phenomena. Nuclear processes are responsible for their energy production in and evolution of stars and for the production of the elements. Understanding the properties of the relatively recently discovered stars such as neutron stars, pulsars, and supernovae require the sophisticated use of the results of nuclear physics research. Some of the challenging problems of general importance at the frontiers of the field of nuclear astrophysics are briefly described. This is followed by a discussion of two frontier areas of nuclear physics proper. One is the incorporation of the structure of the neutrons and protons into our understanding of nuclei. It is indeed a remarkable achievement to have made so much progress understanding nuclear phenomena without the use of a detailed theory of the properties of the particles making up the nucleus. A second area that is at the very beginning of its development is based experimentally on the collision of enormously energetic nuclei. It is hoped that this will result in the formation of a relativistic quark–gluon plasma. If this proves to be the case we shall be able to examine in the laboratory a state of matter which existed very close to the time of the big bang. This review concludes with the description of a phenomenon involving the production of positrons in the collisions of very heavy ions. As this is being written, no viable explanation of these observations exists.

J.D. Bjorken uses the theory of the history of the universe as a framework for the depiction of today's "standard model" of elementary particles and forces. Starting with the quark–gluon (plus leptons) plasma discussed earlier at a time of 10^{-14} seconds after the big bang, we watch the universe cool down. In a series of phase transitions, the universe passes from one state to another until one arrives at the state of matter of today's universe. During the first of these occurring roughly at 10^{-12} seconds, the particles mediating the weak forces acquire a mass. Later, at 10^{-11} seconds, these particles disappear from the plasma, a process referred to as a "freeze out". The quarks now acquire a mass and as the universe cools, the heavier ones "freeze out." At 10^{-5} seconds, the protons, neutrons, pions, etc., the laboratory particles, are formed, and so on. In his final section, Bjorken raises the significant unsettled questions. We are still very far from a satisfactory theory of the elementary particles and forces, but enormous progress has been made in the last twenty years.

Among the critical questions of cosmology, the mean density of the universe occupies a special position. If greater than a critical value, the expansion of our universe will stop and then it will contract. This is simply a consequence of the gravitational forces. If the mean density is smaller than the critical value, the gravitational effects are reduced and the universe will expand indefinitely. George Field discusses this problem examining what is known about the mean density. Is there any hidden matter which is not observable by the light they radiate? The evidence that there must be such matter comes from the study of mass distribution in spiral galaxies, and from inferring the value of the gravitational field of elliptical galaxies from their X-ray emissions. The last has also been used for clusters of galaxies finding that hot gas is distributed between the galaxies. Then there is the issue of matter outside the galaxies that has been observed through the bending of light produced by the gravitational forces exerted by the matter. One believes that "gravitational lenses" are formed by the matter. Field compares this problem with the predictions of the "inflationary" theory leading to the inflationary universe. For that theory to be correct, nine-tenths of the matter must be hidden matter.

Black holes are the subject of the next rather technical two contributions of G.'t Hooft and T.D. Lee. The gravitational field is so strong in the interior of a black hole that light cannot escape. Thus classically black holes do not radiate. This is not the case quantum mechanically. It is with this quantum mechanical radiation that these authors are concerned. The existence of black holes is yet to be demonstrated, although there are several candidates. 't Hooft proposes a transformation of the quantum mechanical state vector from the reference frame of the stationary observer to the reference frame of an accelerated observer which differs sharply from that which occur in other circumstances. He calls this "a new relativity principle in quantum mechanics". It leads to an observable effect. The temperature of the radiation coming from a black hole is double that obtained from the conventional approach and the energy generated by the explosion of a mini-black hole would also be doubled. To experimentally decide between the two theories, all we need is a certified black hole!

T.D. Lee's contribution deals with two questions. Can one infer the properties of the system in the unreachable region beyond the relativistic horizon by examining that component of the system within the region of observability? The answer is yes. It is not that information is sent from one region to the other. Rather the properties of the system within the region of observability reflect the properties of the system outside of that region. This, by the way, is an example of what is referred to as the Einstein, Rosen, Podolsky paradox through which Einstein and his co-authors attacked quantum mechanics, describing it as an incomplete theory. Lee's second point deals with black-body radiation. The latter term refers to the spectrum of radiation that emerges from a black body (a system which absorbs all the light that falls upon it). Lee asks, "Do black holes emit black-body radiation?" The answer is, not always, as it depends on the quantum states of the matter within the black hole.

What lies beyond quantum theory? Is it the final framework for quantitative understanding of all physical phenomena or will there eventually be a generalization for which quantum theory will hold in some limit, as classical mechanics can be obtained from quantum mechanics in the limit, properly taken, that the Planck constant approaches zero? Stephen Adler in his paper begins such a generalization. Quantum theory uses complex numbers. There is one alternative, the case of quaternions whose properties Adler describes. The consequences of a quaternion based theory are explored by Adler and as he points out there are many aspects of the conventional quantum theory that are not present. Thus, it is not a generalization as only a *part* of complex number quantum theory generalizes to the quaternion quantum theory.

The volume now turns to the philosophical issues associated with quantum theory.

For Bohr, complementarity was a general principle that could be applied to situations outside of physics. A trivial example is formed by the two variables clarity and accuracy. Complementarity suggests that the greater the precision of a statement the less its clarity. Bohr in his writing strove for precision with the result that his sentences were long, involuted and opaque. In 1962 he applied the principles of complementarity to biological investigations. His conclusions turned out to be incorrect but his discussion inspired Max Delbruck to leave physics and found the subject of molecular biology. Gunther Stent in his essay discusses Bohr's 1932 talk "Light and Life", then goes on to record Delbruck's role, and Bohr's reprise, "Light and Life Revisited" in 1963. Stent concludes with his own application of complementarity to ethics and biology, with an emphasis on the complementarity variables free will and evolution. This consideration is not entirely philosophical in view of the claims of sociobiology.

The epistemological issues raised by Einstein, Bohr and Podolsky generally referred to as the EPR paradox are the subject of the next three papers by John S. Bell, Abner Shimony, Alain Aspect and Phillipe Grangier. As was mentioned earlier because of the EPR paradox, Einstein *et al.* concluded that the "quantum mechanical description of reality" is not complete. Their analysis is based on an experiment referred to as "gedanken" (thought) because the authors believed that it could not be performed. This was certainly true at the time (1935) that EPR was published. The objection of these authors to quantum mechanics lies in the correlations quantum mechanics predicts for states of a system even though its components may be far apart spatially. John Bell via the theorem that bears his name provided an analysis of the EPR paradox that permits a quantitative test to determine if EPR's objections to quantum mechanics are valid. In a delightful paper he describes this analysis as well as his own summary of the situation as of 1981.

However, by the date of the symposium, experiments, which were thought to be "gedanken" in 1935, because of advances in technology could be and have been done. One of these experiments is reported in the paper by Aspect and Grangier. In this experiment, a single photon impinges on a beam splitter. Quantum theory predicts perfect anti-correlation for detection of the photon on either side of the beam splitter. It is important that one is dealing with a single photon as Aspect and Grangier emphasize. In summary, anti-correlation is observed so that if the photon is detected in one detector, one knows that there will be no photon at the other detector. Quantum mechanics describes this situation perfectly.

The consequences of the experiments of Aspect and his colleagues are considered by Abner Shimony. Bell's analysis is reformulated and extended and the notions of "parameter independence" and "outcome independence" are explained. Parameter independence is not violated by quantum mechanics but "outcome independence" is. This discussion, which deals with the interpretation of quantum mechanics, leads naturally into a discussion of Bohr's philosophy, its lacuna and its relationship to the philosophy of Kant.

Ideological opposition to Bohr's philosophy came from the Soviet Union. According to Loren R. Graham, the work "complementarity" was banished from Soviet publications from 1948 to 1960. Much of the criticism by Soviet philosophers arose simply because they did not understand quantum theory. For example, they thought quantum theory was not causal. Of course, Soviet physicists did understand and used quantum theory effectively. They eventually won the battle. These events as well as the more recent history of the attitude of Soviet philosophers and scientists to "Bohr's quantum mechanics" are the subject of Graham's contribution.

There may still be scientists who believe they they need not be concerned with the uses society makes of their discoveries. In today's high technology where so many decisions are science related, this attitude to say the least is bizarre. As Weisskopf put it, "Compassion without curiosity is ineffective. Curiosity without compassion is inhuman." Curiosity is Weisskopf's synonym for science. Bohr was a compassionate man. He took a deep interest in his fellow colleagues. In the pre-World War II days, his institute in Copenhagen was a place of refuge for the many physicists who had been driven from their positions and homes by Nazis. As one of those who led the study of fission of nuclei, he was greatly concerned with the impact of the existence of atomic bombs on the post-war world. He saw clearly the possibility of a nuclear arms race and, on the other hand, the possibility that because of the existence of nuclear weapons the use of war as an instrument of national policy would be regarded as obsolete. He felt that a necessary condition for the avoidance of the first of these and the achievement of the second is an "Open World" as described in a letter to the United Nations written in 1950. Martin Sherwin's article describes Bohr's attempt to convince Roosevelt and Churchill to open negotiations with Stalin on the control of nuclear weapons and more generally on the international control of atomic energy. As we know all too well, he was not successful and a post-war nuclear arms race ensued. There are lessons to be learned from this attempt of a scientist to influence public policy with regard to nuclear weapons as Sherwin makes clear. Bohr inspired many physicists to enter into the debate on nuclear arms, a debate which continues to this day.

The American attitude towards arms control is the subject of John

Steinbruner's incisive article. "Fear, distrust and political antagonism toward the Soviet Union has prevailed over other international political sentiments. Security has been based on immediate preparations of war; nuclear weapons have been established as a central element in our theory of defense." In these two trenchant sentences, Steinbruner has encapsulated the frightening prospect that faces us today.

There is one positive element that Bohr emphasized in his work in Copenhagen—namely, science is international. The discoveries of science, the laws of science do not recognize the existence of nations or their boundaries. The attempt to hoard scientific advances is bound to fail. Every advance in the effectiveness of nuclear weapons by one nation has been met by a similar development in the other. In this era of high technology can one exploit this fact to bring nations to the negotiating table? The international character of science is the subject of Edoardo Amaldi's paper. The formation of international research organizations such as CERN in which Amaldi has played a leading role is recounted. He also discusses the European attitudes toward nuclear arms control decrying the focus on number of weapons. He reminds us that the non-proliferation treaty, soon to come up for re-ratification, requires the two superpowers to "pursue negotiations in good faith" leading to a "cessation of the nuclear arms race at an early date" and to "nuclear disarmament". During the fifteen years since the ratification of this treaty, there has been little evidence that either power has paid attention to this part of the treaty. There has recently been some movement in this direction, since a treaty eliminating the intermediate-range missiles in Europe seems possible. We can all hope that this is the first step of several to come, stopping the nuclear arms race and eliminating most, if not all, nuclear weapons.

In composing this preface, I was struck by the rich fare this volume offers. On the one hand, there is the life of a truly great man of physics, Niels Bohr. There is the authoritative description of various fields of physics by leading physicists. There is philosophy and epistemology and finally the impact of one of the developments of physics on world society. Of course, these essays cannot provide a complete picture, but a detailed study of any one of them would be deeply rewarding.

Herman Feshbach

Overview

Victor Weisskopf

MASSACHUSETTTS INSTITUTE OF TECHNOLOGY

Mr. President and friends . . . I am honored and pleased to be able to open this Symposium for a man that everybody considers as one of the great men of science who made a tremendous impact on the development of physics. I also must express a deep personal gratitude to this man. Whoever worked with him was strongly influenced by his thinking, by his life style and character. Furthermore, I am deeply grateful to his country because I met my wife when I came to Copenhagen fifty two years ago.

Niels Bohr was born a hundred years ago. What hundred years! I'd like to use Dickens' quotation: "It was the best of times, it was the worst of times . . ." Truly these past hundred years were in some respects the best of times. Think of the tremendous development of science, technology, medicine, and of human culture such as painting, music, and many other expressions of human creativity. How many new ideas and modes of expression appeared in all forms of art and science during this period! I don't need to tell you why it was the worst of times: two world wars, fascism, perversion of socialism, a terrifying increase in the ability and use of new and more effective weapons to annihilate human beings.

Just this contrast, I believe, had a decisive influence on the life, on the style, on the thinking, on the actions, and on the character of Niels Bohr. He was deeply aware of this dichotomy; he was no "ivory-tower" scientist. His ideal was to help to mitigate the worst by the very spirit engendered by science.

He grew up in a highly intellectual milieu. Together with his brother Harold, he became deeply involved in the scientific and cultural problems of these days. They grew up at the end of the last century and the beginning of this century. It was an extremely positive time; quite apart from great successes in science and culture, there was a strong feeling of political stability. Of course the stability was only apparent, but it shaped the thinking of people at that time. A long time of peace in Europe, an optimistic climate, social legislation and technology was applied in a positive way. It was a relatively stable time where people thought "Now we are going to get a better world." But this optimistic mood ended in 1914 when the thunderbolt of the First World War struck.

Bohr's scientific life started in 1905. He was twenty years old at that time. It was the year in which the first paper on relativity was published by Einstein, the existence of the photon was recognized by Einstein; five years earlier Planck started quantum theory. It was a turning point of our science. What a time to begin a career! Bohr had the great luck to be born at the right time. I should have said that mankind had the good luck to have him born at that turning point.

I would like to divide Bohr's life into four periods. In each of them he exerted a tremendous impact on our science. The first period extends roughly from 1910 to 1922. He met Ernest Rutherford, at the beginning of that period in Manchester. Rutherford was an experimental physicist who had little taste for theory in contrast to the theoretical physicist Bohr. Just the clash of those different attitudes turned out to be extremely productive. Bohr published his famous paper on the so called Bohr model of the atom, based on Rutherford's discovery of the atom as a planetary system with the nucleus at the center and the electrons seemingly revolving around the nucleus like planets around the sun. In this paper he introduced two revolutionary ideas: the concept of quantum states, postulating that the electron revolves around the nucleus only on selected orbits, so that it takes a considerable energy to change the orbits. This explained the previously unexplained stability of atoms. The second idea was that the light emission of the atoms comes from transitions ("jumps") of the atom from one quantum state to another, emitting a photon of an energy corresponding to the difference of the two quantum states. He also could express the size and the energy of atoms, in terms of fundamental constants like e, the charge of the electron; h, the Planck constant, and the mass of the electron. It represented a tremendous advance since atomic size and energy were completely unexplained before. His model was based on very intuitive methods. It was not logically clear how he came to his conclusions in his early papers. "These are provisional hypotheses," he said, "that needed a basis, a real theory that was unknown at that time." His contemporaries fell into two groups: the ones that thought this is *the* answer and the final theory, and the other half that did not believe it, that thought it is the craziest thing ever invented. There is an interesting story of Max von Laue and Otto Stern, who made a vow that if this theory is correct, they will give up physics. Well, fortunately, they didn't hold their promise.

It is interesting that Bohr himself was aware, in 1913, what a great step he took. He wrote a letter to his brother, with whom he was in extremely close contact: "It could be that I have found out just a little bit about the behavior of atoms, which perhaps may be a very small piece of true reality." It was a big one.

At the end of this first period, in 1921, Bohr published a paper, the so-called "Aufbau-principle" paper, in which he not only treated the simple atoms, such as hydrogen and hydrogen-like atoms as he did in his first paper, but in which the general properties of the other atoms could be understood on the basis of Bohr's intuitive ideas. The so-called periodic system of atoms, discovered by D.I. Mendelejew in 1869, became a consequence of Bohr's quantum states of the atom.

He became very famous, and was awarded the Nobel prize in '22. He received ample financial support from Denmark and from America—mainly Rockefeller money. This enabled him to lay the foundation to the famous Institute for Theoretical Physics in Copenhagen, that still is a center for research today. Indeed the old building is still standing, although surrounded with new construction.

Then came the second period roughly from 1922 to 1930. I would call it the "heroic" period. It was the time of the birth of quantum mechanics, the logical basis of his earlier intuitive ideas. It is interesting that at that time no paper by Niels Bohr as the single author appeared, except some of his talks. He found a new way of collaborative working. He assembled around himself the most active, gifted, perceptive young physicists of the world. Here are a few names: O. Klein, H. Kramers, W. Heisenberg, W. Pauli, P.A.M. Dirac, L. Landau, G. Gamov, F. Bloch, H. Casimir, J. Slater, C.F. von Weizsacker, etc. A truly international crowd. It was the first really international institute. Whoever joined this Institute, as I had the privilege later on, the ambiance of the place shaped the lives and the thinking of most people who came there: it made it clear that science *is* an international enterprise, that nationality and origin do not count.

In these years the foundations of quantum mechanics were laid; they were conceived, discussed, and the deepest problems of the structure of matter brought to light. Imagine what atmosphere, what life, what intellectual activity reigned there! And Bohr was at his best. He created his style, the so-called *Kopenhagener Geist*. He was the greatest among colleagues: acting, talking, living as an equal in a group of young, optimistic, jocular, enthusiastic people; approaching the deepest riddles of nature with a spirit of attack of freedom from conventional bonds, of joy, that hardly can be described. That was the period when Bohr and his men—yes, men—touched what I would like to call the "nerve of the universe." There were no women among the theorists at that time, with the exception of Miss L. Meitner but she was an experimental physicist, a guest of the group. Well, that's the spirit of that time.

In the course of a few years the basis was laid for the science of atomic phenomena that grew into the vast body of knowledge that is known today. I'd like to quote, slightly changed, a statement by Churchill, actually addressed to the Royal British Airforce in World War II: "Never before have so few done so much in such a short time." Every Ph.D. thesis at that time was an opening of a new field. Today the fundamental ideas of quantum mechanics are not in dispute. All the predictions, even the most surprising and odd predictions, have been verified. But the formulation is still in dispute. Most Bohr disciplines say that they know Bohr's interpretation. I don't pretend this. But still, I would like to use one or two minutes to give mine, without pretending that Niels Bohr would approve it.

The behavior of atom and molecules cannot be described by the classical concepts such as particles, position, movement, velocity, or waves in the case of light. A new way has been found, called quantum mechanics, which describes the atoms to be in a series of discreet quantum states. Quantum mechanics describes the properties of these quantum states, their dynamics, how they change, how they radiate light and how they interact with other atoms or molecules. To some limited extend, the quantum states can be described by classical concepts. But this can only be done in an apparently contradictory way: sometimes by using a wave picture for the electron, and sometimes a particle picture. Both pictures are justified but must be applied to different conditions. Bohr calls these two pictures "complementary."

The quantum states cannot be directly observed. Direct observation has an influence on the state, it changes or even destroys it.

The limits of the application of classical concepts are codified by the Heisenberg uncertainty principle. It is a sort of a signpost which says, "Up to here no further can you apply classical concepts." And if you go further, you are asking inappropriate questions, and you will get inappropriate answers. For example, if want to measure exactly the position of the electron and the atom, you intrude upon the quantum state which you really wanted to investigate. Such inappropriate experiments, change or destroy the object. This is why one must be careful with experiments. All experiments are intrusions, and influence the object. But we can minimize this, and that's what we do when we study, for example, the physics of metals. We are very careful, we only look at the surfce and not try to locate every electron in the metal; if we did, the metal would explode. So we try the least possible intrusion.

Under these conditions, the question arises as to whether quantum states are real in the same sense as a large object is real. Yes, they are real indeed, but not directly observable because that would change and destroy them. They are indirectly observable by observing the consequences of their existence. They are real, although in a new and different sense. The different properties cannot be observed directly and simultaneously. For example, depending on the mode of observation, either the wave properties show up, or the particle properties. One of them evanesces when the other one is observed. And that is what Bohr called complementarity.

He applied his concepts to ideas and concerns outside physics, for example, to the question of free will. We have a clear experience what we mean by "free will," but it evanesces when we analyze causally what happens. There is a complementary relation between reasoning and acting, between justice and compassion, between clarity and truth. That's why Niels Bohr was a very bad speaker, because he was too much concerned with truth. In order to illustrate the situation of how quantum mechanics describes reality, Bohr often mentioned a cubist picture of a person in which we see the object simultaneously from different contradictory angles; Bohr would say these are complementary ways describing the person. You see the object from many seemingly contradictory aspects, but altogether it gives you a deeper impression of the object than you would have gotten from a realistic painting.

Bohr also applied the complementarity to the problem of life versus physics and chemistry. He said that the life phenomena are complementary to physics and chemistry, they evanesce when you perform a thorough chemical analysis, since the object is killed. Now that turned out to be wrong; Bohr went too far. Life phenomena seems to be in principle explainable by physical-chemical processes. We will hear more from Gunther Stent on this topic.

I'm now coming to the third period from 1930 to 1940. At that time his main interests centered upon nuclear physics and quantum field theory. I don't want to say too much about quantum field theory. It was actually produced and developed by Dirac, Wigner, Jordan, Pauli, and Heisenberg, and others. I would like to mention a fundamental seminal paper he wrote with Leon Rosenfeld, which is the corresponding paper to Heisenberg's uncertainty relation, namely the uncertainty relations regarding electromagnetic fields. There is a certain complementary exclusiveness between electric and magnetic fields on the one hand, and the number of photons contained in them on the other. You cannot determine both at the same time.

He had a very essential influence in nuclear physics. It started with the famous "golden year" of 1932, when the neutron, the positron, and anti-matter were discovered. The neutron provided the clue for the understanding of nuclear structure. The nucleus consists of protons and neutrons and the force that keeps the protons and neutron together is much stronger, than the force that holds the electrons within the atom. He therefore developed a theory of nuclei and of nuclear reactions based on strong interaction in a paper with F. Kalckar. It is interesting that the overwhelming personality and influence of Bohr did not prevent, but postponed the discovery of similarities between nuclei and atoms, namely the shell structure. It was discovered only after the war by Maria Mayer on suggestion of E. Fermi, and by J. Jensen in Germany, that there are shells in nuclei like the ones in the atom, in spite of the strong interaction, a fact that was explained later. It is interesting how a strong personality sometimes may even slow down the progress of science.

The fission of uranium was discovered when Bohr was deeply involved in his studies of nuclear structure. Obviously this phenomenon captured Bohr's interest and he wrote fundamental papers with J. Wheeler on this process that had a decisive influence on the development of nuclear energy.

The work on uranium fission inevitably brought Bohr into a realm where physics and human affairs are hopelessly intertwined. He was unusually sensitive to the world in which he lived. Before many others, he was aware that science could not be separated from the rest of the world. The events of world history brought home this point earlier than expected. By the 1930's, the ivory tower of pure science had already been broken. It was the time of the Nazi regime in Germany, and streams of refugee scientists came to Copenhagen and found help and support from Bohr. Bohr's institute was the center for everybody in science who needed help, and many a scientist found a place somewhere else—in England, in the United States—through the help of Bohr's personal actions. He collected money—Danish, American, and British funds—to support these. It was incredible how much energy he could spend on getting the money, on discussing scientific developments—he participated whenever he could at the discussions—and at the same time travelling to America and to England to "sell" his refugees, which he did very successfully. He sold me to the University of Rochester.

We now come to the fourth period: 1940 to his death in 1962. That was the time of the second World War, of the occupation of Denmark, his flight to Sweden. Here is the place to tell the story of the rescue of the Danish Jews, a story which is not enough known. It was an extraordinary action, how the Danish people saved their Jewish compatriots. After the occupation by the Nazis, Denmark was a model protectorate. But soon after, because of the resistance activities, the Nazis decided to use their usual methods, and planned to gather all Danish citizens of Jewish extraction and bring them to concentration camps. Fortunately there was one decent German official, a Mr. Duckwitz—his name should not be forgotten—who told the underground movement a week before that this measure will be taken. And then an action took place which is unique in the history of persecutions. The whole people of Denmark collaborated secretly to bring the Jewish population, to the coast and shipped about ten thousand souls to the Swedish coast in fishing boats, at night after midnight. The Nazis could only catch four hundred of them. Bohr and his family, who were half-Jewish, were among those who went over.

Then a completely new life started for him. He went to England from Sweden, and then to the United States where he joined a large group of scientists in Los Alamos, who at that time were working on the exploitation of nuclear energy for war purposes. He did not shy away from the most problematic aspects of scientific activity. He faced it as squarely as the others, as a necessity. But at the same time, it was his idealism, his foresight, and his hope for peace that inspired many people at that place of war to think about the future, and to prepare their minds for the task ahead. He saw not only death and destruction but also hope in the atomic bomb. He hoped that this weapon would make wars unthinkable and would induce nations to work together solving their problems with methods other than war. He believed that in spite of death and destruction there was a positive future for this world, transformed by scientific knowledge.

At that time Bohr actively engaged in a one-man campaign to persuade leading statesmen of the West of the danger, and the hope, that might come from the atomic bomb. He wanted the men of power to make use of this momentous development, of the terrific consequences of a nuclear war. It should bring East and West together. He was working for an internationalization of atomic energy, peaceful and military. It was Bohr's principle, when you face an impossible situation as he did in the search of atomic structure, completely *new* approaches are necessary. He wanted to apply this principle to the post-war political situation. He talked to Roosevelt, Churchill, and other important statesmen and he experienced the difficult pitfalls of diplomatic life. Although he was able to convince a number of important statesmen, including Roosevelt at the beginning, his meeting with Churchill was a complete failure. Churchill wanted none of this sharing of secrets with the Soviet Union, and went even so far as to accuse Bohr of being too friendly with Russians.

Bohr's great political concept did not come to any fruition. Neither did other attempts succeed to raise nuclear technology to an international level in order to avoid a nuclear armament race between powerful nations. Bohr ended his efforts of international understanding on nuclear weapons with his famous letter to the United Nations, written in 1950, in which he laid down the thoughts about the necessity of an open world. What he meant was not only openness in the sense of free exchange of people and ideas, but also open in the sense of a mutual understanding of the problems of the other side. He predicted that, if no agreement is reached on international cooperation in nuclear matters, the world would be plunged into an accelerating armament race which might explode one day into mutual annihilation.

In the last decade of his life, Bohr spent much time in the organization of international activities in science. He participated actively in the founding of Nordita, a Scandinavian international research organization, and of a new laboratory in Europe, as suggested first by I.I. Rabi, whose purpose should be to bring European fundamental science back to the significance it had before the war. With his help the European Center of Nuclear Research in Geneva (CERN) was founded and has developed into one of the leading laboratories in particle physics. In many ways CERN is based on the same ideas as Bohr's Institute in Copenhagen fifty years earlier, on the idea of international scientific collaboration. But now it was executed on the largest scale, in experimental and theoretical physics.

Physics became a large enterprise; large numbers of people and large machines were necessary to carry out physical research. Bohr recognized this as a logical continuation of what he and his friends have started. He was not afraid of big science if it is imbued with the spirit of the search of truth. He saw the necessity of physics on a large scale, on an international scale. In no other human endeavor are the narrow limits of nationality or politics more obsolete and out of place than in the search of more knowledge about the universe.

He spent his last years following the results of new research, helping to get support for science from governments wherever he could, annd reformulating his philosophy of complementarity. He was proud of the rebirth of European science after the ravages of war and he enjoyed his life as the grand old man of Physics.

When Niels Bohr died, an era ended. The era of great personalities who created modern science. For some other reason our era does not produce any more great personalities. But it was Bohr himself who helped to shape the spirit, and the institutions for the continuation of the scientific endeavor. Much of the external aspects of physics have changed, but the spirit, the driving idea, remains the same. And it is up to us to realize his ideals in the future. He, more than anybody else, knew that to succeed in this, two tasks have to be fulfilled: to continue the quest for deeper insights into the riddles of nature with the same insistence, and the same idealism, as he did; and to avert the threatening catastrophies engendered by today's gross abuse of technical and military applications of science. He did not succeed in the latter, but he showed us the way to proceed.

Niels Bohr and the Development of Physics

A. Pais

ROCKEFELLER UNIVERSITY

On 24 October 1957, Robert Oppenheimer and I took an early train from Princeton to Washington. We were on our way to the Great Hall of the National Academy of Sciences, where, that afternoon, the first Atoms for Peace Award was to be presented in the presence of President Eisenhower. It was a festive event. I recall greeting Supreme Court Justice Felix Frankfurter before the start of the proceedings. We briefly chatted about how bittersweet a moment in Bohr's life this occasion had to be. (More about that later.) Then the meeting was called to order. James Killian, president of MIT read the citation from which I quote:

'You have explored the structure of the atom and unlocked many of Nature's other secrets. You have given men the basis for greater contributions to the practical uses of this knowledge. At your Institute at Copenhagen, which has served as an intellectual and spiritual center for scientists, you have given scholars from all parts of the world an opportunity to extend man's knowledge of nuclear phenomena. These scholars have taken from your Institute not only enlarged scientific understanding but also a humane spirit of active concern for the proper utilization of scientific knowledge.

'In your public pronouncements and through your world contacts, you have exerted great moral force in behalf of the utilization of atomic energy for peaceful purposes.

'In your profession, in your teaching, in your public life, you have shown that the domain of science and the domain of the humanities are in reality of single realm.'

These words eloquently describe that combination of qualities we find in Bohr and only in Bohr: creator of science, teacher of science, and spokesman not only for science *per se* but also for science as a potential source for the common good.

As a creator he is one of the three men without whom the birth of that uniquely twentieth century mode of thought, quantum physics, is unthinkable.

The three are, in order of appearance: Planck, the reluctant revolutionary, discoverer of the quantum theory who did not at once understand that his quantum law meant the end of an era now called classical.

Einstein, discoverer of the quantum of light, the photon, who at once realized that classical physics had reached its limits, a situation with which he never could make peace.

And Bohr, founder of the quantum theory of the structure of matter, also immediately aware that his theory violated sacred classical concepts, but who at once embarked on the search for links between the old and the new, achieved with a considerable measure of success in his correspondence principle.

How different their personalities were.

Planck, in many ways the conventional university professor, teaching his courses, delivering his Ph.D's.

Einstein, rarely lonely, mostly alone, who did not really care for teaching classes, and who never delivered a Ph.D.

And Bohr, always in need of other physicists, especially young ones, to help him clarify his own thoughts, always generous in helping them clarify theirs, not a teacher of courses nor a supervisor of Ph.D.'s but forever giving inspiration and guidance to postdoctoral and senior research.

I first met Bohr when I came to Copenhagen in 1946 as the first of the post-war postdoctoral crop from abroad. Some months later Bohr asked me whether I would be interested in working together with him day by day for the coming months. I was thrilled and accepted. The next morning I went to Carlsberg. The first thing Bohr said to me was that it would only then be profitable to work with him if I understood that he was a dilettante. The only way I knew to react to this unexpected statement was with a polite smile of disbelief. But evidently Bohr was serious. He explained how he had to approach every new question from a starting point of total ignorance. It is perhaps better to say that Bohr's main strength lay in his formidable intuition and insight rathar than in erudition. I thought of his remarks of that morning some years later, when I sat next to him during a colloquium in Princeton. The subject was nuclear isomers. As the speaker went on, Bohr got more and more restless and kept whispering to me that it was all wrong. Finally he could not contain himself and wanted to make an objection. But after having half raised himself he sat down again, looked at me with unhappy bewilderment and asked: 'What is an isomer?'

These little stories may help to convey what it was like to work with Bohr. We learned from him to take nothing for granted and, to use one of his favourite phrases, to be prepared for a surprise every new day. The substance of the work was solid and serious, the style lighthearted and highly informal. One day a visitor, unacquainted with the Copenhagen spirit, remarked to Bohr: 'It seems that at your Institute nobody takes anything seriously.' To which Bohr replied: 'That is true and even applies to what you just said.'

Niels Henrik David Bohr was born in Copenhagen on 7 October 1885, the son of Christian Bohr, an internationally known physiologist, and of Ellen Adler who hailed from a wealthy Danish-Jewish banker family. He had a two-year-younger brother Harald, who was to become a mathematician of great eminence.

Niels was a good pupil in school, though apparently not a driven personality. He was good at handcrafts and sports. He was always a physical man, loving to ski and sail, and, in his youth, to play soccer. Both brothers were good at that; in fact Harald played half-back in the Danish Olympic team that won the silver medal in 1908.

Bohr entered the University of Copenhagen in 1903. In 1906 he won a gold medal of the Royal Danish Academy of Sciences and Letters for a theoretical and experimental investigation of ripples on vibrating liquid jets, work which led to two papers in Royal Society journals. In May 1911, his thesis *Studies on the electron theory of metals* earned him the Ph.D. degree. This very thorough work is based on Lorentz' electron theory. Already then Bohr was aware of the limitations of the classical description. To these difficulties he added one he himself had discovered: 'It does not seem possible, at the present stage of development of the electron theory, to explain the magnetic properties of bodies from this theory,' I have not found in the thesis any reference to the question of spectra.

In October 1911 Bohr went to Cambridge, hoping to work with J.J. Thomson. As I know from Bohr, his first meeting with J.J. went about as follows. Bohr entered, opened J.J.'s book *Conduction of electricity through gases* on a certain page, pointed to a formula concerning the diamagnetism of conduction electrons, and politely said: 'This is wrong'. Some of the subsequent encounters went similarly, until J.J. preferred to make a detour rather than meet Bohr. Rosenfeld later told me a rather similar story. Shortly before his death, Bohr said: 'I considered Cambridge as the centre of physics, and Thomson as a most wonderful man. It was a disappointment to learn that Thomson was not interested to learn that his calculations were not correct. That was also my fault. I had no great knowledge of English, and therefore I did not know how to express myself. . . . The whole thing was very interesting in Cambridge, but it was absolutely useless' [1]. The lack of contact was especially disappointing to Bohr because, then and later, he looked up to J.J. as a great man. He kept busy in Cambridge, however, attending lectures, writing a short paper on the electron theory of metals, and reading *The Pickwick Papers* (a book he always remained fond of) in order to improve his English.

In March 1912 Bohr moved to Manchester for a three months' stay at Rutherford's laboratory. Only a few months before Bohr's arrival Rutherford had discovered that an atom has a nucleus of small dimensions in which nearly all the atomic mass is concentrated. Inspired by Rutherford's discovery Bohr was able, in 1913, back in Copenhagen meanwhile, to decode the Balmer formula for the spectrum of hydrogen and thereby to unlock the secret of the structure of atoms.

It was a triumphant moment in the history of thought. Awareness of spectra predates recorded history, since primitive man must have worshipped the rainbow. Aristotle had produced a theory for this phenomenon, elaborated in learned discourse during medieval times [2]. First Descartes then Newton had grasped the true origins of the rainbow effect. Newton, experimenting with his prisms, had decomposed the sun's light into a continuous spectrum of colours and so had founded spectrum analysis. Newton had also wondered about the dynamical origin of spectra:

'Do not all fix'd Bodies, when heated beyond a certain degree, emit Light and shine; and is not this Emission perform'd by the vibrating motions of their parts?' [3]

It was the twenty-eight-year-old Bohr who in 1913 answered Newton's query. His inspiration came from Rutherford, as said, but in addition from the great experimental advances in quantitative spectroscopy which began in the 1850s. For our purpose the main event in this fascinating period [4] is Balmer's divine guess concerning the line spectrum of atomic hydrogen. Balmer, a Swiss school master from Basle, was mainly interested in biblical architecture. At some point his attention was drawn to data obtained by Angström (1868)

[5] on the frequencies of four hydrogen lines, H_α, H_β, H_γ, and H_δ (now known as the first four Balmer lines). Comparison with modern results shows that Angström's values were accurate to one part in ten thousand. Nothing more, nothing less than these four lines led Balmer to conjecture that the hydrogen spectrum has infinitely many lines with frequencies v_{mn} ($n = 1, 2, 3, \ldots$, $m > n$ and integer) given by (in modern notation)

$$v_{mn} = R[(1/n^2) - (1/m^2)], \qquad R = 3.2916 \times 10^{15} \text{ s}^{-1}.$$

That is the formula published in 1885 by the sixty-year-old Balmer in the first physics paper [6] he ever wrote. Thus in 1885 the constant R (later called the Rydberg) was known to one part in ten thousand— to the great benefit of Bohr. (Fits of many more lines to the formula followed rapidly.) I once asked Bohr what people thought of this formula and of spectra in general between 1885 and 1913 when he, Bohr, saw what it meant. I do not remember exactly what Bohr replied but I can guess that it must have been something similar to what he said some years later.

'One thought [spectra are] marvelous, but it is not possible to make progress there. Just as if you have the wing of a butterfly then certainly it is very regular with the colours and so on, but nobody thought that one could get the basis of biology from the colouring of the wing of a butterfly [7].

What was Bohr like in 1913? According to Richard Courant: 'Somewhat introvert, saintly, extremely friendly, yet shy' [8]. According to the Danish physicist Rud Nielsen: 'He was very friendly, an incessant worker, and seemed always in a hurry. Serenity and pipe smoking came later' [9].

At that time Bohr was also a newlywed. In the summer of 1912 he married Margrethe Nørlund. It was a superb marriage, as I know especially well since during the summer of 1946 I lived with the Bohr family in their country house in Tisvilde, one of the happiest periods in my life.

Now to Bohr's paper [10] on the hydrogen atom, dated 5 April 1913. At its beginning, Bohr refers, for the first time, I believe, to the fact that, according to the classical theory, 'the electron will no longer describe stationary orbits' but will fall inward to the nucleus due to energy loss by radiation. Then he plunges into the quantum theory. His first postulate: An atom has a state of lowest energy (he calls it the permanent state, we the ground state) which by assumption does not radiate, one of the most audacious hypotheses ever introduced in physics. His second postulate: Higher 'stationary states' of an atom will turn into lower ones such that the energy difference E is emitted in the form of a light quantum with frequency f given by $E = hf$ (h is Planck's constant).

I shall not discuss in detail Bohr's quantum constraint on the hydrogen orbits. Suffice it to say that it is equivalent to the one we learned in school: The orbital angular momentum L is restricted to the values

$$L = n(h/2\pi), \qquad n = 0, 1, 2, \ldots .$$

Not only does the Balmer formula follow at once, but R now appears in terms of fundamental constant

$$R = (2\pi^2 e^4 m/h^3)\text{s}^{-1}.$$

This prediction of R, 'inside the experimental errors in the constants entering in the expression for the theoretical value' [10] is the first triumph of quantum dynamics.

Bohr's ideas were rapidly given serious consideration, but the initial response was mixed. Success and logic appeared to have parted ways. Why should an atomic ground state not radiate? There was also the issue of causality, first raised by Rutherford even before Bohr's paper [10] on the hydrogen atom had appeared. On 6 March 1913, Bohr had sent his paper to Rutherford, requesting [11] that it be submtted to *Philosophical Magazine*. In his reply (20 March) [12] Rutherford remarked: 'There appears to me one grave difficulty in your hypothesis which I have no doubt you fully realize, namely how does an electron decide what frequency it is going to vibrate and when it passes from one stationary state to another? It seems to me that you would have to assume that the electron knows beforehand where it is going to stop.' These conflicts with causality, classical causality, characteristic for the old quantum theory, the period which begins with Planck and ends with quantum mechanics, would remain unresolved until after 1925 when that new mechanics brought clarity. Numerous other serious difficulties arose in the years 1913–1925. Just one more example: attempts to apply the Bohr theory to the spectrum of helium, the next simplest atom, led to disaster—not surprisingly, in the absence of spin and the exclusion principle.

Bohr was of course well aware of all these problems. As was written of him at the time of his death: 'The tentative character of all scientific advance was always on his mind, from the day he first proposed his hydrogen atom, stressing that it was merely a model beyond his grasp. He was sure that every advance must be bought by sacrificing some previous certainty and he was forever prepared for the next sacrifice' [13].

The days of the old quantum theory are among the most unusual in all of the history of physics. As said, paradoxes were plentiful, yet, now here, now there, additional evidence showed that there had to be something right about the Bohr theory. There were Bohr's own prediction [14] on the ratio of Rydberg constants for singly ionized helium and hydrogen, agreeing with experiment to five significant figures; Sommerfeld's theory of fine structure; the introduction of selection rules, beginning with the rule $\Delta L = \pm 1$ for electric dipole transitions, first stated by Bohr [15] and independently by Rubinowicz [16]; and Bohr's treatment of the periodic table of elements, according to which atomic electrons are arranged in shells such that the chemical properties are largely determined by the electron configuration in the outermost shell.

Bohr's most important contribution to the development of physics is his pioneering work, only briefly sketched in the foregoing, on the quantum dynamics of the atom. This is perhaps best appreciated by recalling the situation at the turn of the century as described [17] by d'Andrade:

'It is perhaps not unfair to say that for the average physicist at the time, speculations about atomic structure were something like speculations about life on Mars—very interesting for those who like this kind of thing but without much hope of support from convincing scientific evidence and without much bearing on scientific thought and development.'

It was Bohr who made atomic structure into a subject of scientific inquiry, building on the wealth of information on spectra amassed in the nineteenth and early twentieth century. He himself created many tools of the old quantum theory, most notably the correspondence principle. In 1923 Hans Kramers, Bohr's closest collaborator through most of the period under discussion, wrote [18] of this principle: 'It is difficult to explain of what it consists, because it cannot be expressed in exact quantitative laws and it is, on this account also difficult to handle.' [Roughly speaking, the principle states that for large wavelengths (slow oscillations) the theory should be in formal accord with classical mechanics and electrodynamics.] Another contemporary characterized [19] the old quantum theory like this: 'It is all mysterious, but [one] cannot deny that all the reasoning is sound, and that is a fruitful mysticism.'

The best characterization of Bohr's activities during those years was given [20] in 1949 by the seventy-year-old Einstein: 'That this insecure and contradictory foundation was sufficient to enable a man of Bohr's unique instinct and tact to discover the major laws of the spectral lines appeared to me as a miracle—and appears to me as a miracle even today. This is the highest sphere of musicality in the sphere of thought.'

Those years of struggle in the *clair-obscur* left an indelible mark on Bohr's style, once again best expressed [21] by Einstein: 'He utters his opinions like one perpetually groping and never like one who believes to be in the possession of definite truth.' As Bohr himself often used to say, never express yourself more clearly than you think. That admonition goes a long way toward explaining why Bohr was such a divinely bad lecturer. He was bad because he spoke too softly and because he would every now and then skip an argument well thought out beforehand (as I know well, having helped him prepare lectures on a few occasions and listening to their subsequent delivery). He was divine because he would be actively struggling with the concepts under discussion even as he was delivering his lecture. Rabi has told me the following story. During the First International Conference on the Peaceful Uses of Atomic Energy, Bohr gave one of the evening lectures [22], in English. As usual, simultaneous translation from a prepared text was provided. Playing with his head set Rabi noticed that one of the translations was in English. He now had the choice, either to listen to the translation and understand what Bohr was saying; or to listen to Bohr directly. He chose the latter.

Bohr's role as teacher started shortly after the appearance of his 1913 papers on the quantum theory of the atom. In April 1916 he was appointed to the newly created chair in theoretical physics in Copenhagen. On 3 March 1921 his Institute for theoretical physics (the later Niels Bohr Institute) was formally opened. Soon physicists from far and wide came to work at his Institute, arguably the world's leading centre in theoretical physics during the twenties and thirties. The international character of the enterprise was manifest from the start. By 1930 some sixty physicists hailing from Austria, Belgium, Canada, China, Germany, Holland, Hungary, India, Japan, Norway, Poland, Roumania, Switzerland, the United Kingdom, the United States, and the USSR had spent time in Copenhagen [19].

In 1925 quantum mechanics arrived. In March 1927 Heisenberg stated his uncertainty principle. On 16 September 1927, at the Volta

Meeting in Como, Bohr enunciated for the first time [23] the principle of complementarity, which embodies the physical interpretation of the uncertainty relations:

'The very nature of the quantum theory . . . forces us to regard the space-time coordination and the claim of causality, the union of which characterizes the classical theories, as complementary but exclusive features of the description, symbolizing the idealization of observation and definition, respectively.'

Already then he emphasized that we have to treat with extreme care our use of language in recording the results of observations that involve quantum effects. 'The hindrances met with on this path originate above all in the fact that, so to say, every word in the language refers to our ordinary perception.'

Bohr's deep concern with the role of language in the appropriate interpretation of quantum mechanics never ceased. In 1948 he put it as follows:

'Phrases often found in the physical literature, as "disturbance of phenomena by observation" or "creation of physical attributes of objects by measurements", represent a use of words like "phenomena" and "observation" as well as "attribute" and "measurement" which is hardly compatible with common usage and practical definition and, therefore, is apt to cause confusion. As a more appropriate way to expression, one may strongly advocate limitation of the use of the word phenomenon to refer exclusively to observations obtained under specified circumstances, including an account of the whole experiment' [24].

This usage of *phenomenon*, the one to which nearly all physicists now subscribe, was inacceptable to Einstein. In contrast to the view that the concept of phenomenon *irrevocably* includes the specifics of the experimental conditions of observation, Einstein held that one should seek for a deeper-lying theoretical framework which permits the description of phenomena independently of these conditions. That is what he meant by the term *objective reality*. It was his almost solitary position that quantum mechanics is logically consistent but that it is an incomplete manifestation of an underlying theory in which an objectively real description is possible.

I have written elsewhere [25] about the great debate between Bohr and Einstein concerning the foundations of quantum mechanics. No agreement between them was ever reached. There exists a masterful account of the Bohr-Einstein dialogue published [26] by Bohr in 1949. There can be no doubt that Einstein's lack of assent to complementarity was a very deep frustration for Bohr. It is our good fortune that this led him to keep striving at clarification and better formulation, and not only that. It was Bohr's own good fortune too. Einstein appeared forever as his leading sparring partner—even after Einstein's death he would argue with him as if he were still alive. Bohr never explicitly said so to me, but from my many conversations with him on quantum physics I know that complementarity, the elucidation of quantum mechanics, was his contribution most precious to him. There lay his inexhaustible source of identity in later life. In retrospect I find it remarkable how very rarely the subject of the old quantum theory came up in our discussions.

In this hour I can of course not do justice to all of Bohr's contributions to physics. With particular regret I must pass by his fundamental work in nuclear physics. I shall devote the remainder of this talk to Bohr's 'great moral force in behalf of the utilization of atomic energy for peaceful purposes'.

On 16 January 1939, Lise Meitner and Otto Frisch submitted a letter [27] to *Nature* in which they interpreted data obtained by Hahn and Strassmann [28] on the interaction of neutrons with uranium in terms of fission, a term they coined in that letter. On that same day Bohr arrived in New York, on board the *Drottningholm*, for a few months' stay in the United States. Shortly before leaving Copenhagen Bohr had been told by Frisch about fission. On 26 January Bohr made this news public at the Fifth Washington Conference on Theoretical Physics, at the Carnegie Institution. On 7 February the fifty-four-year-old Bohr submitted to the *Physical Review* an important contribution [29] to the theory of fission. He had deduced that only the rare isotope uranium-235 undergoes fission when a natural uranium sample is irradiated by slow neutrons. Thus Bohr's involvement with fission began in terms of pure physics.

We move to the winter of 1943. The Bohrs are in Copenhagen. It was more than three years since the Germans had occupied Denmark, formally leaving a Danish government in place, however. At that time Bohr received via underground channels a message from Chadwick suggesting that he come to England. Chadwick intimated the importance of such a move: 'You will have a very warm welcome and an opportunity to serve in the common cause'*. In his reply [31] Bohr expressed his sense of duty to help where he could in Denmark.

---

* Details of this exchange of messages between Bohr and Chadwick, and, of course, more information concerning the British participation in atomic bomb projects are described in a book by Gowing [30].

In August 1943, the Germans took over the Danish government. Hostages were being taken. Bohr was now in acute danger. In late September he and his family escaped to Sweden. On 6 October he was flown to England, where his son Aage joined him a week later. The rest of the family settled in Stockholm.

Almost at once upon his arrival in England, Bohr was briefed by Sir John Anderson on a very special aspect of the British war effort: research and development of an atomic bomb, a project code named Tube Alloys. In December Bohr, accompanied by Aage, left for the US to serve as roving ambassador for the British in regard to atomic matters. His official title was 'Consultant to the British Directorate for Tube Alloys'. Lord Halifax, the British ambassador in Washington was to be his main contact person.

Already then it had become clear to Bohr that there was not just the technical—military issue of producing a bomb but also more fundamental problems regarding the post-War world. He formulated his views in a memorandum [32] dated 3 July 1944, a document to which I shall come back, and from which I quote:

'[We must seek] an initiative aimed at forestalling a fateful competition about the formidable weapon, [an initiative which] should serve to uproot any cause of distrust between the powers on whose harmonious collaboration the fate of coming generations will depend.'

Later, in 1950, he was to expand on this idea in these words [33]:

'Humanity will . . . be confronted with dangers of unprecedented character unless in due time measures can be taken to forestall a disastrous competition in such formidable armaments and to establish an international control of the manufacture and use of the powerful materials.'

In one phrase, already in 1943–4 Bohr had the idea: We must tell the Russians at once. Because, he argued, if we don't then, right after the War, serious distrust will disrupt relations between wartime Allies, with potentially grave consequences.

Bohr discussed his ideas with his friend Felix Frankfurter, who was so impressed that he promised to convey these ideas to President Roosevelt. After having done so, Frankfurter reported back to Bohr that the President intended to take up these issues with Prime Minister Churchill.

Bohr had also talked with Anderson and Halifax who decided that Bohr should talk directly to Churchill. Others added their voice after Bohr returned to England, in April 1944. His old friend Sir Henry

Dale, President of the Royal Society, wrote to Churchill: 'I cannot avoid the conviction that science is even now approaching the realization of a project which may bring either diaster or benefit on a scale hitherto unimaginable to mankind.'

Meanwhile an event had occurred which was to complicate matters. On 28 October 1943, Kapitza had sent a letter to Bohr in Stockholm, inviting him to join scientists in the Soviet Union. The letter reached Bohr in London, from where he sent a cordial but non-committal reply [34], seeing to it that copies of both letters were transmitted to the British authorities.

On 16 May 1944 Bohr was received by Churchill whose adviser, the physicist Lord Cherwell, was also present.

Let us briefly interrupt the story in order to catch a glimpse of the status of the bomb project at that time.

On the preceding 4 February, General Groves, administrative head of the Manhattan Project, had written to the President that it was still unknown how much fissionable material was needed to produce a bomb [35].

During the following June/July the discovery of spontaneous fission of plutonium-240 caused serious reconsideration of detonation methods for plutonium bombs [36].

The point of all this is that at the time of the Bohr-Churchill meeting scientists did not yet have the means for constructing a testing device, let alone a deliverable weapon. Yet already then Bohr was urging the case for an open world.

The meeting was a disaster. Here was the War Lord, desperately preoccupied with immense responsibilities, used to incisive decisions, three weeks before D-day. There was Bohr, hard to follow as always, failing to make any impact. 'We did not speak the same language', Bohr said later [37]. It would be most inapppropriate, it seems to me, to blame either man for a missed opportunity.

In June 1944 Bohr returned to the United States.

On 19 August, at Quebec, Roosevelt and Churchill signed, 'Articles of Agreement governing collaboration between the authorities of the USA and the UK in the matter of Tube Alloys.'

On 26 August Bohr met in complete privacy with Roosevelt for a meeting that lasted an hour and a half. It was for that occasion that Bohr had prepared the memorandum [32] mentioned earlier. This time Bohr got his points across and the President sounded promising [38].

On 18 September, Churchill and Roosevelt held a private meeting at the President's residence in Hyde Park, New York. An *aide-mémoire* of that encounter contains three points [39]. First, the bomb project shall be kept strictly secret for the time being. Secondly, the US and the UK shall continue their collaboration on these projects after the war. Thirdly, 'Enquiries should be made regarding the activities of Professor Bohr and steps should be taken to ensure that he is responsible for no leakage of information, particularly to the Russians.' The next day Churchill sent a note [40] to Cherwell: 'The President and I are much worried about Professor Bohr. How did he come into this business? He is a great advocate of publicity. He says he is in close correspondence with a Russian professor, an old friend of his in Russia to whom he has written about the matter and may be writing still. The Russian professor has urged him to go to Russia in order to discuss matters. What is all this about? It seems to me Bohr ought to be confined or at any rate made to see that he is very near the edge of mortal crimes. I had not visualized any of this before, though I did not like the man when you showed him to me, with his hair all over his head, at Downing Street. Let me have by return your views about this man. I do not like it at all.'

Bohr had gotten nowhere.

There is an aftermath to this tale.

Bohr spent the spring term of 1950 at the Institute for Advanced Study in Princeton. During that period he was deeply preoccupied with the preparation of an Open Letter to the United Nations, from which I have already quoted [33]. I helped him with the preparation of several drafts.

The following story dating from that period was told to me by Robert Oppenheimer.

At one point during this period Bohr called on Secretary of State Dean Acheson to discuss with him the content of his planned Letter. The meeting began at, say, two o'clock, Bohr doing the talking. At about two thirty Acheson spoke to Bohr about as follows. Professor Bohr, there are three things I must tell you at this time. First, whether I like it or not, I shall have to leave you at three for my next appointment. Secondly, I am deeply interested in your ideas. Thirdly, up till now I have not understood one word you have said. Whereupon, the story goes, Bohr got so enraged that he waxed eloquent for the remainder of the appointment.

Bohr's Letter, dated 9 June 1950, was delivered to the Secretary General of the United Nations on 12 June.

On 24 June the Korean war broke out.

On 1 November the United States exploded in the Pacific its first thermonuclear device.

Once again Bohr was defeated by overwhelming historical forces.

And that is why Frankfurter and I chatted about how bittersweet a day in Bohr's life the 24th of October 1957 had to be.

Bohr's rich and full life came to an end on 18 November 1962. He was seventy-seven years old.

As I try to find a single phrase expressing what Bohr meant to me I am reminded of what Bohr once wrote [41] of Rutherford: 'He was a second father to me.' That is just what Bohr himself was to all those who worked with him for some length of time.

In October 1937 Bohr was in Bologna, attending the Galvani Conference, when word came of Rutherford's death. Then and there Bohr made a brief speech in memory of that great man. What he said then [42] I would like, in conclusion, to apply to Bohr himself.

'His untiring enthusiasm and unerring zeal led him on from discovery to discovery and among these the great landmarks of his work, which will for ever bear his name, appear as naturally connected as the links in a chain.

Those of us who had the good fortune to come in contact with him will always tresure the memory of his noble and generous character. In his life all honours imaginable for a man of science came to him, but yet he remained quite simple in all his ways. When I first had the privilege of working under his personal inspiration he was already a physicist of the greatest renown, but nonetheless he was then, and always remained, open to listen to what a young man had on his mind. This, together with the kind interest he took in the welfare of his pupils, was indeed the reason for the spirit of affection he created around him wherever he worked. . . . The thought of him will always be to us an invaluable source of encouragement and fortitude.' (Reprinted by permission from *Nature*, Vol. 140, pp. 752. Copyright © 1937 Macmillan Magazines Ltd.)

References

[1] Interviews of N. Bohr by T.S. Kuhn, 1 and 11 Nov. 1962, Archives of the History of Quantum Physics, Niels Bohr Library, American Institute of Physics, New York.

[2] R.C. Dales, *The scientific achievement of the middle ages* (Univ. of Pennsylvania, Philadelphia, 1978), Chapter 5.

[3] I. Newton, *Opticks* (Royal Society, London, 1704), Query 8.

[4] Described in some more detail in A. Pais, *Inward bound* (Oxford University Press, 1986), Chapter 9, section (b).

[5] A. Angström. *Recherches sur le spectre solaire* (Uppsala Press, 1868), p. 31.

[6] J. Balmer, *Verh. Naturf. Ges. Basel* **7**, 548 (1885).

[7] Interview of N. Bohr by T.S. Kuhn, 7 Nov. 1962, Archives of the History of Quantum Physics, Niels Bohr Library, American Institute of Physics, New York.

[8] R. Courant, in *Niels Bohr* (Ed. S. Rozental) (Wiley, New York, 1967), p. 159.

[9] J.R. Nielsen, Phys. Today **16**, 22 (1963).

[10] N. Bohr, *Philos. Mag.* **26**, 1 (1913).

[11] N. Bohr, letter to E. Rutherford, 6 March 1913, repr. *in Niels Bohr, Collected works* (North Holland, Amsterdam, 1981), vol. 2, p. 111.

[12] E. Rutherford, letter to N. Bohr, 20 March 1913, repr. *in Niels Bohr, Collected works,* vol. 2, p. 112.

[13] *The New York Times*, 19 November 1962.

[14] N. Bohr, *Nature* **92**, 231 (1913).

[15] N. Bohr, *Dansk. Vid. Selsk. Skrifter* **4**, 1 (1918); repr. *in Niels Bohr, Collected works,* vol. 3, p. 67.

[16] A. Rubinowicz, *Naturwissenschaften* **19**, 441, 465 (1918).

[17] E.C. d'Andrade, *Proc. R. Soc. London Ser. A,* **244**, 437 (1958).

[18] H.A. Kramers and H. Holst, *The atom and the Bohr theory of its structure* (Knopf, New York, 1923), p. 139.

[19] P. Robinson. *The early years, the Niels Bohr Institute 1921–1930* (Akademisk Forlag, Copenhagen, 1979), p. 51.

[20] A. Einstein, in *Albert Einstein, philosopher-scientist* (Ed. P.A. Schilpp) (Tudor, New York, 1949).

[21] A. Einstein, letter to B. Becker, 20 March 1954.

[22] N. Bohr, Physical science and man's position, Proc. first Int. Conf. on the Peaceful Uses of Atomic Energy (United Nations, New York and Geneva, 1956), vol. 16, p. 57.

[23] N. Bohr, *Nature* **121**, 580 (1928).

[24] N. Bohr, *Dialectica* **2**, 312 (1948).

[25] A. Pais. *Subtle is the Lord* (Oxford University Press, 1982), Chapter 25.

[26] N. Bohr. in *Albert Einstein, philosopher-scientist,* p. 199.

[27] L. Meitner and O.R. Frisch, *Nature* **143**, 239 (1939).

[28] O. Hahn and O. Strassmann, *Naturwissenschaften* **26**, 756 (1938); **27**, 11, 19 (1939).

[29] N. Bohr, *Phys. Rev.* **55**, 418 (1939).

[30] M. Gowing, *Britain and atomic energy, 1939–1945* (McMillan, London and New York, 1964).

[31] R. Moore. *Niels Bohr* (Knopf, New York, 1966), p. 298.

[32] N. Bohr, memorandum to F.D. Roosevelt, dated 3 July 1944.

[33] N. Bohr, Open Letter to the United Nations, dated 9 June 1950, published as a pamphlet by J.H. Schultz Forlag, Copenhagen, 1950.

[34] N. Bohr, letter to P. Kapitza, 29 April 1944, repr. in R. Moore, *op. cit.*, p. 336.

[35] R.G. Hewlett and O.E. Anderson, *A history of the United States Atomic Energy Commission* (Pennsylvania State University Press, University Park, Pa., 1962), vol. 1, p. 243.

[36] R.G. Hewlett and O.E. Anderson, *op. cit.*, vol. 1, p. 251.

[37] M. Gowing, *op. cit.*, 1939–1945, p. 355; R. Moore, *op. cit.*, p. 343.

[38] M. Gowing, *op. cit.*, 1939–1945, p. 357.

[39] R. Moore, *op. cit.*, p. 353.

[40] See M. Gowing, *op. cit.*, 1939–1945, p. 358; R. Moore, *op. cit.*, p. 352.

[41] N. Bohr, *Proc. Phys. Soc. London* **78,** 1083 (1961).

[42] N. Bohr, *Nature* **140,** 752 (1937).

On the Origins of the Copenhagen Interpretation

Arthur I. Miller

UNIVERSITY OF LOWELL AND HARVARD UNIVERSITY

What is impressed indelibly into the memories of all physicists who interacted with Niels Bohr is, as Léon Rosenfeld (1967) recollected, Bohr's "unrelenting effort to attain clarity [of foundations]—true as ever to his Schiller aphorism, 'only fullness leads to clarity/ And truth lies in the abyss'." In a similar vein Victor Weisskopf distinguished between research at the three major centers of atomic physics: "In Munich and Göttingen you learned to calculate . . . In Copenhagen you learned to think" (in Stuewer, 1977).

Who was Niels Bohr? By this question I mean the following: How did this man so presciently go right to the heart of the problems that beset atomic physics during 1913–1927? How did he come to exercise such a marked influence on a generation of physicists? What are the roots of the Copenhagen interpretation of atomic physics? This essay focuses on the last question, but it necessarily touches the others. We shall see that resolution of the ambiguities and paradoxes of atomic physics at the moment of its invention in 1925, and then in its interpretative period of 1926–1927, required the turn of mind of someone whose world-view concerned physics, philosophy, linguistics and, as we shall speculate here, art too.

By 1927 atomic physics had fallen into an abyss of ambiguities. And this occurred for reasons that had little to do with considerations of empirical data. Rather, by mid-1925 the core problem in atomic physics was the failure of physicists to extend into the atomic domain intuitive concepts and their visualizations of phenomena that had been assumed essential to understanding nature. I will discuss how this realization came slowly and with great reluctance.

Then I will turn to the critical period of 1926 through 1927 when Bohr realized the fullness of the concepts required to raise atomic physics from the abyss. More than anyone else, during those years he interacted with Werner Heisenberg who had, as Weisskopf put it so well, "learned to think" in Copenhagen. Bohr and Heisenberg grappled with questions that cut to the very core of how we construct knowledge in the world in which we live, questions such as: How thoroughly connected are intuition and visualization? How is the intuition that leads to one sort of visualization replaced by another?[1] Their replies to questions such as these comprise the heart of the Copenhagen Interpretation, which altered our view of physical reality in ways that are still not completely understood.

I will proceed as follows:

(1) to set the stage I will review the state of Bohr's atomic theory in 1922.

(2) I shall then use the principal problem of atomic physics during 1923 through 1925—namely, dispersion, as an Ariadne's thread to trace the demise of Bohr's theory and then the invention of matrix mechanics by Heisenberg in June of 1925.

(3) I shall then turn to the problems of interpreting the new atomic physics during Fall 1926 through Autumn 1927 which led to Heisenberg's invention of the uncertainty principles and then Bohr's formulation of complementarity.

(4) I will then discuss some early reactions to complementarity, in particular those of Heisenberg and P.A.M. Dirac.

(5) I will then explore certain of the non-scientific per se roots of complementarity in psychology and notably cubist art.

First with levity and then with ever-increasing concern, Bohr, Heisenberg and Wolfgang Pauli referred to the mixture of half-classical and half-quantum concepts in the Bohr theory as a "swindle." By allowing physicists to play both ends against the middle, this swindle served as a useful guide into the atomic domain. The reason is that the perception-laden meaning of mathematical symbols from the violated classical mechanics permitted a visualization of the atom in its stationary states as a miniscule solar system. Bohr's concern with extrapolating language carrying meanings from the world of sense perceptions into the atomic domain is a central theme in what turned out to be a series of papers that he wrote from 1913 through 1927.

Figures 1 and 2 show what a magnificent edifice the Bohr theory had become by 1922. Bohr's atomic models held great appeal for many physicists. For example, in (1923) Max Born waxed poetic: "the thought that the laws of the macrocosmos in the small reflect

MAIN LINES OF THE ATOMIC STRUCTURE
OF SOME ELEMENTS

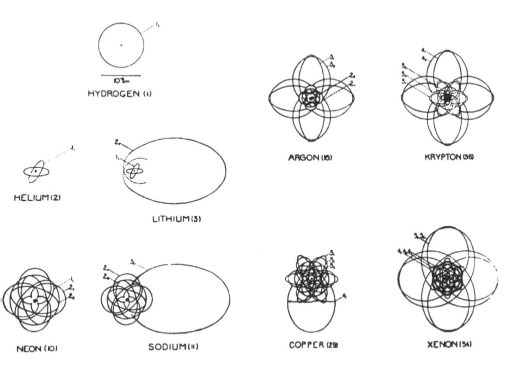

Figure 1. (a) and (b) (facing page) depict what Born (1923) wrote of as the "alluring result of Bohr's theory . . . that the atom is a small planetary system." (From Kramers and Holst, 1923a)

the terrestrial world obviously exercises a great magic on mankind's mind."

In the summer of 1922 Bohr lectured on his atomic theory at the so-called Bohr Festspiele in Göttingen. There he met the 20 year old *Wunderkind* Heisenberg. As a result of Heisenberg's incisive questions on the problem of dispersion, Bohr invited him for private conversations.

In a letter of 27 November 1933, Heisenberg reminisced to Bohr of their conversations at the Festspiele[2]: "[Until that meeting] I could do physics only in the Sommerfeld style [i.e., emphasis on mathematical techniques]. Yet without understanding every detail from your lectures suddenly I had almost the impression of understanding the real context of atomic physics." Similarly had Bohr

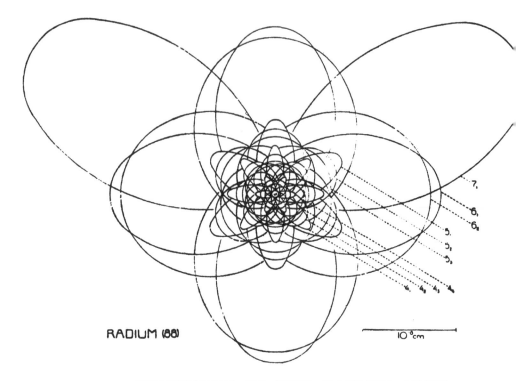

RADIUM (88)

10^{-8}cm

STRUCTURE OF THE RADIUM ATOM

Figure 2

been impressed with Heisenberg. "He understands everything," Bohr was reported to have said after their conversations.

One fundamental difficulty that Bohr discussed with Heisenberg was that the finite width of spectral lines implied an indeterminancy in the time duration of the emitted radiation and electron transition. Bohr's assistant Hendrik Kramers wrote in (1923b) that Bohr was greatly concerned over this indeterminancy because it "leads to the difficult paradox of the quantum theory [namely], that an adequate description of elementary processes in space and time is not possible." According to classical physics such a description is assumed possible. To Bohr ferreting out a paradox was essential to clarification.

By the end of 1922 the honeymoon of the Bohr theory was over. Besides its failure to produce the stationary states of the simplest three-body systems, there were models of the anomalous Zeeman effect which introduced half-integral quantum numbers and defied visualization. Bohr concluded that the fundamental difficulties which beset atomic physics all had as their common denominator the problem of the interaction of light with atoms. Bohr's principal reason was that atoms do not at all respond to incident light as a solar system atom should.

But in order to deal with the problem of dispersion Bohr had to confront a concept that he had avoided right from the start in atomic physics—namely, Albert Einstein's light quantum. Most physicists were critical over light quanta and their criticisms had little to do with empirical data. Rather, they were opposed to the light quantum's counterintuitiveness. In the German-language scientific literature the term "intuition" or *Anschauung* has a particularly loaded connotation that is rooted in Kantian philosophy. By "intuition" physicists like Bohr, Born, Heisenberg, Pauli and Erwin Schrödinger, meant the mental imagery that is abstracted from phenomena that we have witnessed in the world of sense perceptions. For example, the mental imagery or customary intuition of light as a wave phenomenon is abstracted from the behavior of water waves. But light quanta were counterintuitive because no visualizable model could be constructed for how they produced interference. Then there was the mind-boggling wave-particle duality of light which led one physicist to wonder how something could behave "as though it possessed at the same time the opposite properties of extension and localization" (Richardson, 1916). Light quanta were to be avoided right from the start. Bohr agreed and, in fact, had referred in his seminal (1913) paper to the atom emitting "*homogeneous* radiation" (italics in original).

In the face of the withering away of the solar system atom, Bohr turned to a foundational examination of his theory in a major paper that he completed in November 1922 entitled, "On the Application of the Quantum Theory to Atomic Structure: The Fundamental Postulates of the Quantum Theory (1923)." He focused on dispersion. So he necessarily discussed the nature of light and so too another customary intuition of classical physics—namely, the assumed possibility of tracing the continuous development of physical systems in space and time. To Bohr the key point in "forming a consistent image of phenomena" was to reconcile the discontinuities of atomic physics with the inherent continuity of classical physics. One approach that Bohr suggested involved the light quantum and maintaining conservation of energy and momentum in individual processes. But this was an unsatisfactory solution, he continued, because the "image of light quanta precludes explaining interference." Yet the undeniable usefulness of the light quantum hypothesis for explaining certain phenomena led Bohr to conclude that a contradiction-free description of atomic processes could not be arrived at by "use of conceptions borrowed from classical electrodynamics." Since in classical physics the conservation laws of energy and momentum are linked with a continuous space-time description then, continued Bohr, these laws may "not possess unlimited validity." Presently, however, he was not prepared to take this step. Bohr's guide in the atomic domain would be the correspondence principle, upon which rested his second method for solving "fundamental difficulties"; namely, Bohr proposed that atoms respond to light waves like an ensemble of oscillators whose frequencies are those of the possible atomic transitions.

In summary, Bohr attempted to toe the line concerning the basic connected assumptions that heretofore had to be made in order to understand nature: continuity, tracing the development of physical systems in space and time, and the conservation laws.

In (1923) Rudolf Ladenburg and Fritz Reiche proposed a mathematical formulation for Bohr's proposal. They used the correspondence principle to translate the classical equation for dispersion into one that could be interpreted "formally" as the atom responding to radiation like "*Ersatzoszillatoren.*"

Their work permitted Bohr, with Kramers and John C. Slater, in (1924) the means to avoid Arthur Holly Compton's "formal interpretation" of his experiments on the scattering of x-rays from atoms in terms of light quanta. For although there may be essential discontinuities in individual atomic processes, wrote Bohr, our "customary intuition" demands that light be a continuous phenomenon. And Bohr was willing to pay a high price to maintain a wave image of light.

Bohr's physics of desperation essentially combined the most extreme consequence of the first method of 1922 (renouncing energy and momentum conservation in individual atomic processes) with the *Ersatzoszillatoren*, which Bohr, Kramers, and Slater referred to as virtual oscillators. In the 1924 version of Bohr's atomic theory the bound electron became unvisualizable because an electron in a stationary state is represented by as many oscillators as there are transitions to and from this state.

As support for these radical moves Bohr noted that they were a step toward resolving the paradox of an entity that is both extended and localized. While subsequent work on dispersion by Born, Heisenberg, and Kramers utilized the virtual oscillators, neither the violations of energy nor momentum were well received.

The Winter term of 1924/1925 was Heisenberg's first extended stay at Bohr's three-year old Institute in Copenhagen. Kramers was Bohr's assistant and the problem at hand was dispersion. Kramers had just published a note in which he extended Ladenburg's and Reiche's 1923 result to absorption processes. Kramers' procedures and results did not refer to solar system atoms. Rather, the virtual oscillator representation permitted him to deal with only measurable quantities which entered through the amplitude of the Fourier decomposition and frequencies for possible electron transitions.

Suffice it to say that Heisenberg's research on dispersion with Kramers convinced Heisenberg of the far-reaching advantages of the virtual oscillator metaphor.

The Kramers-Heisenberg paper, completed in December 1924 with their famous dispersion relation, turned out to be the high-water mark of the Bohr theory. No further progress was made. By spring 1925 a lull fell across the landscape of atomic physics. Born (1925) wrote that he hoped to live to see the "true physics of the atom." In a letter of 21 May 1925 to Kramers, Pauli wrote that he considered taking up "film making" (Pauli, 1979).

In papers published during the late spring and summer of 1925 Bohr reviewed the grim situation. He described the June 1924 Bothe and Geiger experiment that disconfirmed the Bohr Kramers and Slater theory as "forcing" the concept of light quanta onto quantum theory, thereby further muddying the waters. His reason was that the light quantum's "unavoidable fluctuations" makes it especially difficult to use "intuitive pictures" to discuss the structure of atoms and collision processes. Bohr lamented the "essential failure of pictures in space and time." But here a new twist entered—namely, Bohr conceded that since in individual atomic processes energy and momentum are conserved, and since in individual processes the light quantum is "forced" on quantum theory, then the laws of conservation of energy and momentum must be separated from a space-time description of atomic processes. In August of 1925 Bohr welcomed Heisenberg's new atomic physics, subsequently dubbed quantum mechanics by Born. Heisenberg was the sort of physicist who thrived in periods of flux.

Heisenberg's quantum mechanics was the final link in a chain that reached back to the Bohr Festspiele in 1922 which piqued Heisenberg's interest in dispersion, and where he met Ladenburg who was beginning to formulate an oscillator representation for bound electrons. Then there was the Bohr, Kramers, and Slater theory that drove home to Heisenberg the "kind of price" required for progress—namely, that there would be no "cheap solutions (AHQP, 13 February 1963)." And finally there was the Kramers-Heisenberg (1925) paper. In June (1925) Heisenberg represented the unvisualizable bound electron with virtual oscillators whose coordinates did not necessarily commute. Almost immediately Born realized that Heisenberg's new kinematics was more general than for representing bound electrons as harmonic oscillators.

Although the renunciation of a picture of the bound electron had been a necessary prerequisite to Heisenberg's invention of the new quantum mechanics, the lack of an "intuitive" [anschauliche] interpretation was of great concern to Bohr, Born, and Heisenberg. This concern emerges from their scientific papers of the period 1925–1927.

With the publication in early 1926 of Schrödinger's version of atomic physics with its wave imagery and assumed continuity, the search for some sort of visualization of atomic processes intensified and took a subjective turn in the published scientific literature. Schrödinger wrote (1926) that he formulated the wave mechanics because he "felt discouraged not to say repelled, by the methods of the transcendental algebra which appeared very difficult to me and by lack of visualizability"'of the quantum mechanics. Needless to say, wave mechanics appealed to classical realists such as Einstein who had nothing but praise for it.

Heisenberg thought otherwise. On 8 June 1926 he wrote to Pauli (Pauli, 1979): "The more I reflect on the physical portion of Schrödinger's theory the more disgusting I find it . . . What Schrödinger writes on the visualizability of his theory . . . I consider trash. The great accomplishment of Schrödinger's theory is the calculation of matrix elements." And it was to this end that Heisenberg exploited what he insisted was only the mathematical equivalence of the two theories.

On the time scale of atomic physics thus far the period late 1925 through 1927 is the equivalent of at least a decade. The floodgates were opened. Problems that had resisted treatment in the old Bohr theory were solved one right after the other. For example, in Fall 1925 Heisenberg and Pascual Jordan solved the anomalous Zeeman effect. And, in another virtuoso performance of 1926, Heisenberg cracked the problem of the ortho and para helium spectra while almost discovering Fermi-Dirac statistics. Heisenberg's terse comment in a letter to Pauli of 19 November 1921 on the dazzling methods which became his trademark in theoretical physics was that "success sanctifies the means" (Pauli, 1979).

But despite these prodigious feats of problem solving, severe fundamental difficulties lurked in the new theory because until the fall of 1927 the new atomic physics lacked unambiguous interpretation of its mathematical symbols. This situation was rooted in the total failure of physicists to extend the visual imagery of classical physics—that is, customary intuition with its perception-laden language, into the domain of the atom.

But they had nothing else, and continued reliance on customary intuition was a stumbling block in Bohr's and Heisenberg's struggles at Bohr's Institute during the fall of 1926 into 1927 to interpret the syntax of the new atomic physics. As Heisenberg reported from Copenhagen in his November (1926) paper, "Quantum Mechanics," the "electron and the atom possess not any degree of physical reality as the objects of daily experience." Then, continued Heisenberg, there was the "light quantum hypothesis" that contradicted the "known laws of optics"—that is, the light quantum's reality was still not entirely understood. The probing of foundations in this extraordinary paper of Heisenberg reflects what Pauli wrote of him to Kramers on 27 July 1925—"Heisenberg has learned a little philosophy from Bohr in Copenhagen" (Pauli, 1979).

By late 1926 Bohr accepted the wave-particle duality of light and matter and, recalled Heisenberg (AHQP, 25 February 1963), "wanted to take this dualism [as the] central point." Consequently, Bohr could deal effectively in *Gedanken* experiments with pictures. Heisenberg persisted in focusing on his own quantum mechanics with its essential discontinuities and unvisualizable particles.

On 23 November 1926 Heisenberg wrote to Pauli that after reading Dirac's transformation theory he understood better how "things are related"—that is, waves and particles. However, Heisenberg went on, "That the world is continuous I consider more than ever as totally unacceptable. But as soon as it is discontinuous, all our words that we apply to the description of facts are so many c numbers. What the words 'wave' or 'corpuscle' mean we know not any more" (Pauli, 1979). Heisenberg recalled of this period that "we couldn't doubt that [quantum mechanics] was the correct scheme but even then we didn't know how to talk about it [these discussions left us in] a state of almost complete despair'' (1975).

Requiring a respite from their intense interactions, in early February of 1927 Bohr went on a skiing trip to Norway. During this break Heisenberg recalled Einstein's statement to him in Spring 1926 that "it is the theory that decides what we can observe" (Heisenberg, 1969). Although Heisenberg had initially disagreed, in 1927 he moved beyond Einstein's suggestion in a paper that is the sequel to the November 1926 paper "Quantum Mechanics"—namely, Heisenberg's paper, "On the Intuitive Content of the Quantum-Theoretical Kinematics and Mechanics," that he submitted for publication in March (1927). How important was the concept of intuition to Heisenberg is indicated by its inclusion into the title of this classic paper in the history of ideas. In this paper Heisenberg demarcated boldly between "to be understood intuitively" and the visualization of atomic processes. Focusing exclusively on the quantum mechanics with its unvisualizable particles, and taking support from the redefinitions of physical reality in the large required by the special and general theories of relativity, Heisenberg proposed that in the atomic domain a revision of our usual kinematical and mechanical concepts "appears to follow directly from the fundamental equations of the quantum mechanics." That is, Heisenberg permitted the mathematics of the quantum mechanics to determine the restrictions on such perception-laden symbols as position and momentum. These restrictions are the uncertainty relations. Consequently, Heisenberg redefined the concept of intuition through the theory's mathematics, and separated intuition from visualization.

Focusing exclusively on the particle aspect of light and matter led Heisenberg to the following conclusions:

(1) Revision of customary concepts of physical reality is rooted in the "typical discontinuities" of atomic processes.

(2) Quantum mechanical probability follows from the mathematics of the quantum theory "[in which] all quantum-theoretical quantities 'in reality' are matrices." However, cautioned Heisenberg, we should not conclude that the quantum mechanics is "an essentially statistical theory in the sense that only statistical conclusions can be drawn from specified data." The reason is that the conservation laws permit accurate prediction of certain quantities.

(3) Since the uncertainty relations placed limits on the accuracy to which initial conditions could be determined, then the "invalidity of the causal law is definitely established in quantum mechanics."

In later years Heisenberg recalled Pauli's response to the uncertainty principle paper as (1960): "*Es wird Tag in der Quantentheorie.*"

Bohr thought otherwise. In the interim he had realized that the need to redefine physical concepts was the wave-particle duality of light and matter, and not the essential discontinuities. Bohr pressed his own view relentlessly with Heisenberg and a tense atmosphere developed.

Although by May of 1927 relations improved between Bohr and Heisenberg, all differences of opinion were not yet settled. For example, in a letter to Pauli dated 16 May 1927, Heisenberg wrote that there are "presently between Bohr and myself differences of opinion on the word 'intuitive'" (Pauli, 1979). This is reasonable because to Bohr a mathematical formalism that stressed discontinuities and unvisualizable particles could not decide what was intuitive. Heisenberg went on to mention that Bohr had been discussing with him a "general work on the 'conceptual foundations' of the quantum theory from the viewpoint, 'There are waves and particles'." As is well known, Bohr often spun out ideas which were further clarified through intense discussion. So, very likely, their analysis of Heisenberg's notion of the quantum measurement process served also to crystallize and sharpen Bohr's own thoughts toward complementarity.

On 16 September 1927 at the International Congress of Physics, Lake Como, Italy, Bohr read a version of this "general work" which would be the final published installment in the series of papers that reached back to 1913. It is a Bohr tour-de-force, dense with a labyrinthine web of arguments that lead essentially to the two conclusions which comprise the principle of complementarity (1928):

(1) In the atomic domain Planck's constant links the measuring apparatus to the physical system under investigation in a way that "is completely foreign to the classical theories." This is how the unavoidable intrinsic statistics enter the quantum theory. Since, in the atomic domain the notion of an undisturbed system developing in space and time is an abstraction, then "there can be no question of causality in the ordinary sense of the word," that is, the strong causality of classical physics. Instead Bohr linked causality with the predictive powers of the conservation laws, and not with space-time pictures. We recall that already in 1925 Bohr had realized the necessity to separate space-time pictures from the conservation laws. In 1927 the space-time pictures became restricted metaphors.

(2) Bohr's masterstroke was to realize that the wave-particle duality of light and matter was only paradoxical because of limitations in the atomic domain on our language. Rather, in the atomic domain both horns of the dilemma are connected. That is, Planck's constant links quantities that characterize a localized entity like a particle (energy and momentum) with quantities that characterize an extended entity like a wave (frequency and wavelength). Bohr reasoned that just as the large value of the velocity of light had prevented our realizing the relativity of time, the minuteness of Planck's constant rendered paradoxical the wave-particle duality of matter and light. Since Planck's constant places restrictions on the use of our language in the atomic domain, then so too on our customary intuition or visual imagery, which enables us to describe only things that are either continuous or discontinuous but not both. Yet subatomic particles are simultaneously localized and extended. So, stressed Bohr, the wave and particle modes of light and matter are neither contradictory nor paradoxical, but complementary. Both modes or sides are required for a complete description of the atomic entity.

It is an understatement to say that Bohr's Lake Como lecture was not completely understood. The recorded comments by Born, Enrico Fermi, Heisenberg, Kramers and Pauli were orthogonal to Bohr's lecture and to each other. The deep meaning of complementarity would be explored in earnest during the legendary Bohr-Einstein dialogues that began in October of 1927 at the Solvay Conference.

To the best of my knowledge Bohr discussed versions of the Como lecture principally with Pauli and Dirac. Pauli was in overall agreement with Bohr's view. I was somewhat surprised to learn that Dirac was not. Dirac's disagreements serve to further explicate Bohr's view of quantum theory.

In a letter to Dirac of 14 March 1928[2], Bohr tried to clarify points from their recent discussion in Cambridge: "Although I realise the tentative character of the formulation [i.e., most likely regarding inclusion of relativity into quantum mechanics] I still believe that the point of view of complementarity is suited to describe the situation. I think that we cannot too strongly emphasize the inadequacy of our ordinary perception when dealing with quantum problems."

In his characteristically terse manner Dirac spelled out his disagreements in a letter to Bohr dated 9 December 1929[2]: "Although I believe that quantum mechanics has its limitations and will ultimately be replaced by something better, (and this applies to all physical theories) I cannot see any reason for thinking that quantum mechanics has already reached the limit of its development. I think it will undergo a number of small changes, mainly with regard to its method of application; and by that means most of the difficulties now confronting the theory will be removed."

A letter to Bohr from Dirac of 9 June 1936[2], shows that by this time their differences of opinion turned about their different and inarticulable notions of aesthetics. Dirac wrote: "[You] emphasize the beauty and self-consistency of the present scheme of quantum mechanics, but I do not think they provide an argument against the possible existence of a still more beautiful scheme in which, perhaps, the conservation laws play an entirely different role."

What about Heisenberg's view of complementarity? Just as Heisenberg came to accept the wave-particle duality of light and matter when the appropriate mathematical apparatus became available, his writings from the period 1928–1929 in conjunction with his reminiscences reveal that he came to understand what he referred to as the "symmetry of the complementarity picture" from what he took to be the mathematical statements of the complementarity principle—the methods of the quantization of wave fields formulated by Dirac in 1927 for the electromagnetic field and then for particles with mass by Jordan, Oskar Klein and Eugene Wigner during 1927–1928. This is a form of the quantum theory in which the wave and particle aspects of matter can be transformed mathematically into one another and yet remain mutually exclusive. As Heisenberg wrote in a review article of (1929) entitled, "The Development of the Quantum Theory: 1918–1927": "[Second quantization] demonstrates that in the formalism of the quantum theory the particle and wave pictures appear only as two different manifestations of one and the same physical reality."

Thus far we have explored the philosophical-scientific origins of complementarity.[3] Could there have been other roots of this far-ranging concept?

As had been the case in Einstein's invention of special relativity, the roots of complementarity run deeper than considerations of physics or even philosophy per se. For example, in 1913 there was Bohr's concern over the problem of assigning semantics or meaning to the mathematical symbols in his first atomic theory paper. Well known is Bohr's life-long interest in how meaning is assigned to words; how meaning is affected by context; and how context depends on the mental imagery constructed from the world in which we live. As Bohr put it in the published 13 April 1928 *Die Naturwissenschaften* version of his Como lecture, "every word in the language refers to our mode of intuition."

The linguistic root of complementarity has a psychological part which could have been triggered by or at least gained support from Bohr's awareness of William James's book *Principles of Psychology*, Søren Kierkegaard's writings and/or their interpretation by the Danish philosopher Harald Høffding, who had been a university teacher of Bohr's and then life-long friend. Bohr broached the psychological dimension in the April 1928 printed version of the Como lecture, and it echoes James and Kierkegaard. For example, wrote Bohr, the failure of our "customary intuition" in the atomic domain can be traced to "the general difficulty in the formation of human ideas, inherent in the distinction between subject and object." Suffice it say here that the origins of Bohr's psychological component of complementarity is still a puzzle for the historian of ideas, owing mainly to mutually disconfirming historical data. This complex issue has been explored in some depth by Gerald Holton (1970) and Max Jammer (1966).

What I should like to add to the mosaic that is the background of complementarity is Bohr's interest in art, especially cubism. We might expect, therefore, to see in his study a painting by one of the acknowledged masters of this genre—for example, a Braques, a Gris, a Duchamp, or maybe a Picasso. Instead Bohr exhibited Jean Metzinger's 1924 painting *L'Écuyère*. This choice indicates a quite special interest in cubism, and perhaps a clue to yet another path to complementarity—that is, assuming that Bohr had known about Metzinger prior to 1927. It may be the case that after 1927 Bohr found in Metzinger's writings one more example of complementarity. Let's make the first assumption and attempt to find what it was in Metzinger that interested Bohr. Most art historians consider Metzinger to have been a minor cubist painter, but everyone agrees that he was a major theorist of the cubist school. In (1912), Metzinger and Albert Gleizes published a systematic exposition of cubist methods in their widely-read book *Du Cubisme*. A cubist painting, they wrote, represented a scene as if the observer were "moving around an object [in order to] seize it from several successive appearances. . . ." Cubists achieved this motif through the interpenetration of figure and space in order to free the artist from a single perspective in favor of multiple viewpoints. And this was what impressed Bohr about cubism. Mogens Anderson (1967), a Danish artist and friend of Bohr, recollected Bohr's pleasure in giving "form to thoughts to an audience at first unable to see anything in [Metzinger's] painting—They came with a preconceived idea of what art should be." Such had been the case in 1913, when atomic physicists had a preconceived visual image of the atom. By 1925 atomic physicists had come to realize the inadequacy of visual perception, as had the cubists. In 1927 Bohr offered a motif for the world of the atom with striking parallels to the motif of multiple perspectives offered by cubism for glimpsing beyond and behind visual perceptions: According to complementarity the atomic entity has two sides—wave and particle—and depending on how you look at it, that is, what experimental arrangement is used, that is what it is.

Here we have explored how Niels Bohr was led into ever-deeper levels of analyses of atomic phenomena that moved from examining physics per se into an analysis of perceptions and then into an analysis of thinking itself. In this way, one that squares with Schiller's aphorism, Bohr recognized the fullness contained in the wave-particle duality of matter and light, thereby raising atomic physics out of the abyss.

I will conclude with a quote from Bohr's letter of 24 August 1927 to Heisenberg that captures so well the emotional intensity of their research in atomic physics, research that defined a heroic age[2]: "The kind words you wrote about your stay in Copenhagen were a great pleasure to read. Also for me this has been an unforgettable time."

References

1. In my (1984) I have addressed these quotations in the wider context of cognitive science.
2. Letter on deposit at the American Institute of Physics, New York City.
3. Most recently the philosophical roots of complementarity have been investigated by Holton (1960), Jammer (1966), and Miller (1984), among others.

Bibliography

Quotations from AHQP (Archive for History of Quantum Physics) are taken from the interviews of Werner Heisenberg by Thomas Kuhn. The abbreviation *Zsp* is for *Zeitschrift für Physik*.

Andersen, Mogens, "An Impression," in Rozental (1967), pp. 321–324.

Bohr, Niels, "On the Constitution of Atoms and Molecules," *Phil. Mag.*, **26**, 1–25 (1913).

―― "On the Application of the Quantum Theory to Atomic Structure: Part I. Postulates of the Theory," *Proc. Camb. Phil. Soc.* (Supplement) (1924). Published in *ZsP*, **13**, 117–165 (1923).

―― "The Quantum Theory of Radiation," *Phil. Mag.*, **47**, 785–802 (1924) (with H. Kramers and J. Slater).

―― "The Quantum Postulate and the Recent Development of Atomic Theory." Versions are in *Atti del Congresso Internazionale dei Fisici* (Bologna: Zanichelli, 1928), pp. 565–598; *Nature* (Supplement) 580–590 (14 April 1928); *Die Naturwissenschaften*, **15** (13 April 1928).

Born, Max, "Quantentheorie und Störungsrechnung," *Die Naturwissenschaften*, **27**, 537–550 (1923).

―― *Vorlesungen über Atommechanik* (Berlin: Springer, 1925).

Heisenberg, Werner, "Über quantentheoretische Umdeutung kinematischer und mechanischer Beziehungen, *ZsP*, **33**, 879–893 (1925).

―― "Quantenmechanik," *Die Naturwissenschaften*, **14**, 899–904 (1926).

―― "Über den anschaulichen Inhalt der quantentheoretischen Kinematik und Mechanik," *ZsP*, **43**, 172–198 (1927).

―― "Die Entwicklung der Quantentheorie, 1918–1928," *Die Naturwissenschaften*, **26**, 490–496 (1929).

―― "Errinerungen an die Zeit der Entwicklung der Quantenmechanik," in Fierz, M. and V.F. Weisskopf (eds.), *Theoretical Physics in the Twentieth Century* (New York: Interscience, 1960), pp. 40–47.

―― *Der Teil und das Ganze: Gespräche im Umkreis der Atomphysik* (München: Piper, 1969), translated by A.J. Pomerans as *Physics and Beyond: Encounters and Conversations* (New York: Harper, 1971). [1969]

―― "Discussion with Professor Werner Heisenberg, in O. Gingerich (ed.), *The Nature of Scientific Discovery: A Symposium Commemorating the 500th Anniversary of the Birth of Nicolaus Copernicus*, (Washington: Smithsonian Institution Press, 1975), pp. 556–573.

Holton, Gerald, "The Roots of Complementarity," *Daedalus*, Fall 1970, 1015–1055; reprinted in Gerald Holton, *Thematic Origins of Scientific Thought: Kepler to Einstein* (Cambridge: Harvard University Press, 1973), pp. 115–161.

Jammer, Max, *The Conceptual Development of Quantum Mechanics* (New York: McGraw-Hill, 1966).

Kramers, Hendrik, *The Atom and the Bohr Theory of Its Structure* (London: Gylendal, 1923a), translated from the first Danish edition of 1922 by R.B. and R.T. Lindsay (with H. Holst).

―― "Das Korrespondenzprinzip und der Schalenbau des Atoms," *Die Naturwissenschaften*, **27**, 550–559 (1923b).

Ladenburg, Rudolf, "Absorption, Zerstreuung und Dispersion in der Bohrschen Atomtheorie," *ZsP*, **4**, 451–468 (1923) (with F. Reiche).

Metzinger, Jean, *Du Cubisme* (Paris: Figuiere, 1912) (with A. Gleizes).

Miller, Arthur I., *Imagery in Scientific Thought: Creating 20th-Century Physics* (Boston: Birkhäuser, 1984; Cambridge: MIT Press, 1986).

Richardson, O.W., *The Electron Theory of Matter* (Cambridge: Cambridge University Press, 1916).

Rosenfeld, Léon, "Niels Bohr in the Thirties," In Rozental (1967), pp. 114–136.

Rozental, S., *Niels Bohr: His Life and Work as Seen by His Friends and Colleagues* (New York: Wiley, 1967).

Schrödinger, Erwin, "Über das Verhältnis der Heisenberg-Born-Jordanschen Quantenmechanik zu der meinen," *Annalen der Physik, 70*, 734-756 (1926).

Stuewer, Roger (ed.), *Nuclear Physics in Retrospect: Proceedings of a symposium on the 1930's* (Minneapolis: University of Minneapolis Press, 1977).

Niels Bohr and Atomic Physics Today

Daniel Kleppner

MASSACHUSETTS INSTITUTE OF TECHNOLOGY

Today, more than seven decades since he first laid the groundwork for quantum mechanics, Niels Bohr's spirit continues to animate science. Bohr extended and enormously enriched the vocabulary of physics, but his imprint on modern thought is to be found far beyond the enclaves of science. The little sketch of the Bohr atom—electrons whirling around a nucleus in ring-like trajectories—is universally recognized as a symbol of knowledge and scientific progress. It is not the symbol's arcane connotations of stationary states and quantum mechanics to which the public mind resonates, however, but to its metaphor of a solar system, a microcosm of planets encircling the sun, an abstraction of the visible world that has been an article of faith since the Copernican revolution. In large part it was this ability of Bohr to provide the affective underpinnings for quantum mechanics—the planetary model, the correspondence principle, complementarity and the construct of the "Copenhagen interpretation"—that has made his role in physics so enduring.

I propose to describe here some recent developments in atomic physics that one way or another are related to ideas in Bohr's earliest papers on quantum theory. Bohr used hydrogen as the principle stepping stone for the new physics; I shall follow his example by focusing on contemporary developments that involve hydrogen.

The family of hydrogen has grown since Bohr's time. Today its members include muonium and positronium, discovered by Vernon Hughes and Martin Deutch, respectively. "Geonium"—hydrogen without a nucleus, a free electron in a trap—to adapt the name coined by its inventor, Hans Dehmelt, could also be considered one of the family. It is particularly fitting to cite these three scientists here since they, along with Norman Ramsey, who pioneered the hydrogen maser, have just been awarded the Rumford Premium of the A.A.A.S. at this Symposium in honor of the Bohr Centenary. The family of hydrogen will also be taken to include Rydberg atoms—a species of hydrogen-like atoms—and the very highly charged single-electron atoms that have recently been created. However, at the risk of a family feud, I will omit the heavy hadronic atoms that have been created by particle physicists, and anti-hydrogen, whose time is coming, but is not yet here.

To set the stage, Fig. 1 shows the familiar energy levels and some of the spectral sequences of hydrogen. The Balmer sequence, terminating on $n = 2$, is the only sequence in the visible regime. The rich magenta glow of a hydrogen discharge tube arises from the Balmer-α line. The Lyman sequence, terminating on the ground state, lies in the ultraviolet, and the higher lying sequences lie in the infrared or beyond. For very high values of n the states are known

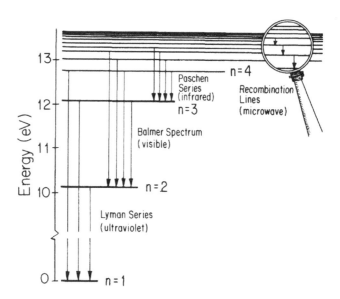

Figure 1. Energy levels and spectrum of hydrogen according to the Bohr model. The red transition $n = 2 \rightarrow 1$ (the Balmer-α line) was first observed in the sun's absorption spectrum by Fraunhofer in 1817. Balmer identified the optical series of hydrogen in 1883; the near infrared sequence terminating on $n = 3$ was observed by Paschen in 1908, and the ultra violet sequence terminating on the ground state was identified by Lyman in 1916. Radio recombination lines, arising from transitions in the vicinity of $n = 100$, were discovered by Mezger and Hoglund in 1965.

as *Rydberg* states: transitions among Rydberg states usually occur in the millimeter wave regime, and at longer wavelengths.

Optical Spectroscopy of Hydrogen

The agreement between the theoretical and observed values is inside the uncertainty due to experimental errors . . .

    Niels Bohr, on the numerical value of the Rydberg constant [1]

 Bohr's first paper on hydrogen [1] provided tentative explanations of two problems beyond the scope of classical physics: why atoms are stable and why they radiate sharp spectral lines. However, his arguments, resting on such implausible concepts as stationary states and "jumps" between energy levels, could hardly be justified. As Bohr recognized, his proposals were heretical to physics as it then existed. Nevertheless, one had to take the arguments seriously for in spite of his unorthodox reasoning, Bohr obtained not only the correct form for the hydrogen spectrum,

$$\frac{1}{\lambda} = R \left(\frac{1}{n^2} - \frac{1}{m^2} \right),$$

but he also derived the following value for the Rydberg constant R, (more precisely, R multiplied by the velocity of light, c)

$$\frac{2\pi^2 m e^4}{h^3} = 3.1 \times 10^{15}.$$

"The agreement between the theoretical and observed values" that Bohr claimed follows from comparison of this number with the experimental value of 3.290×10^{15}. To derive R from the fundamental constants was a glorious achievement. Only once before had physical theory asserted itself in such a fashion—when Maxwell demonstrated that the speed of light is related to the electric and magnetic force constants.

 Spectroscopy provided the raw materials for Bohr's theory—the Balmer spectrum and the Rydberg constant—but progress in the optical spectroscopy of hydrogen was slow and not particularly rewarding for over fifty years following Bohr's first papers. Until the era of lasers and modern spectroscopy, the resolution of optical spectroscopy was limited by the Doppler effect due to thermal motion of the atoms. Doppler broadening is particularly serious for

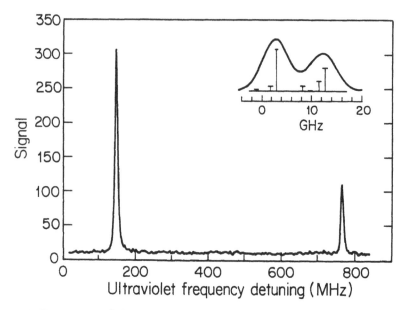

Figure 2. Spectrum of the ls → 2s two-photon "forbidden" transition, observed by Doppler-free spectroscopy. The two lines are due to hyperfine interactions. The line width is 10 Mhz. This transition, which has an intrinsic linewidth of about 1 Hz, is expected to play a principle role in the development of ultra-high resolution spectroscopy in the coming decade. The inset shows the spectral profile of the Balmer-α line by conventional spectroscopy. The fine structure interval, 10,000 Mhz, is visible, but none of the underlying hyperfine structure, nor the Lamb shift, is resolved. (Courtesy of T.W. Hänsch, ref 2.)

hydrogen because of its low mass and relatively high speed for any given temperature. The best resolution that one could obtain after decades of effort is shown in the inset of Fig. 2. The inset is a profile of the Balmer-α line taken by absorption spectroscopy in a roughly collimated atomic beam. A doublet structure due to fine structure is clearly resolved, but the enormously important underlying structure due to the Lamb shift is totally unresolved, as are the yet smaller hyperfine structure effects.

In the early 1970's, the barrier of Doppler broadening was breached by laser saturation spectroscopy, largely through the pioneering work by T.W. Hänsch. Since then, spectroscopic resolution has continued to increase at an extraordinary rate. Laser techniques provide a precision thousands or millions times greater than before. Transitions that were previously inaccessible can now be observed. For instance, Fig. 2. shows the spectrum of the "forbidden" two-photon transition, $1s \rightarrow 2s$ [2]. The potential resolution of this spectral line is more than ten million times higher than for other optical lines in hydrogen. This is because the ultimate resolution of a spectral line is limited by the natural linewidth due to spontaneous emission, and the $2s$ state is remarkably long-lived; 0.15 s, compared to 2×10^{-9} s for the $2p$ state. Consequently, the natural linewidth for the $1s \rightarrow 2s$ transition is a few Hz, whereas the natural linewidth for the Lyman series is typically a few hundred Mhz. The potential resolution of the $1s \rightarrow 2s$ transition is approximately 10^{15}, an astronomical figure compared to the resolution of allowed topical transitions. Achieving this resolution will require not only substantial advances is laser technology, but the development of techniques for "holding" a hydrogen atom without perturbing it for times approaching one second. There is no fundamental reason why these problems cannot be solved. On the contrary, there is reason to be optimistic about advances in the coming decade.

The Rydberg constant is now known to a few parts in 10^{10} [3]. For the moment, this is about as precise as it can be measured, for the accuracy is limited not by the experimenter's skill in controlling lasers or studying hydrogen, but by our fundamental ability to define distance. (The Rydberg constant is essentially a measure of wavelength—actually inverse wavelength—and can be regarded as an atomic unit of distance.) To achieve higher spectroscopic accuracy we need techniques for measuring the *frequency*, not the *wavelength*, of an optical transition. In principle, optical frequencies can be measured by the same procedures that are used for radio frequency signals, for which precision of 1 part in 10^{15} is readily achieved. Converting this principle to practice is perhaps the most important challenge facing precision spectroscopy today.

What is to be gained by measuring the Rydberg constant to a few more significant figures? For the moment, no physical theory is critically dependent on R. However, by comparing different optical transitions in hydrogen it will be possible to obtain more precise values for the Lamb shift and for the proton structure effects that currently limit comparison of experiment with QED theory. The measurements can also be expected to improve our knowledge of other fundamental constants, for instance the electron-proton mass ratio. The most likely reward, however, is that the optical techniques that must be developed will find important applications in other areas of spectroscopy, as well as in atomic physics, and metrology. Scientific surprises seem to occur whenever the precision with which we can study nature is increased substantially: it would be strange if the next generation of spectroscopy were an exception.

The Fine Structure of Hydrogen

The doublet structure of the Balmer-alpha line of hydrogen (illustrated in Fig. 2, inset) was discovered in 1888 by A.A. Michelson in his search for narrow spectral lines for use as a wavelength standard. Bohr suggested that the splitting was relativistic: electrons moving in orbits with identical total energy but different instantaneous kinetic energy would have slightly different effective masses. Some twelve years later Dirac proved that Bohr's guess was fundamentally correct. Among the results of the Dirac theory is the prediction that all states of hydrogen having the sample principle quantum number n and the same angular momentum j will have exactly the same energy. The fine structure of hydrogen provides a natural testing ground for the Dirac theory. Conventional optical spectroscopy of the fine structure is not equal to this task because, as Fig. 2 shows, its resolution is simply not good enough: Doppler broadening obscures all details of the fine structure.

The challenge of studying the hydrogen fine structure was met by Willis Lamb who developed a technique for studying transitions within the manifold of $n = 2$ states by microwave spectroscopy [4]. (Doppler broadening is negligible because of the low frequency of the microwave transition). Lamb found that the $2S_{1/2}$ level is shifted upward compared to the $2P_{1/2}$ level by approximately 1060 MHz. This shift, the Lamb shift, together with the experimental observation of the anomalous magnetic moment of the electron, provided an experimental imperative for superseding the Dirac theory. The creation of quantum electrodynamics (QED) by Tomonga, Schwinger and Feynman, successfully answered this challenge. The Lamb shift of hydrogen and the anomalous magnetic moment of the electron have continued to provide the primary arena for low energy tests of QED.

The principle experimental impediment to measuring the Lamb shift arises from the short lifetime of the $2p$ state: the result is that the natural line width for the $2s \rightarrow 2p$ transition is enormous, approximately 100 MHz. Because the Lamb shift is only 1060 MHz, the intrinsic resolution of the measurement is low. Nevertheless,

Lamb managed to "split" the natural line width by about a factor of 1000, achieving a final precision close to 100 kHz.

The most recent Lamb shift measurement has improved on this precision by using a method that appears to defy the uncertainty principle: the experimental linewidth is reduced below the linewidth predicted from the uncertainty principle.

The reduction of the line width is achieved by a technique based on an argument that is very much in the spirit of the Copenhagen School. The natural line width is proportional to the *mean* decay rate, or inverse lifetime, of an ensemble of identically prepared atoms. However, atoms radiate statistically: in any ensemble of excited atoms, some will "live" for times much greater than the mean lifetime. The distribution of lifetimes is generally described by an exponential decay curve. By selecting only those atoms whose lifetime happen to lie in the "tail" of the curve, an arbitrarily narrow linewidth can be achieved. There is as one might expect, a price to pay: the number of long-lived atoms decreases exponentially with the time. However, if the signal is sufficiently strong, the result is a net improvement in the measurement. Using this method, Pipkin and Lundeen [5] have measured the Lamb shift to an accuracy of 9 kHz, or to a fractional accuracy of 1 part in 10^5!

The theoretical challenge of calculating the Lamb shift to an accuracy of 1 part in 10^4 is every bit as formidable as the experimental challenge, but this, too, has been met [7]. The results are in reasonable agreement at least for the present:

Experiment	1 057.845(9) MHz	[5,6]
Theory	1 057.875(10) MHz	[7]

The major obstacle to probing QED further via the Lamb shift is the theoretical uncertainty in the correction required by the structure of the proton. It seems unfair that hydrogen, the prince of atoms, should be plagued by a nucleus that we do not understand, but in fact we need to learn more about the shape of the proton if the Lamb shift in hydrogen is to be pushed further as a test of QED.

The Hyperfine Structure of Hydrogen

A small splitting in the ground state of hydrogen gives rise to hydrogen's well known 21 cm transition, a line that radio astronomers describe as the best known line in the universe. The splitting originates in the magnetic interaction between the electron and proton, the hyperfine interaction. The hyperfine frequency, $\Delta\nu_H \approx 1420$ MHz, is far too small to be detected optically. It was first measured by Rabi using atomic beams [8]. The accuracy, about 1 part in 10^4, was crude compared to later measurements with the hydrogen maser [9], but it was high enough to reveal that the electron had an anomalous magnetic moment. Today, $\Delta\nu_H$ is one of the best determined frequency intervals in physics. The value is [10]

$$\Delta\nu_H = 1{,}420{,}405{,}751.766\ 7(10)\ \text{Hz}.$$

This figure is accurate to twelve figures: in contrast, theory is only accurate to six figures. The theoretical problem, once again, is due to uncertainties in the proton structure. Because of this, the hyperfine separation of hydrogen has little to say about QED or any other urgent theoretical problem. Nevertheless, $\Delta\nu_H$ plays an important role in science.

Any phenomenon that can be measured to high precision can, in principle, be employed as a metrological standard. The hyperfine separation of hydrogen serves that purpose as an atomic frequency standard in the form of the hydrogen maser. As the world's most stable clock, the hydrogen maser has been used to measure the effect of the Earth's gravity on time—thhe Earth's gravitational red shift [11]. The effect amounts to approximately 1 second in ten thousand years, (2 parts in 10^{11}) but using a rocket-borne atomic clock, as sketched in Fig. 3, the gravitational red shift has been measured to a precision of about 1 part in 10^4.

Hydrogen masers play an important role in radio astronomy by serving as the local frequency references needed to keep the worldwide very-long-baseline-interferometry system synchronized. They are also employed in the global positioning system for navigation. These examples provide one illustration of the assertion made earlier that the development of high precision measurement techniques generally leads to interesting new science and unexpected applications.

Muonium—Hydrogen Without a Hadron

Muonium is a hydrogen-like atom consisting of a positive muon and an electron. Muonium is ideal for studying QED because the muon, like the electron, is a structureless lepton, free from all the com-

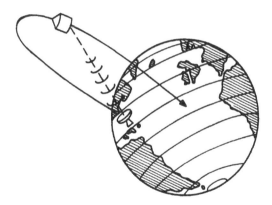

Figure 3. The earth's gravitational red shift has been measured by comparing a rocket-borne atomic clock, a hydrogen maser, with a similar clock at a ground station. The difference in rate of time is approximately 1 part in 10^{11}, about 1 second in 3,000 years, but this difference was measured to a precision of about 1 part in 10^4. (Courtesy of R.F.C. Vessot, ref. 11.)

plications of hadrons and their underlying quark structure. In principle, muonium is the best understood atomic system.

Muonium is one of the so called "exotic" atoms, atoms that must be created in the laboratory because they are never found in Nature. It was first created by Vernon Hughes and his colleagues [12]. Muons are created by the decay of pions. The experimenter then has two microseconds—the time for a muon to decay into a positron and a neutrino—to "cool" the muon from hundreds of MeV to below 1 eV, allow it to pick up an electron and radiate to a low-lying atomic state, and carry out a spectroscopic measurement. Such an experiment is necessarily complex, but Nature is helpful in providing muons that are naturally spin polarized due to the parity violating weak interaction by which they are created. Furthermore, the direction of the decay products points out the final direction of the muon's spin. Muonium research has been revolutionized in recent years by the constructions of high intensity pion sources thhat have enormously increased the useful flux of atoms.

The muonium hyperfine structure $\Delta\nu_M$ provides one of the most sensitive tests of QED. Because the hyperfine interaction involves the muon and electron magnetic moments, both of these moments must be known to compare experiment and theory. The electron magnetic moment is known precisely: the muon's moment is determined as part of the hyperfine measurement.

The most recent measurement of $\Delta\nu_M$ is accurate to 3.6 parts in 10^8 [13]. Although this is less than the precision of $\Delta\nu_H$, the theory for muonium can in principle be carried out to this same high precision. In practice, theory is limited by uncertainty in the muon's magnetic moment, which is about 3.6 parts in 10^7.

Measuring the hyperfine separation of muonium is a formidable experimental enterprise, but not nearly so demanding as measuring muonium's Lamb shift. The reason is that muonium is normally created in a high pressure gas that serves to thermalize the muons and provide them with electrons. However, in a gas, the metastable 2s state is immediately quenched by collisions. To observe the Lamb shift, the muonium must be in a vacuum. Nevertheless, two groups [14,15] have observed the Lamb shift, and measured it to the 1% level. This is not accurate enough to provide a critical test of QED, but it is an impressive step in that direction.

Positronium

Positronium is composed of an electron and positron. Like muonium, positronium is hadron-free. However, electrons and positrons are antiparticles, and that introduces new physics because antiparticles like to annihilate each other. Furthermore, although the reduction of a two particle system into a single particle system is trivial non relativistically, the relativistic treatment presents a theoretical challenge, particularly when the masses are identical.

Positronium was first created in 1951 by Martin Deutch [16], but in spite of efforts over the years, only recently has it become possible to observe an optical transition [17]. The experiment is tortuous: 0.5 MeV positrons from a radioactive source are cooled to a few hundred eV in a copper crystal moderator. They are then stored in a magnetic bottle accumulator, and eventually dumped into a second copper crystal moderator, where some of the positrons pick up electrons near the surface and are ejected as positronium atoms in the ground state. At just the right moment a laser is flashed to induce the $1s \rightarrow 2s$ transition. The excited atom is photoionized by the same laser light, and detected.

To observe an optical transition in positronium is a considerable feat: to simultaneously measure the transition to high precision must

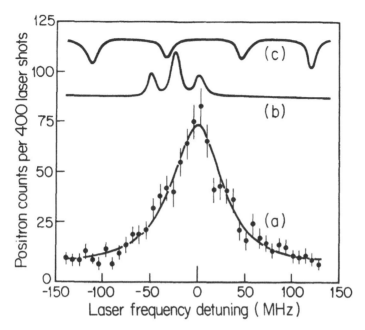

Figure 4. Spectrum of the $1s \rightarrow 2s$ transition in the $e^- - e^+$ atom, positronium. Curve *a* shows the data and the fit to the resonance line. Curve *b* is the spectrum of a nearby calibration line in tellurium, and curve *c* shows the frequency markers of an etalon used for interpolation and control. In spite of the formidable problem of creating thermal positronium from high energy positrons, the experiment achieved a spectral resolution of 10^7, and a precision that approaches 10^9. (Courtesy of S. Chu, ref. 17.)

be counted as a *tour de force*. Figure 4 shows the data from such a measurement. The result [17] is

$$\nu(1s \rightarrow 2s) = 1,233,607,185(15) \text{ MHz.}$$

This result is accurate to 2 parts in 10^9, which approaches the accuracy of the Rydberg constant for hydrogen itself! The theoretical prediction is not quite as accurate as experiment, but theory and experiment agree.

High-Z Hydrogen

The gross features of single electron atoms with nuclear charge number Z can be found from hydrogen by changing the energy scale by a factor of Z^2. Thus the transition wavelength for He^+ are essentially the same as for hydrogen, except for a scaling factor of $\frac{1}{4}$, (plus a small reduced mass correction) a fact that Bohr recognized in his earliest work on hydrogen. Over the years techniques have been developed for studying higher and higher charged state ions, but only in the past few years has it beeen possible to observe hydrogen- like ions near the end of the periodic table. Now, hydrogen-like uranium, U^{91+} has been created [18]. Its Lyman-α line has a wave- length is 0.015 nm, in the hard x-ray regime.

QED is a perturbative theory; the results generally emerge as a power series in the parameter is $Z\alpha$. (α is the fine structure constant, $\approx\frac{1}{137}$.) For hydrogen, $Z\alpha$ is less than 1% and high order terms rapidly vanish. For hydrogen-like uranium, however, its value is approxi- mately 2/3, and high order terms are important. This behavior is shown in Fig. 5, where various contributions to the Lamb shift are plotted. Because of this scaling behavior, even a low resolution mea-

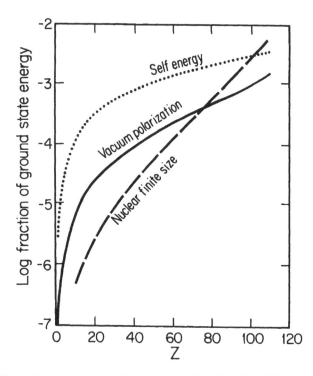

Figure 5. Plot of various contributions to the Lamb shift as a function of nuclear charge. Because of the enormous enhancement of these terms at high Z, relatively low precision experiments on highly charged ions can provide precise measurements of these terms.

surement of the Lamb shift at high-Z can provide a relatively precise check on the high order terms.

Hydrogen-like uranium has been observed, but the Lamb shift measurement turns out to be much easier if one uses helium-like uranium, U^{90+}. The Lamb shift is found by measuring the lifetime of a $2p$ state. The lifetime of the $2\,^3P_0$ state is 10^{-10} second, long enough to be determined by time of flight. Approximately 30% of its decay rate is due to radiation to the short-lived $2\,^3S_1$ state: this rate is sensitive to the energy splitting, which depends on the Lamb shift.

The Lamb shift in U^{90+} is about 70 eV. The measured value is 70.4(8.1) eV [18], in good agreement with the theoretical value, 75.3(0.4) eV [19]. This, however, is only the beginning: future measurements are expected to be good to one part in a thousand.

Geonium—The World's Simplest "Atom"

If one is willing to stretch the definition of "atom" to include systems with less than two particles, then geonium, not hydrogen, ranks as the simplest atom of all. "Geonium" is the term coined by Hans Dehmelt to describe a free electron in a trap. A truly free electron would be unmeasurable, and one can simply regard the trap as the measuring apparatus. However, the trap also confines the electron, playing a more active role than one usually allows for a measuring apparatus. The word "geonium" is suggestive of an electron bound in a trap that is fixed to the earth. Presumably the results would be the same for a trap in a space station or on some other planet.

Confining clouds of electrons or ions in static and dynamical fields is easy: what puts geonium in a class by itself is the ability to study a *single* electron in a trap. The trapping lifetime achieved by Dehmelt and his colleagues appears to be effectively infinite: one electron was confined for over nine months. By coupling the electron's motion to a tuned circuit, the electron's energy can be thermalized to cryogenic temperatures, the micro-eV regime [20].

Geonium was created to measure the anomalous magnetic moment of the electron. By the Dirac theory, the g-factor of the free electron is exactly 2; by QED, it differs from 2 by approximately $\alpha/2\pi$, approximately 1 part in a thousand. Over the years, the g-factor anomaly, $a_e = (g - 2)/2$, has been the center of one of the most grueling battles between theory and experiment on QED.

Today, the battle is being fought at a level of precision of 40 parts per billion.

The theoretical value for a_e is generally calculated as a power series in $\alpha/\pi \approx 2.3 \times 10^{-3}$:

$$a_e = 0.5 \left(\frac{\alpha}{\pi}\right) + A_2 \left(\frac{\alpha}{\pi}\right)^2 + A_3 \left(\frac{\alpha}{\pi}\right)^3 + A_4 \left(\frac{\alpha}{\pi}\right)^4 + \cdots$$

The experimental result is [21]

$$a_e(\exp) = 1,159,652,193(4) \times 10^{-12},$$

while the theoretical result is [22]

$$a_e(\text{the}) = 1,159,652,302(128) \times 10^{-12}.$$

The principal contribution to the theoretical error is uncertainty in the fine structure constant α. One would, of course, reverse the argument and use a_e to determine α. The principle contribution to the experimental error comes from small inhomogeneities in the magnetic field that are introduced to detect the spin motion. A new generation of geonium experiments is being planned in which this obstacle is removed.

Where will the battle end? Like all perturbation theories, if QED is carried out to high enough order in α/π, the calculated result will start to diverge from the physical value. Because QED is the model for QCD and essentially all other modern gauge theories, one would very much like to know the limit where "high enough" lies. For the moment, however, that limit appears to be over the horizon.

Rydberg Atoms

According to the theory the necessary condition for the appearance of a great number of lines is therefore a very small density of the gas: for simultaneously to obtain an intensity sufficient for observation the space filled with the gas must be very great.

(Niels Bohr [1] explaining why emission lines from high-lying states of hydrogen are not observed in laboratory spectra.)

According to the Bohr theory, the radius of an excited atom increases with n^2, where n is the principle quantum number. The reason that radiation from high-n states should not be expected in laboratory discharge sources is that the cross sections of such atoms, which scale as n^4, are gigantic: the atoms would be destroyed by collisions before they could radiate. Bohr's prescription for observing was prophetic: to find low density and large volume one needs to look to the heavens. Interstellar space satisfies both criteria, and radiation from Rydberg atoms—for such is the name given to atoms in high-lying states—was first detected by radio astronomers.

Using lasers, Rydberg atoms can now be prepared and studied with ease in the laboratory. Figure 6 shows a spectrum of atoms in the region of $n \approx 450$. The signal is essentially an excitation spectrum using light from a precisely controlled laser. The data are generated by ionizing the Rydberg atoms, not by detecting their radiation.

Although the Bohr model played a crucial role in the development of quantum mechanics, attempts to apply the model to problems other than the hydrogen atom were a failure. Bohr's own unsuccessful attempt to understand helium was the forerunner of a now largely forgotten literature on the structure of even more complex atoms. Nevertheless, there were a few useful elaborations of the Bohr theory, and one of these was to the problem of hydrogen in an applied electric field—the Stark effect of hydrogen. Sommerfeld solved the problem classically, and then imposed periodic quantum conditions. (The Bohr-Sommerfeld quantization law emerged from this effort.) His results are in reasonably good agreement with calculations from quantum mechanics for fields not too close to the ionization limit.

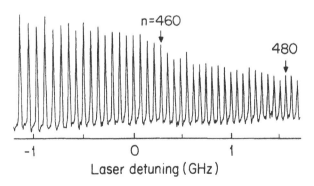

Figure 6. Portion of an excitation spectrum of Rydberg states of barium observed by laser spectroscopy with an atomic beam. For $n = 460$, the atomic diameter is approximately 10 μ, and the binding energy is 70 μeV. Such atoms will spontaneously ionize in a field of about 10 mV/cm. (Courtesy of H. Rinneberg.)

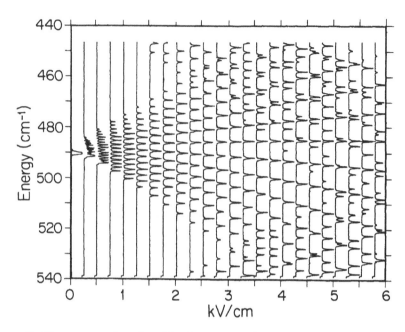

Figure 7. The energy levels of hydrogen, and most Rydberg atoms, form simple fan-like patterns in an applied electric field. These data are for the $n = 15$ manifold of lithium. Levels from the $n = 14$ and 16 manifolds can be seen approaching from below and above, respectively.

The Stark structure of Rydberg states of hydrogen is dramatic: the energy levels split into fan-like manifolds that form intricate geometrical patterns. Figure 7 shows one of these manifolds in Rydberg states of lithium.

Rydberg atoms have such remarkable properties that they form almost a separate atomic species. They have been used to study strong field phenomena, collisions, energy transfer, atom-surface interactions, multiphoton processes, non linear dynamics, and a host of other processes [23].

Stationary States, At Last

. . . we are led to assume that these configurations will correspond to states of the system in which there is no radiation of energy; states which consequently will be stable as long as the system is not disturbed from the outside.

                         Niels Bohr [1], introducing the concept of stationary states

Bohr's description of radiation as occurring by "jumps" between stationary states was self contradictory: truly stationary states should not radiate. The resolution of this difficulty lies in the fact that radiation is inherently a slow process: stationary states are *almost* stationary. However, the dynamics of radiation was a problem beyond the scope of the Bohr theory.

According to QED, atoms in excited states will spontaneously radiate by a process that can be pictured as an interaction between the atom and the vacuum fields of the space in which it is embedded. Recently it has been shown that in cavities the vacuum fields can be suppressed so that an excited atom cannot radiated: its state is truly stationary, at least with respect to spontaneous emission. Early evidence for such effects was found in the fluorescence of atoms near mirrors [24], and in the radiation of free electrons in a cavity-like trap [25]. However, with Rydberg atoms it is possible to control the effect and measure it precisely.

Rydberg atoms in a beam pass between two plates which behave like a waveguide near cutoff. The wavelength is slowly varied across cutoff, at which point the atoms abruptly stop radiating and the number of excited atoms transmitted by the system jumps. (See Fig. 8.) The radiative lifetime of an excited atom has been shown to increase by a factor of at least twenty [26].

What is the significance of "turning off" spontaneous emission? From one point of view the effect could be regarded simply as a stunt. However, that such an elementary process could go unrecognized for so long suggests that although quantum electrodynamics may indeed be the most highly developed theory in physics, a body of basic phenomena may have been overlooked. In particular, one might naturally expect that if a cavity can change the lifetime of an atom, it can also change the energy of the atom. That turns out to be the case. The effects are small, but they could contribute a measurable shift to measurements of a_e in the next generation of geonium experiments [27], and produce a measurable correction to the Lamb shift [28]. In any case, the existence of such a "cavity shift" illustrates an interesting point: One cannot isolate an atom from the vacuum by putting it into a cavity, for the atom will then interact with the cavity. Thus, the idea of a truly isolated atom is fundamentally non-physical: the atom together with its surroundings—vacuum, cavity, or whatever—must be viewed as a single system. Such a point of view would undoubtedly have been congenial to Niels Bohr.

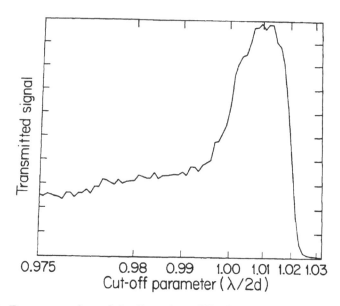

Figure 8. Demonstration of the "turning off" of spontaneous emission. The signal represents the intensity of Rydberg atoms after they travel a time comparable to their natural radiative lifetime. The atoms travel between parallel plates: if the plates are moved to than one-half wavelength apart, spontaneous radiation ceases, and the intensity rises. (Ref. 26.)

References

1. N. Bohr, *Philosophical Magazine* **26**, 1 (1913).
2. E.A. Hildun, U. Boesl, D.H. McInture, R.G. Beausoleil and T.W. Hänsch, *Phys. Rev. Lett.* **56**, 576 (1986).
3. P. Zhao, W. Lichten, H. Layer and J. Berquist, *Phys. Rev. Lett.* **58**, 1193 (1987).
4. W.E. Lamb, *Phys. Rev.* **72**, 241 (1947).
5. S.R. Lundeen and F.M. Pipkin, *Phys. Rev. Lett.* **46**, 232 (1981).
6. G. Newton, D.A. Andrews and P.J. Unsworth, *Phil. Trans. Roual Soc. London* **290**, 373 (1979).
7. T. Kinoshita and J. Sapirstein, *Atomic Physics 9*, ed. R.S. Van Dyck, Jr. and E.N. Fortson, World Scientific (Singapore), 1, 984, p. 83.
8. J.E. Nafe, E.B. Nelson, and I.I. Rabi, *Phys. Rev.* **71**, 914 (1947).
9. H.M. Goldenberg, D. Kleppner and N.F. Ramsey, *Phys. Rev. Letters,* **5**, 361 (1960).
10. H. Hellwig, et al., *IEEE Trans. Instr. Meas.* **IM19**, 200 (1970).
11. R.F.C. Vessot et al., *Phys. Rev. Lett.* **45**, 2081 (1980).
12. V.W. Hughes, C.W. McColm, D.W. Ziock and R. Prepost, *Phys. Rev. Lett.* **5**, 63 (1960).

13. F.G. Marian et al., *Phys. Rev. Lett.* **49,** 993 (1982).
14. A. Badertscher, et al., *Atomic Physics 9*, ed. by R.S. Van Dyck, Jr., and E.N. Fortson, (World Scientific, Singapore, 1984) p. 38.
15. C.J. Oram, *Atomic Physics 9,* ed. by R.S. Van Dyck, Jr., and E.N. Fortson, (World Scientific, Singapore, 1984) p. 83.
16. M. Deutsch, *Phys. Rev.* **82,** 455 (1951).
17. S. Chu, A.P. Mills, Jr. and J.L. Hall, *Phys. Rev. Lett.* **52,** 1689 (1984).
18. H. Gould and C.T. Munger, *Atomic Physics 10*, ed. H. Narumi and I. Shiramura, (N. Holland, Amsterdam, 1987) p. 95.
19. P.J. Mohr, *Phys. Rev. Lett.* **34,** 1050 (1975); P.J. Mohr, *Phys. Rev.* **A26,** 2338 (1982).
20. H. Dehmelt, *Atomic Physics 7*, ed. D. Kleppner and F.M. Pipkin, (Plenum, N.Y. 1981) p. 337.
21. P.B. Schwinberg, R.S. Van Dyck, and H.G. Dehmelt, *Phys. Rev. Lett.* **47,** (1981) 1679; R.S. Van Dyck, P.B. Schwinberg and H.G. Dehmelt, *Atomic Physics 9*, ed. by R.S. Van Dyck, Jr., and E.N. Fortson, (World Scientific, Singapore, 1984) p. 53.
22. T. Kinoshita and W.B. Lindquist, *Phys. Rev. Lett.* **47,** (1981) 1573; *Phys. Rev.* **D27** (1983) 853, 867,877, 886.
23. D. Kleppner, *Atoms in Unusual Situations*, ed. J.P. Briand (Plenum, NY, 1986) p. 57.
24. K.H. Drexhage, *Progress in Optics*, ed. E. Wolf (North Holland, Amsterdam, 1974), V XII, p. 165.
25. G. Gabrielse and H. Dehmelt, *Phys. Rev. Lett.* **55,** 67 (1985).
26. R. Hulet, E.S. Hilfer, and D. Kleppner, *Phys. Rev. Lett.* **55,** 2137 (1985).
27. L.S. Brown and G. Gabrielse, *Rev. Mod. Phys.* **50,** 233 (1986).
28. P. Dobiasch and H. Walther, *Ann. Phys. Fr.* **10,** 825 (1985).

Condensed Matter Physics: Does Quantum Mechanics Matter?

Michael E. Fisher

CORNELL UNIVERSITY*

Introductory Words

Herman Feshbach, the organizer of this Symposium in honor of Niels Bohr, asked me, in his original invitation, for a review of the present state of condensed matter physics, with emphasis on major unsolved problems and comments on any overlap with Bohr's ideas regarding the fundamentals of quantum mechanics. That is surely a difficult assignment and, indeed, goes well beyond what is attempted here; nevertheless, I will take the liberty of raising one issue of a philosophical or metaphysical flavor.

Does Quantum Mechanics Matter?

Condensed matter physics is a multifaceted subject—a beautiful diamond. The aim of this lecture is, in part, to hold this diamond up to view and to turn it around, alas rather rapidly, so that a few flashes may catch the eye. In doing this I will, however, concurrently address the question, "Does quantum mechanics matter for condensed matter physics?" This is not meant facetiously—it is not a rhetorical question. Some of my colleagues at Cornell were, I believe, a bit horrified on learning that I could consider this question seriously or even pose it! One of them indeed, said, "That seems to me as crazy as discussing planetary motion and asking 'Does Newtonian mechanics matter?'" However, I want to suggest that the question for

* Now at Institute for Physical Science and Technology, University of Maryland, College Park, Maryland 20742

condensed matter physics is not really quite like that. Consider the analogous but hopefully less controversial question, "Does human biology matter to sociology?" Obviously there *is* some connection. But suppose we go further towards more fundamental levels of science, asking the same question:

> *Sociology*—Does *human biology* matter?
> Does *cell biology* matter?
> Does *biochemistry* matter?
> Does *molecular physics* matter?
> Does *nuclear physics* matter?
> Does *quantum chromodynamics* matter?

Surely no one can seriously maintain that quantum chromodynamics matters to sociology!

For every area in science, there are other areas of direct relevance, further areas of less relevance, and also relevant interconnections between what may properly be regarded as *different levels* of discourse. I believe that this remark reflects, to some extent, Bohr's idea of *complementarity*; the issue is, however, interesting and subtle. I would like to explain my view of it in further detail.

To a very good degree of approximation, many disciplines in the sciences are "essentially" self-contained in the sense that there is a natural language which "fits" the phenomena and a corresponding "calculus", which, according to the case at hand, may be more-or-less mathematical. It is perhaps easiest to grasp this within the discipline of physics itself. It has long been granted, for instance, that thermodynamics has a wide domain of validity within which one does not have to worry about the atomic constitution of matter or about statistical mechanics. The situation as regards most applications of Maxwell's equations is really very similar, i.e., they are used to discuss electrical circuits on appropriate scales, to design antennae, etc., without worrying about their more fundamental basis in quantum electrodynamics or condensed matter physics (the latter as regards any magnetic, dielectric or conducting properties). This situation seems even clearer as regards the discipline of chemistry where, unfortunately, a mathematically complete calculus is, in practise, absent in most cases: Schrödinger's equation really does have *almost no* relevance to what one actually does in vast areas of chemistry.

The issue is truly the same in condensed matter physics. The basic problem which underlies the subject is to understand the many, varied manifestations of ordinary matter in its condensed states and to elucidate the ways in which the properties of the "units" affect the overall, many-variable systems. In that enterprise it has become increasingly clear that an important role is and, indeed, should be played by various special "models." It is, nowadays, a triviality that the behavior of dilute gases is universal and in most regards quite independent of the chemical constitution of the molecules or atoms. We no longer regard the universal ideal gas laws as peculiar or mysterious. One should not, therefore, be so surprised that other theoretical models that have been abstracted from complicated physical systems, like the basic Ising and Heisenberg models originally devised for magnetism, take on a life of their own quite regardless of the underlying atomic and molecular physics (for which quantum mechanics does matter).

This situation serves to illustrate another point, namely, the picture of *levels* or *strata* in science. One might, as a theorist, argue that more attention should be paid to the connection between models, like the ideal gas or the Ising and Heisenberg models, and fundamental physical principles as embodied in quantum mechanics. The various models so important in modern condensed matter physics, often appear to the outsider as inspired guesses, whose success is judged solely by their ability to explain the phenomena which motivated them. Would it not be an equally great success for theory to *derive* any of these models from atomic theory? Is that not where quantum mechanics enters modern condensed matter physics?

Certainly, the connection between levels of description deserves clarification in all science. However, the mere fact that such an 'artificial,' 'nonfundamental' model as the Ising model provides insight into a wide range of *contrasting* examples of condensed matter such as anisotropic ferromagnets, gas-liquid condensation, binary alloys, structural phase transitions, etc., shows that the question of "connecting the models with fundamental principles" is *not* a very relevant issue or central enterprise. I stress again, that in any science worth the name, the important point is to gain *understanding*. The language in which the understanding is best expressed cannot be dictated ahead of time, but must, rather, be determined by the subject as it develops. Accordingly, it really would *not* be a "great success" to derive the Ising model from atomic theory or quantum mechanics. Indeed, for which physical system would one be "deriving" it? For a ferromagnet where the electronic spins are crucial?

For a liquid-gas system where the atoms or molecules are the basic unit? For an alloy where the metallic ions are the "variables"? Or for a ferroelectric material in which the rotation of some large unit of correlated atoms in a lattice is the basic dynamical variable? Of course, in each of these cases, one *does* want to understand what the important variables actually are; but, in many ways, it is more important to understand how these variables interact together and what their "cooperative" results will be.

On the other hand, it is certain that all the sciences are linked together. Furthermore, some disciplines are more "fundamental" than others. However, it is not so easy to identify "fundamental principles" *per se*. Does Schrödinger's equation constitute a truly "fundamental principle"? Surely not! It must merely be some sort of low-energy approximation to a more basic theory. The same goes for quantum chromodynamics: evidently that is also only a limiting theory! And so on. Personally, I find the elucidation of the connections between various disciplinary levels a topic of great interest. Typically, delicate matters of understanding appropriate asymptotic limits, both physical and mathematical, are entailed. The issues involved are often very subtle; nevertheless, one must admit, they seldom add significantly to one's understanding of either the "fundamental" starting theory or of the target discipline. Sad but true! Every now and then, when one is lucky, one can uncover some special corner of the "higher" discipline where more basic or fundamental principles reveal some limitations that have not been suspected before or where "natural assumptions" can be proved false; but that is a bonus not to be expected generally. Nevertheless, I am myself interested in and will continue to encourage those braver souls who really try to understand the interconnections between different disciplines, or subdisciplines, or between layers within a discipline. Many, however, who would claim to be working from fundamental principles do not, in my view, really approach the task appropriately. As indicated, the aim should be, for instance, to justify the Ising model for some specific system on the basis of, say, Schrödinger's equation. To often, however, individual theorists have, under what one may fairly characterize as the prevailing absolutist philosophy in physics, attempted to derive *all* the properties of a real system from "fundamental principles," often using gross approximations, rather than attempting to justify the intermediate language that one has discovered is appropriate.

Clearly there are elements of Bohr's "complementarity" in these points of view. However, I think it is a mistake to view complementarity as merely a two-terminal black box. In real science there are many layers and interleavings, some of which we may hope to see disappear, but many of which will remain, both complementary and supplementary. In a word, a fully reductionist philosophy, while tenable purely as philosophy, is the wrong way to practice real science!

The Task of Condensed Matter Physics

With that over-lengthy preamble let me return to the main subject. My aim is, first of all, to explain what I think condensed matter physics does, or should be doing, and what defines condensed matter physics, and thence to approach the question, "Does quantum mechanics matter?"

The key fact about condensed matter physics is the existence of many states of matter—and we keep discovering more! In fact, one of the things that makes condensed matter physics exciting is that every year or two, something new turns up.* We will consider some of these new things.

So the first question is, "What are the states of matter?" Can we elucidate them? When we have discovered them, can we characterize them? What is their nature? More specifically, this usually means determining the type of order, and the particular spatial and temporal correlations that build up. Then, we may ask, "How do the various states transform into one another? How are they interrelated?" These latter questions demonstrate that the study of phase transitions, my own love, is one of the fundamental aspects of condensed matter physics.

The Stability of Matter

Now there is one most basic way in which quantum mechanics truly matters. We would not even have condensed matter if we did not have quantum mechanics! But even here it is worth enquiring as to

---

* Since this lecture was presented verbally the startling discovery of "high T_c" superconductors,[1] which lose their resistance already above 90°K, has borne out this assertion!

which aspects of quantum mechanics are actually important. The first issue to appreciate is that the very term "condensed matter" implies a large number of elementary units, atoms, electrons and nuclei, spins, etc., say N in number. Then the so-called "thermodynamic limit" in statistical mechanics, namely $N \to \infty$, is an important and basic idealization. (Nevertheless, both theoretically and experimentally, one is increasingly able to study large but still finite systems. And 'large', amusingly enough, sometimes turns out to be not much bigger than 10, 20 or 30. From the viewpoint of interconnections with wave mechanics and the more basic theories, this is quite a puzzle: Why is 'large' so small?)

There is a second, related issue which underpins all statistical mechanics. Statistical mechanics relates microscopic mechanics to macroscopic thermodynamics. Thermodynamics postulates—or embodies the deep physical insight—that there is a bulk free energy function which is proportional to the amount of matter or, one says, is an "extensive" quantity. At low temperatures the free energy becomes just the energy. Statistical mechanics must yield, or explain, the extensive nature of the free energy and energy and thus that matter is *stable*. More concretely if $E_0(N)$ is the ground state energy of a system of size N, say N electrons and N protons, stability means that

$$E_0(N) \geq -\varepsilon N \qquad \text{for all} \qquad N, \tag{1}$$

where ε, a finite constant independent of N, is essentially the limiting binding energy per degree of freedom (one might also well work on a per unit volume basis).

Now one of the fundamental things Bohr did with his theory of the atom was to explain why it was stable—why it did not collapse, the electrons falling into the nucleus to yield a ground state energy unbounded below. However, the finite energy of the Bohr atom, which is naturally measured in terms of the Rydberg*

$$\varepsilon_0 = me^4/2h^2, \tag{2}$$

is *not* enough to explain why, granted Bohr mechanics and Coulomb interactions, bulk matter is stable with, as one observes, ε not so different from ε_0.

Indeed, some years back, F.J. Dyson and A. Lenard[2] proved the converse! Suppose one has the standard Schrödinger wave equa-

---

* Here m and e denote, as usual, the mass and charge of an electron.

tion—kinetic energy in the usual form—and Coulomb forces: What is the ground state energy of a neutral system of N such particles? The answer is provided by the *upper* bound

$$E_0(N) < -\varepsilon_1 N^{7/5} \qquad \text{for some fixed} \qquad \varepsilon_1 > 0; \qquad (3)$$

note the exponent $\frac{7}{5}$ which exceeds unity. So although one single hydrogen atom is stable, according to Bohr, many hydrogen atoms would seem to collapse; and, indeed, to collapse quite catastrophically!

It should be mentioned here, that we leave gravity aside. We will choose to work on scales where it is unimportant—of course, all real matter will eventually collapse into a black hole if N gets large enough!

But what then is the answer to the puzzle posed by Eq. (2)? Why is matter stable? The missing piece, is in fact, provided by Fermi-Dirac statistics. Dyson and Lenard in their very beautiful paper,[2] proved that if the negative particles, i.e., the 'electrons', are fermions, and if there are no more than q species of them (with masses $m_j^- \leq m$ and charges $|e_j| \leq e$ for $j = 1, 2, \cdots, q$), then there is, in fact, a lower bound of the required form, namely,

$$E_0(N) > -\varepsilon_2 q^{2/3} N, \qquad (4)$$

with ε_2 proportional to ε_0. Thus wave mechanics *with* Fermi statistics is enough to stabilize matter.*

This result echoes a theme running through modern condensed matter physics: "Fermions are vital." They are hard to deal with; we do not really understand them well yet despite a half-century's acquaintance. But we know they are vital, and in this sense quantum mechanics certainly matters.

On the other hand, there is "another end" to the question of the existence of the free energy for Coulomb forces—namely, what happens to the N-particle interactions at long distances. One must prove that the forces are not too repulsive. More concretely the free energy, $F(N)$, must satisfy a "tempering condition",[3] $F(N) < \varepsilon_x N$. How does this come about? Coulomb forces can be strongly repul-

---

* In retrospect one recalls that the ground state of the unrestricted many-particle Schrödinger wave equation will have a nodeless wave function symmetric under the interchange of particle coordinates. Such a wave function is acceptable for bosons but, of course, is unacceptable for fermions which must, therefore, have a higher energy.

sive at long distances; but Coulomb forces screen! This is a crucially important property, but one that is completely classical! The delicate proof[4] that screening tempers the repulsions uses only the ideas of Newton, specifically the fact that when there are spherical distributions of charges, only the total charge outside a given region matters—a reflection of the fact that Coulomb's law is a true inverse square law (or, more appositely, an inverse $(d - 1)$-power law where d is the spatial dimensionality).

Now I will discuss several interesting forms of matter, describing in each case the experimental situation and then giving some idea of our theoretical understanding—and concurrently pointing out the degree to which quantum mechanics enters.

Polymeric Matter

The first sort of condensed matter I would like to consider is polymeric matter. Polymers are long molecules with some finite thickness: see Fig. 1. the importance of their thickness has been stressed by many researchers, most notably P.J. Flory, and is realized to be a crucial feature determining their behavior. It means that a polymer chain of length N units (or monomers) cannot be thought of as a free random walk: such a walk may revisit any region of space. But a real polymer chain interacts with itself and different segments cannot approach too closely. This poses the so-called *excluded volume problem*.

The mean size of a polymer molecule is properly described by its *radius of gyration*, R_G, which is the square root of the average

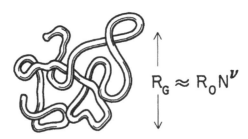

Figure 1. Depiction of a polymer molecule of N units and its mean size or "radius of gyration" R_G.

squared distance between all the monomers that compose it. Because of the excluded volume, R_G grows in a nontrivial way with polymer length, namely, according to the law

$$R_G \approx R_0 N^\nu, \tag{5}$$

where the exponent, ν, differs from the value $\nu_0 = \frac{1}{2}$ which characterizes a free random walk. Indeed, we know that $\nu > \frac{1}{2}$ (for spatial dimensionalities $d < 4$) so that the rate of growth exceeds that for an unimpeded random walk; in fact, one has $\nu \approx 0.588$ in $d = 3$ dimensions.

Now, thanks to renormalization group ideas and other theoretical ways of looking at the problem, the exponent ν is fairly well understood: but other challenging problems remain beyond the size of a single isolated molecule. One might say, "Well, aren't polymers for chemists?" The answer is that many of their properties have proved too difficult for traditionally trained chemists! Consider a solution of polymer molecules of overall concentration c. It has been the condensed matter physicists who have finally had to think hard about this problem and to solve it. To give a feel for the theory, consider the simplest thermodynamic quantity of such a solution, namely, the osmotic pressure, π, regarded as a function of the concentration, c. This must also depend strongly on the parameter, N, the length of the polymer chains: but how? It turns out that one can construct a universal theory—actually a "scaling theory"[5]—for the osmotic pressure and other solution quantities.

To be more explicit, suppose one has n polymeric molecules in a total volume V so that the concentration is

$$c = n/V = \rho_1/N, \tag{6}$$

where ρ_1 is the overall density of the *monomers*. Consider the *reduced* osmotic pressure

$$\tilde{\pi} = \pi(c,N)/ck_B T; \tag{7}$$

in the limit of extreme dilution, $c \to 0$, one expects this to approach the ideal gas value, $\tilde{\pi} = 1$; but interactions between different molecules must change $\tilde{\pi}$ drastically already at low polymer concentrations, especially when N is large. One concludes theoretically[6] that the combined dependence on c and $N \gg 1$ can be written

$$\tilde{\pi} = \frac{\pi(c,N)}{ck_B T} \approx Z(X), \tag{8}$$

where $Z(X)$ is a *universal function* of the *scaled concentration*

$$X = c/c^* \qquad \text{with} \qquad c^* = c_0/N^{d\nu}. \qquad (9)$$

Note the appearance here of the exponent ν which describes the excluded volume of a single chain via Eq. (5). The parameter c_0 depends only on the nature of the individual monomers.

At low concentrations one can expand in powers of X and so finds

$$\bar{\pi} = 1 + B_2(N)c + O(c^2) \qquad (10)$$

where the second virial coefficient obeys $B_2 \approx b_2 N^{d\nu}/c_0$, in which b_2 is a numerical coefficient. As anticipated, B_2 depends strongly on N. However, for high concentrations the theory shows[6] that

$$Z(X) \approx Z_\infty X^{1/(d\nu - 1)} \qquad (X \to \infty) \qquad (11)$$

where Z_∞ is a constant. This, in turn, implies the surprising result

$$\pi/k_B T \approx C_0 \rho_1^\delta \qquad \text{with} \qquad \delta = d\nu/(d\nu - 1), \qquad (12)$$

in which C_0 depends only on the properties of a monomer. Thus at high concentrations the osmotic pressure becomes *independent* of the length, N, of the polymer chains! Furthermore π then increases with monomer density according to a universal power law with exponent $\delta \approx 2.31$ (for $d = 3$).

How well does all this theory work? As an example, regard Fig. 2: this shows the osmotic compressibility, $\partial\pi/\partial c$ vs. X, which, by Eq. (8), just depends on the scaling function Z and its derivative. The scaled concentration, $X = c/c^*(N)$, runs over four decades, the compressibility over two—and the data come from solutions in different solvents and for a range of N. The only physical parameter used to draw the theoretical line in Fig. 2 is c_0 which is fitted to the second virial coefficient, B_2. Equally good agreement with theory[6] is found for the pressure itself, for the monomer density correlation length, etc.

Thus we see in polymeric matter new, subtle and universal behavior which we have succeeded in understanding theoretically. But quantum mechanics has had essentially nothing to say about the problem! Indeed, one feels that if some of the giants of the past, like Boltzmann or Gibbs or Rayleigh, were able to rejoin us today, they would be able to engage in research at the cutting edges of condensed matter physics without taking time off to study quantum mechanics first! No doubt they would *want* to learn about quantum mechanics—who would not? But while learning quantum mechanics, they could all be doing front-line research using their knowledge of clas-

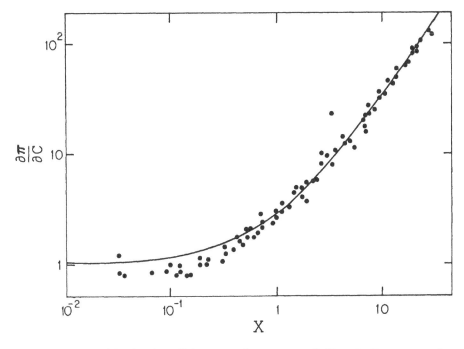

Figure 2. Logarithmic plot of the osmotic compressibility, $\partial\pi/\partial c$, versus the scaled concentration, $X = c/c^*$, for polystyrene solutions in toluene and in methylethylketone verifying the universal scaling form (solid line) with asymptotic slope $1/(d\nu - 1) \simeq 1.31$. [After Y. Oono, Ref. 6.]

sical physics, Newtonian mechanics, statistical mechanics and electrodynamics, and their mathematical and modelling skills.

Magnetic Matter

Let us now consider magnetic matter, which will yield further insights into modern condensed matter physics. The simplest example of an interesting magnetic material is a ferromagnet, something that can point north and enable us to navigate. Ferromagnetism, of course, ultimately rests on the fact that electrons have spin. An electronic spin is governed by quantal laws, so quantum mechanics is surely relevant. However, we must enquire further.

Historically, the first microscopic model of magnetism that was addressed theoretically was the so-called Ising model: here the individual spins are supposed just to point 'up' or 'down'—the simplest possibility. More concretely, the spins are pictured as sitting at the sites of a space lattice so that at site i, a classical spin variable, s_i, takes on only the two values $s_i = \pm 1$.

Now the fundamental question about the model was "At a given temperature, will all the spins tend to point the same way?" In other words, "Is there any ferromagnetism?" At high enough temperatures the spins will surely be disordered (yielding a paramagnet). But suppose one couples nearest-neighbor spins on the lattice together with an interaction energy $-Js_is_j$ $(J > 0)$, so there is a tendency to align;* what then happens? This was analyzed sixty years ago by E. Ising in his thesis. In an exact calculation he found there was *no ferromagnetism* for a linear chain of spins, i.e., for a one-dimensional system. There was *no* transition from paramagnetism. One could *not* describe real ferromagnets by such a model!

Now one can argue (and, as far as I understand the history, I believe it was argued): "Well, this Ising model is so crude, it does such injustice to the true quantum-mechanical state of affairs, that it is hardly surprising that it fails to describe ferromagnetism. One should improve matters by using proper quantum-mechanical spin variables, namely, Heisenberg spins. These can point in any direction; they are vectors, $\mathbf{S} = \frac{1}{2}\hbar(\sigma_x, \sigma_y, \sigma_y)$; and, furthermore, they are operators, σ_x, σ_y and σ_z obeying the Pauli commutation relations."

Accordingly, let us examine the Heisenberg model of ferromagnetism with couplings $-J\mathbf{S}_{ij}\cdot\mathbf{S}_j$. It turns out, however, that the corresponding chain of Heisenberg spins (again a one-dimensional system) also has no phase transition! It never becomes ferromagnetic which $T > 0$. So quantum mechanics was no help on this problem. Indeed, it turns out to have been an actual hindrance.

We know now that it is the *one-dimensionality* of the spin chain, *not* classical mechanics, that was the reason why Ising found no phase transition in his model. Neither for Ising spins nor for Heisenberg spins does one have spontaneous magnetization (or spontaneous "order") in one dimension.[7] But for Heisenberg spins matters are worse: there is even no ferromagnetism in two dimensions.[7] On the other hand for Ising spins—not withstanding that they represent crude approximations—there is ferromagnetism in two dimensions; and this accords with the observation of ferromagnetism

---

* That is, the pair configurations $s_i = s_j = 1$ and $s_i = s_j = -1$ have energy $-J$, while the configurations in which $s_i = -s_j$ have the higher energy $+J$.

even in films of a few monolayers thickness and in other isolated
magnetic layers.

Finally, in three dimensions there is ferromagnetism for both Heisenberg and Ising spins. So, again, it is not quantum mechanics that
determines whether spontaneous magnetization occurs. Well then,
what does? The answer lies in the interplay of entropy, fluctuations,
and symmetry. What was "wrong" with the Heisenberg model was
its symmetry. Not all real magnets have the full rotational spin symmetry of the Heisenberg model. That symmetry is broken, at the
microscopic level, by further interactions, some of which turn on
the subtleties of quantum mechanics. Ising spins embody only a
discrete, 'up-down' reflection symmetry: but for condensed matter
physics, that is not the prime issue.

The Importance of Dimensionality

The real issue underlying the existence of ferromagnetic order was
pointed out by R.E. Peierls. Let me remind you of the arguments
that Peierls and L.D. Landau developed which give us an understanding of what matters as regards such a phase transition. Consider
Fig. 3(a): at high temperatures, any magnetic material has a smooth
magnetization curve, M versus H, which does nothing very inter-

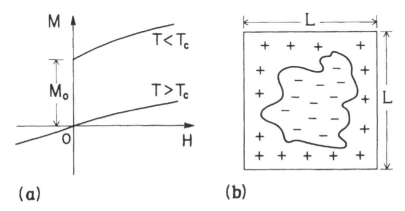

(a)   (b)

Figure 3. (a) Plot of magnetization, M, vs. magnetic field, H, at fixed temperature, T, for a ferromagnetic material, illustrating the existence of a spontaneous magnetization, M_0, below the Curie point, T_c. (b) Example of a
counter-domain or droplet of overturned spins (of linear dimensions L) in
a ferromagnetic domain of 'up' or + spins.

esting; but at low enough temperatures—specifically below the ferromagnetic Curie temperature or critical point, T_c—a jump, measured by the spontaneous magnetization, $M_0(T)$, appears at zero field. What allows or prevents such a spontaneous magnetization?

The way to understanding is found by considering the thermodynamic stability of a counter-domain of, say, 'down' or '$-$' spins embedded in a ferromagnetic domain of 'up' or '$+$' spins, as illustrated in Fig. 3(b). The spins around the outer boundary are held fixed, pointing 'up', and we ask whether overturning a cluster or droplet of spins in the middle yields a state of lower or higher free energy; in other words, is the ferromagnetic 'up' state stable or not? The change in free energy for a counter-domain with linear dimensions of order L may be estimated, neglecting all constants of proportionality, as follows:

$$\Delta F_L = \Delta E - T\Delta S,$$

(i) $\sim M_0 H L^d,$ from the bulk change of
                                       magnetic energy,

(ii) $+ JL^{d-1},$ the surface/interface energy
                                       of the droplet,

(iii) $- k_B T L^{d-1},$ the corresponding surface
                                       entropy,

(iv) $- k_B T \ln L^d,$ the positional entropy of the
                                       droplet center,   (13)

where d is the dimensionality, J the strength of the ferromagnetic couplings, M_0 the magnetization and H the external field. In zero field we can ignore the first term (i). What Peierls showed then matters is the free energy associated with the droplet surface or interface. There is an energetic term (ii) and also an entropic term (iii). In leading approximation both are proportional to the size (or 'area') of the droplet surface. Finally, there is also a positional entropy, (iv), since the counter-domain can occupy many distinct spatial locations.

Now, if the ferromagnetic state is to be stable, the free energy change, ΔF_L must be positive for all L however large: the presence of a large counter-domain (which could reduce the mean magnetization to a small value) must not lower the free energy. By comparing terms, we see that one first needs $T < J/k_B$, so that the temperature must be sufficiently low; secondly, one must have $L^{d-1} > \ln L^d$, which implies $d > 1$. The spatial dimensionality, d, is thus crucial. If $d = 1$ the last term, (iv), dominates and the spontaneous magnetization is destroyed; for $d > 1$ an Ising-like system will display magnetic order for $T < T_c$ (with $T_c > 0$).

To make these arguments rigorous one must allow for "bubbles" within the droplets and for bubbles within the bubbles and so on. The full analysis was first accomplished by R.B. Griffiths[8] but the physical basis of his rigorous proof is just the principle of control by the interface identified by Peierls.

If one looks similarly at a Heisenberg system, it turns out[7(c)] that, because of the continuous symmetry, there is no sharp interface bounding an overturned droplet. Rather, there is a diffuse transition region—an infinitely thick Bloch wall: that reduces the corresponding free energy increment by a factor of L. Consequently, one then needs $L^{d-2} > \ln L^d$ which pushes up the borderline dimensionality by 1, so that $d > 2$ is required for spontaneous magnetization. Evidently the only feature of the quantum mechanics of Heisenberg spins used is the existence of the full rotational symmetry; but that enters equally in the classical limit ($\hbar \to 0$ or $S \to \infty$).

Criticality

Now the actual nature of the transition to spontaneous order can be studied in much greater detail. The two-dimensional Ising model was solved in a wonderful manner by Lars Onsager, now over forty years ago.[9] He discovered that the specific heat should have a divergent singularity of characteristic logarithmic form, namely,

$$C(T) = A \ln |t| + B + \cdots , \qquad (14)$$

where $t = (T - T_c)/T_c$ measures the deviation from the critical point T_c. Figure 4, from Onsager's original paper,[9] shows the exact form of $C(T)$ for the square lattice compared with the best that approximate theories could then do. One sees that the approximations are not only quantitatively wrong; they are *qualitatively wrong* as well, the functional form being simply incorrect—a somewhat shocking outcome!

Now experimentalists these days can perform most precise and

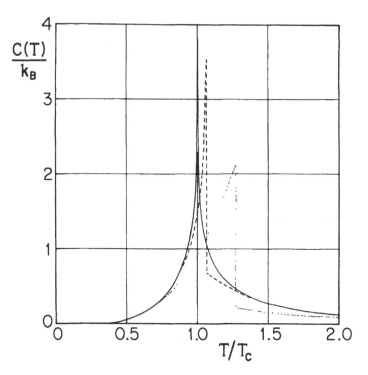

Figure 4. The exact specific heat of the square lattice Ising model (solid curve) and two approximate calculations (dashed and dotted plots). [After L. Onsager, Ref. 9.]

delicate thermodynamic measurements. Figure 5 shows the corresponding curve for a real two-dimensional magnetic material.[10] Actually, it is a layered antiferromagnetic crystal, Rb_2CoF_4 (but Onsager's results apply equally to the square lattice antiferromagnet, with $J < 0$). The real electronic spins act like Ising spins here because there are anisotropic interactions that keep them aligned predominantly parallel (or antiparallel) to a crystalline axis. The interactions between layers are very weak (for special structural reasons) and matter only for T extremely close to T_c. The experimental results in Fig. 5 can be superimposed upon the Onsager solution of Fig. 4 with essentially no adjustments! So again it is the two dimensions and Ising-like symmetry that matters; quantum mechanics does not.

Now there is yet another great discovery by Onsager for two-dimensional Ising models. This concerns the spontaneous magnetization, $M_0(T)$, and, by analogy, the liquid-gas coexistence curve, $\Delta\rho(T)$, for fluids: these both vanish when $T \to T_c-$ as

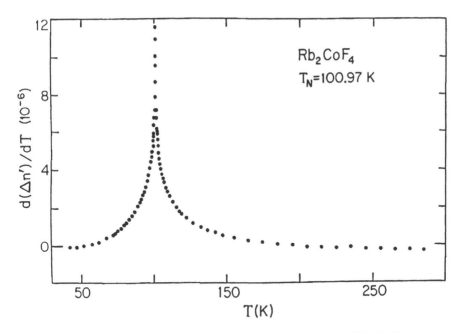

Figure 5. The specific heat of the layered antiferromagnet Rb_2CoF_4 (as measured via the linear birefringence). [After P. Nordblad, et al., Ref. 10.]

$$M_0(T), \quad \Delta\rho(T) \approx B \, |t|^\beta, \qquad t = (T - T_c)/T_c, \qquad (15)$$

where the exponent takes the universal value $\beta = \frac{1}{8}$ for $d = 2$ dimensions. For a long time we did not have any well controlled two-dimensional lattice gases on which to experiment! But recently it has become possible to work with submonolayers of simple gases adsorbed on almost perfect crystalline substrates. Figure 6 shows the data of Kim and Chan[11] for the gas-liquid coexistence curve of methane adsorbed on graphite. One sees a very steep, bullet-nosed curve, implying a rather small value of the exponent β. Now, methane is a fascinating chemical compound. As explained by R.N. Zare, earlier in this Symposium, quantum mechanics has quite a job describing its detailed molecular properties. But we do not need all that here. Only the uniform two-dimensional character of the substrate and the facts that the methane molecules sit adsorbed on the graphite, and feel mutual, short-range attractions are essential. Indeed, the experimental data yield an exponent $\beta = 0.127 \pm 0.020$; despite the relatively large uncertainty—which is not surprising given the difficulties of the experiment—the agreement with Onsager's prediction $\beta = 0.125$ is remarkable.

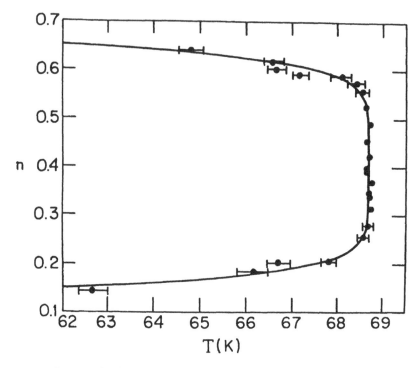

Figure 6. The gas-liquid coexistence curve for methane, CH_4, adsorbed on graphite at submonolayer coverages, n, vs. temperature, T. [After H.K. Kim and M.H.W. Chan, Ref. 11.]

Another interesting question which I would have liked to have gone into is "Why are these two-dimensional critical exponents all rational numbers?". There have been wonderful developments in the last few years coming out of conformal field theory and string theory.[12,13] Perhaps, they are best characterized as mathematical physics rather than condensed matter physics, but, in a nutshell, conformal covariance in two dimensions tells us why two-dimensional critical exponents—not just the few I have mentioned but many others—will, in fact, be rational numbers.[13,14] As an extra bonus, the analysis can also generate detailed information about the correlation functions precisely at two-dimensional critical points.[12,14]

The Irrelevance of $\hbar$ to Critical Phenomena

Allow me to enter a little more technically into the question of quantum mechanics and critical phenomena. One can show that Planck's constant, $\hbar$, is an "irrelevant parameter" for criticality (at nonzero temperatures)! In saying this one refers to calculations performed within a renormalization group framework and the associated, now standard, terminology. One starts by considering the behavior of, say, a spin system for very large values of the spin quantum number, S. Now when $S \to \infty$ the quantal spin variables simply become classical vectors. Then one can ask, "Does the actual value of S matter to the critical behavior? Is finite S different from $S = \infty$?"

Consider also the transition to superfluidity at the lambda point in liquid helium. For describing this, one is inclined, at first, to say that surely quantum mechanics is essential. However, it transpires that for this normal-to-superfluid transition there is really only one feature of quantum mechanics that is essential, namely, that to describe superfluid order one needs, in essence, a wave function. Now quantum mechanics tells us something important about wave functions which we tend to hide from beginning students: specifically, a wave function is a complex number that has both a magnitude *and* a phase. Furthermore, the actual value of the phase cannot be observed owing to the so-called gauge symmetry. Gauge symmetry is an $O(2)$ symmetry corresponding simply to rotations in a plane, say, the complex plane or an (x,y) Cartesian plane.

But the same $O(2)$ or XY symmetry of the basic "order parameter" applies equally to many magnetic systems: their spin vectors are free to rotate into any direction in some crystalline plane; and the analogous result applies to other sorts of materials as well. This XY symmetry, together with the appropriate spatial dimensionality, serves to determine the nature of the critical behavior. For example, the superfluid density, $\rho_s(T)$ of liquid helium, vanishes according to

$$\rho_s(T) \approx D \, | \, t \, |^{\zeta}, \qquad (16)$$

when T approaches the lambda point, T_c, with a characteristic exponent ζ. The same happens for the magnetic analog, the so-called helicity modulus, which measures what happens when one twists the alignment of the spins at the two ends of a ferromagnetic system. For the Bose-Einstein condensation of an ideal Bose gas one has $\zeta_0 = 1$. However, the exponent for real helium, an interacting Bose gas, takes a value $\zeta \approx \frac{2}{3}$; the same goes for XY magnets. Again the value of ζ is the same whether the system is classical or quantal.

It may be instructive for some to see how this insensitivity to quantum mechanics comes about mathematically within a now traditional perturbation-theoretic formulation. For a classical system, say an $S = \infty$ magnet, there will be a 'bare' or unperturbed Green's function behaving like

$$G_0(k,T) \propto \frac{1}{k^2 + t_0}. \tag{17}$$

The square of the momentum or wave vector, k, derives from short-range spin-spin couplings re-expressed in Fourier space, while t_0 depends on T and becomes small as the transition is approached. For a quantum mechanical system matters are, of course, more complicated: e.g., for bosons of mass m the unperturbed Green's function becomes

$$G_0(k,\bar{\omega}_n,T) \propto \frac{1}{k^2 + i\bar{\omega}_n + t_0} \tag{18}$$

where the (rescaled) Matsubara frequencies are given by

$$\bar{\omega}_n = 4\pi(mk_BT/\hbar^2)n \qquad \text{with} \qquad n = 0, 1, 2, \cdots .$$

The effect of quantum mechanics is, thus, to introduce the infinite set of frequencies $\bar{\omega}_n$. But if the temperature is *nonzero* these are discrete nonvanishing frequencies for $n \geq 1$. Thus as t_0 becomes small near the transition only the $n = 0$ component of G_0 can become large when $k^2 \to 0$. All the other components remain finite and harmless. Hence the two expressions (17) and (18) become virtually identical. The terms with $n = 1, 2, \cdots$ serve only to fix details like the precise value of T_c, the magnitude of certain amplitudes, such as D in (16), and the size of the asymptotic corrections to a pure power law as criticality is approached.

Helium in Vycor—Crossover to Ideal-Bose-like Behavior

My colleague at Cornell, John Reppy, has in recent years been measuring very precisely the variation of the superfluid density of helium adsorbed in a porous glass—a sponge-like material called "Vycor." The point of studying helium in Vycor is that by controlling the overall density one can effectively modulate the interactions between helium atoms. Thus the superfluid transition temperature can be reduced far below the usual bulk lambda point, $T_c = 2.17°K$. Figure 7 shows the data for $\rho_s(T)$ obtained by Reppy and collaborators[15] for fillings of the Vycor which yield transition tem-

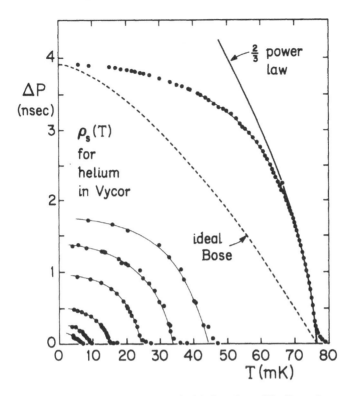

Figure 7. Measurements of the superfluid density of helium-four adsorbed in Vycor glass at various overall densities showing the crossover from a critical behavior characterized by an exponent $\zeta \approx \frac{2}{3}$, to ideal-Bose-like behavior (see dashed curve) as the critical temperatures, T_c, fall below 25 mK. [After Crooker et al., Ref. 15.]

peratures of 80 mK or less. The figure also illustrates a $\zeta \approx \frac{2}{3}$ power law of ρ_s, characteristic of bulk, fully interacting helium, and the ideal-Bose behavior which corresponds to an exponent $\zeta_0 = 1$, i.e., a straight line near the transition. Now, by varying the amount of helium in the Vycor one can control the critical temperature, T_c. For the higher T_c values, one finds good agreement with the bulk $\frac{2}{3}$ law (except very near T_c where imperfections, probably miscellaneous spatial inhomogeneities near the boundary etc., result in a small tail). But notice that by the time one pushes T_c down to 20 mK or less, the plots change shape and approach the transition in rather linear fashion. In fact one seems to be observing a tuning of the system away from bulk helium criticality over towards ideal-Bose-gas behavior; lowering T_c is apparently serving to switch off the interactions.

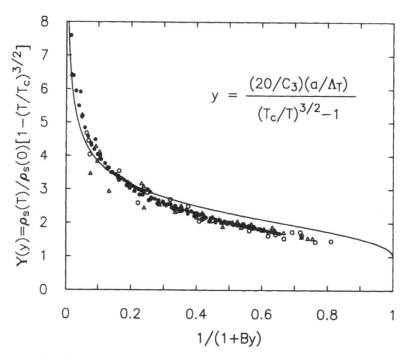

Figure 8. Scaling plot of the data of Crooker et al., (Ref. 15) for the superfluid density of helium adsorbed in Vycor glass, including eleven values of T_c: from 6 to 12 mK (open circles); from 15 to 35 mK (triangles); and from 37 to 77 mK (solid circles). In the definition of the scaling variables (see also the text) $C_3 \simeq 4.2$ is a numerical constant and $\Lambda_T = (h^2/2\pi m^* k_B T_c)^{1/2}$. [After Rasolt, et al., Ref. 16(a).]

Now it turns out that scaling and renormalization group theory serve to give an understanding of this crossover in critical behavior. The data of Fig. 7 (and more) can be rearranged and thence reduced to a single experimental curve—see the dots, circles and triangles in Fig. 8. Furthermore, this plot can be well described by a theoretically calculated curve (solid line) which actually answers an old theoretical question, namely, "How does ideal-Bose criticality change when interactions between the bosons are switched on?" To be more concrete, we regard T_c for a particular filling of the Vycor as the control parameter. Then theory predicts[16] that the ratio

$$Z = \frac{\rho_s(T;T_c)}{\rho_s^{ideal}(T/T_c)} \tag{19}$$

The figure axes: y-axis $Y(y) = \rho_s(T)/\rho_s(0)[1-(T/T_c)^{3/2}]$, x-axis $1/(1+By)$, with equation $y = \dfrac{(20/C_3)(a/\Lambda_T)}{(T_c/T)^{3/2}-1}$.

should, as $T_c \to 0$, be a function only of a scaled variable y given, apart from unimportant details, by

$$y^2 \propto T_c / |\ t\ |^{\phi_T} \qquad \text{with} \qquad t = (T - T_c)/T_c. \qquad (20)$$

The so-called crossover exponent, ϕ_T, is found generally to be[16]

$$\phi_T = 2(4 - d)/(d - 2)^2, \qquad (2 \le d \le 4) \qquad (21)$$

so that for the real three-dimensional system one has $\phi_T = 2$. Figure 8 represents, in somewhat different form, a plot of the scaling function $Z(y)$ and a check that the data do, indeed, obey scaling with $\phi_T = 2$. There is only a single fitting parameter entering the figure: this is $\mathcal{R} = (a^*/a)(m^*/m)^{1/2} \simeq 1.8$ in which a and m refer to the scattering length and mass of isolated helium atoms while the stars denote the effective values resulting from the motion of the atoms in the porous Vycor glass. The calculation of the scaling function, $Z(y)$, is an exercise in purely classical statistical mechanics (and has been carried out to leading order in an $\varepsilon = 4 - d$ expansion[16]): its form depends only on the dimensionality and on the XY symmetry. Quantum mechanics enters in the scale factor required in Eq. (20) to make y dimensionless. This is also where the ratios m^*/m and a^*/a are needed: their calculation certainly represents a quantum-mechanical problem, indeed a rather hard one! But the outcome should serve only to refine the rather reasonable fitted value, 1.8, of the product $\mathcal{R}$.*

Planck's Constant is Worth Only One Extra Dimension

Now consider a situation in which it is established that Planck's constant $\hbar$ enters from the outset: I have already emphasized that, in modern condensed matter physics, the dimensionality, d, is important to both theorists and to experimentalists. Experiments on two- and three-dimensional systems have been described above; and one-dimensional systems are likewise realizable in the laboratory. Indeed, one is almost not a condensed matter physicist these days

---

* This necessarily brief, summary of the experiments and theory overlooks the random character of Vycor and does not address the *onset* of super-fluidity at a finite filling value *at* $T = 0$. The latter problem certainly requires quantum mechanics; but see the next section.

if the dimensionality is not one of your parameters, and, often, one of the most important. My thesis, however, is that in order to calculate the partition function, and other system properties, the cost of including Planck's constant amounts to no more than allowing for an extra spatial dimension! In some ways this is a well known, almost trivial point; but it is still worth emphasizing and I include it for those who have not seen it before. So here is one way of demonstrating the fact.

Suppose the Hamiltonian operator for a system in d dimensions can be written $\hat{\mathcal{H}} = \hat{\mathcal{H}}_0 + \hat{\mathcal{H}}_1$ where the eigenvalues, $U(\sigma)$, of $\hat{\mathcal{H}}_0$ are known, i.e., one can write

$$\hat{\mathcal{H}}_0 \,|\, \sigma\rangle = U(\sigma) \,|\, \sigma\rangle, \tag{22}$$

where the eigenfunctions $|\,\sigma\rangle$ are labelled by the set of quantum numbers, σ. Quantum numbers are just classical variables so that evaluation of the partition function, $Z(T)$, for $\hat{\mathcal{H}}_0$ alone, reduces to a problem in ordinary, classical, d-dimensional statistical physics. But the full partition function is given by

$$Z(T) = \text{Tr}\{e^{-\beta(\hat{\mathcal{H}}_0 + \hat{\mathcal{H}}_1)}\}, \qquad (\beta = 1/k_B T) \tag{23}$$

where, in general, $\hat{\mathcal{H}}_0$ and $\hat{\mathcal{H}}_1$ *do not commute*: that is, of course, precisely where $\hbar$ enters in a crucial way! One can calculate the matrix elements

$$\langle\sigma\,|\,\hat{\mathcal{H}}_1\,|\,\sigma'\rangle \equiv V(\sigma,\sigma'); \tag{24}$$

but how can one deal with the noncommutativity in the partition function?

The answer is to say the magic word "Trotterize"! This means that one uses Trotter's formula for the exponential of two non-commuting matrices (or more general operators) which can be written

$$e^{A+B} \approx \underbrace{e^{A/n}\, e^{B/n}\, e^{A/n}\, e^{B/n} \,\ldots\, e^{B/n}}_{2n \text{ factors}}, \tag{25}$$

as $n \to \infty$. This is a generalization of the familiar definition of the standard exponential function via $e^x = \lim_{n\to\infty}[1 + (x/n)]^n$. Substituting this formulae in Eq. (23) gives rise to

$$Z(T) = \sum_{\{\sigma_0\}} \sum_{\{\sigma_1\}} \cdots \sum_{\{\sigma_n\}} e^{-(\beta/n)[U(\sigma_0) + V(\sigma_0,\sigma_1) + U(\sigma_1) + V(\sigma_1,\sigma_2) - U(\sigma_2) + \cdots]}$$

$$\tag{26}$$

At first inspection, this is a miserable-looking expression for the partition function, perhaps worse than what we started with; but when all is said and done, one finds that what has arisen corresponds merely to a classical system in one more dimension. In other words, d has become $d + 1$!

To understand this in a little more detail consider Fig. 9. Each vertical layer represents a d-dimensional classical system resembling the original one with $\hat{\mathcal{H}}_0$ only but at a temperature nT; the kth layer introduces a factor $\exp[-\beta U(\sigma_k)/n]$. Between adjacent layers, however, there appear interactions in the extra direction, indicated by the dashed lines which represent the factors $\exp[-\beta V(\sigma_k, \sigma_{k+1})]$ in Eq. (26). Note that all the quantities now appearing are ordinary, classical, commuting variables. Of course, the extra dimension and further couplings represent a complication. But who is scared of anisotropic systems? One often deals with them. Incidentally, when the analogous steps are taken in quantum field theory, the extra dimension represents time and then Lorentz invariance comes in to help. The time-like dimension finishes up looking just like the other dimensions so full spatial isotropy is actually restored! Quantum field theory, from this point of view, can be regarded as classical statistical mechanics, in four-dimensional space! These connections

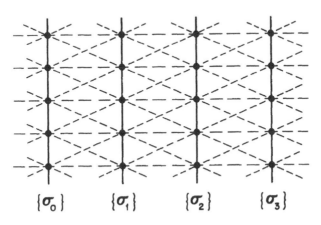

$\{\sigma_0\}$ $\qquad$ $\{\sigma_1\}$ $\qquad$ $\{\sigma_2\}$ $\qquad$ $\{\sigma_3\}$

Figure 9. Schematic representation of the reduction of a quantal statistical mechanical problem in d dimensions which are represented vertically, to a classical statistical mechanical problem in one extra dimension represented horizontally. The sets of quantum numbers $\{\sigma_k\}$ for $\hat{\mathcal{H}}_0$ in the kth vertical layer can be regarded as spin-like variables coupled, between layers, via the noncommuting part, $\hat{\mathcal{H}}_1$, of the original Hamiltonian, $\hat{\mathcal{H}} = \hat{\mathcal{H}}_0 + \hat{\mathcal{H}}_1$.

are now quite well explored and have proved mutually fruitful. Nevertheless, if one wants to deal with quantum-mechanical systems more generally—as one certainly does in condensed matter physics—one is basically faced with a problem of classical mechanics in one more dimension. As we have seen, varying the number of dimensions is of great interest, in any case, so quantum mechanics does not present an intrinsically new challenge.*

Disordered Matter—Ferromagnets in Random Fields

Now I would like to convey the flavor of some recent advances in condensed matter physics. As a first example, I will describe some theoretical progress tied to some challenging experiments. One form of matter which is always there to plague experimentalists is matter that is a bit "mucky" or "dirty": it is disordered by, for example, impurities that are randomly frozen in. Condensed matter physicists (with the aid of chemists and expert crystal growers) have recently gained the ability to prepare materials with controlled disorder and to perform careful experiments on them. There are many different, fascinating aspects to this; I shall focus on just one, which I hope will be found both entertaining and instructive.

Take a ferromagnet—for instance an Ising-like magnet—and imagine that there is disorder in the form of random local magnetic fields. To be specific suppose the total field at site i is

$$H_i = \overline{H} + h_i, \tag{27}$$

where the random piece, h_i, varies from site to site with a zero mean, $\overline{h_i} = 0$, but has a nonzero variance $\overline{h_i^2} \neq 0$: the simplest example is $h_i = \pm h_R$. Now the central question is, do the random fields, h_i, destroy the ferromagnetism? The current theoretical view is that in high enough dimensionalities, typically $d \geq 4$, nothing drastic will happen; the ferromagnetic order will remain intact. So the question becomes one of determining that special dimensionality above which the original phase transition in the pure system is not spoiled by the disorder. This dimensionality is usually called d_{lc}, the *lower critical*

---

* It should, perhaps, be added that dynamics presents a distinct challenge both classically and quantum-mechanically: the former is now fairly well understood; the latter still presents profound technical and conceptual questions.

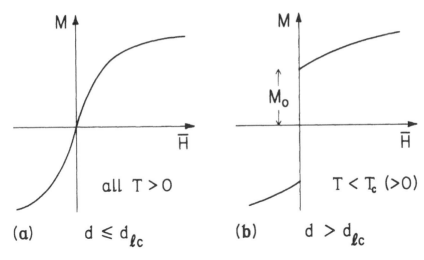

Figure 10. Illustrations of the effects to be expected when random local magnetic fields (of zero mean) are imposed on a ferromagnet: (a) at and below the lower critical dimension, d_{lc}, the spontaneous magnetization for $T > 0$ is destroyed by the randomness; (b) above d_{lc} the ferromagnetic order survives below some nonzero critical point T_c.

dimension. Above d_{lc}, as illustrated in Fig. 10(b), the system exhibits a first-order magnetic transition; but *below* d_{lc} something new can happen: as indicated in Fig. 10(a), one expects the original transition to be destroyed in some way by the randomness when $d \leq d_{lc}$, however low the temperature.

Now, theorists came to hold contrasting views as to the correct answer to this question; one can read the different opinions in the literature. Some said d_{lc} should be 2; others said 3. Both of these conclusions are experimentally significant: two-dimensional magnets have already been mentioned, and one can certainly study three-dimensional examples: as indicated, the randomness is controlled by careful addition of impurities. The first theoretical argument was presented by Imry and Ma[17] in 1975: as we will see, that led to $d_{lc} = 2$. Later others supported this conclusion. But in the meanwhile a number of reputable theorists started to argue for $d_{lc} = 3$. And, indeed—although it is, perhaps, a little unkind of me to record it—some individuals backed both horses! Table I shows the situation reached by 1983–84. It lists the opinions expressed by theorists from various institutions around the world as time progressed. Evidently there was no consensus. Indeed, matters came

Table I. Some theoretical opinions as to the value of the lower critical dimension, d_{lc}, for Ising-like ferromagnets in random fields. Apart from Imry and Ma, only the institutions* of the various authors are recorded.

$d_{lc} = 2$		$d_{lc} = 3$	
1975	(Imry and Ma) UCSD	1976	Tel-Aviv and UCSD
1982	UCSD and IBM	1979	Rome and Paris
1982	ILL, Grenoble	1981	IBM and Weizmann Inst.
1983	Tel-Aviv and IBM	1981	KFA Jülich, Tel-Aviv and IBM
1983	IBM	1981	Edinburgh
		1982	MIT and Helsinki
		1983	UCSB
1983	Geilo vote 4		17

* Some of the abbreviations are: UCSD and UCSB: University of California at San Diego and at Santa Barbara; IBM: T.J. Watson Laboratory, New York.

to such a pass that at the 1983 Spring School in Geilo, Norway, the issue was put to a vote! As I recall, not everybody cared to express an opinion; nevertheless, 17 voted for $d_{lc} = 3$ while only 4 (or 5) voted for $d_{lc} = 2$. So $d_{lc} = 3$ won!

Well, let us look into the issues involved more closely and learn how this story turned out. How can one determine d_{lc}? The first conclusion, $d_{lc} = 2$, actually follows by extending the Peierls' argument outlined above for a pure system. The appropriate picture resembles that in Fig. 3(b) except that now the interface of the counter-domain may, owing to the quenched or frozen-in random fields, want to *wander* in order to incorporate more sites on which the local field already points 'down'. This effect can be represented by adding a new term to the estimate (13) for the total free energy increment, ΔF_L, of the counter-domain. Imry and Ma argued[17] that the typical net total random field in a domain of volume V, should, by the central limit theorem, scale like $\sqrt{V}$. This suggests that the term needed should be

$$\delta F_{\text{random}} \sim -M_0(\overline{h_i^2}L^d)^{1/2}. \tag{28}$$

Minimizing ΔF_L with this term included leads immediately to the conclusion that the order is stable if, for large L, $L^{d-1} > L^{d/2}$; in other words $d_{lc} = 2$. However, the argument, and its later developments, was clearly not absolutely convincing—recall the bubbles-within-bubbles problems with the original Peierls argument.

Now, as mentioned, there are real materials in which one can examine both two- and three-dimensional random magnets. It is not possible to produce directly a local magnetic field that points randomly up or down; but Aharony and Fishman[18] showed that one could, instead, study impure *anti*ferromagnets in *uniform* magnetic fields. Theoretically, that becomes essentially the same problem. So what had seemed a purely abstract model—a theorists' plaything—turned out to apply to some real materials, namely, diluted antiferromagnetic crystals in uniform fields. Some of the best real examples of such systems are[19] $Rb_2Mn_{1-c}Mg_cF_4$ for $d = 2$ dimensions and $Fe_{1-c}Zn_cF_2$ for $d = 3$ dimensions.

Now the experimentalists were having trouble isolating a clear transition to antiferromagnetism in the three-dimensional systems; that fact seemed to favor $d_{lc} = 3$ so the Imry-Ma conclusion was further doubted. And then modern field theory came to our aid! But, as we will see, in this case it was not truly of assistance. There is, now, in field theory the wonderful principle of supersymmetry wherein fermions and bosons are seen to be aspects of the same more basic field. It was further discovered that one could do perturbation theory for a random-field system, starting at $d = 6$ dimensions, writing $d = 6 - \varepsilon$, and treating $\varepsilon = 6 - d$ as an expansion parameter. The formal perturbation expansion for the random systems about six dimensions turned out to be exactly the same as around four dimensions for the corresponding pure systems—and field theory explained why by revealing a supersymmetry connection between the models. Thus the random system in $d = 6 - \varepsilon$ dimensions looked like the pure system in $d = 4 - \varepsilon$ dimensions. Well, we knew what happens in the pure case: for an Ising system one has $d_{lc} = 1$—that is just Peierls' argument again! The change $\Delta d = 2$ from $d = 4 - \varepsilon$ to $d = 6 - \varepsilon$ thus gives $d_{lc} = 1 + 2 = 3$ for the random system. In other words, Ising-like order should survive random fields only for $d > 3$. Many people examined this argument, and it seemed to hold up well: see Table I.

Here was a serious impasse! It was one of those situations where experiment really needed guidance from theory. Sometimes experiment serves as a true guide to theory; but this was not such a case. In fact, it finally took some really hard and careful mathematics—truly "mathematical physics," with all its epsilons and deltas—to settle the matter.[20] Sad to say, physics is not a democratic subject: the Geilo vote was wrong; democracy was dismissed! John Z. Imbrie,[20] at Harvard, succeeded in showing that d_{lc} was strictly less than 3 (in fact, less than $2 + \delta$ for any positive δ). So we can be convinced that $d_{lc} = 2$ after all!*

Now I dislike criticising colleagues—but what tends to happen when somebody really proves something that was not previously obvious, is that all those good theorists who guessed right beforehand, say, "Well, why do we need a rigorous proof? It was really physically obvious!" And all those who did not guess correctly keep discreetly quiet; or, perhaps, they refer grudgingly to Imbrie's work saying, "He has a reasonable argument." What Imbrie actually proved, with full rigor, was that in more than two dimensions there is long-range ferromagnetic order at $T = 0$ even in the presence of small random fields. So we must conclude that Ising-like order certainly survives in $d > 2$ dimensions. This is really a striking result, but I fear that Imbrie's name will be forgotten except, I trust, by the historians of science, because those who were proved right will march on with little reflection while the others will turn to fresh problems. Maybe I am too pessimistic but I do feel that this episode has something to teach us about Science.

Modulated Matter

My next topic concerns an unusual type of phase, or set of phases of matter that are observed in certain alloys and other systems as well. Consider the alloy Ag_3Mg: other like examples are Au_3Zn, Cu_3Pd, Au_3Cu, Al_3Ti, etc. As illustrated in Fig. 11(a), this forms a face-centered lattice structure with the silver ions (open circles) on the cube faces and the magnesium ions (closed circles) on the corner sites. However, this same alloy or ones with closeby composition (say $Ag_{3-x}Mg_{1+x}$ with x small) can form a quite different phase, the so-called $\langle 2 \rangle$ phase, wherein alternate planes of Mg atoms normal to one crystal axis are displaced slightly—actually by half a basic cube diagonal—with respect to one another. As shown in Fig. 11(b), two planes shifted "up" are followed by two planes shifted "down", and so on: the two-fold repeat justifies the terminology $\langle 2 \rangle$. So here

---

* It is still not, at this point in time, clear precisely *how* the supersymmetry field-theoretic arguments fail. However, in another simpler but analogous case it is seen that there is no true renormalization group fixed point on which to base a super-symmetric perturbative expansion. [D.S. Fisher, Phys. Rev. Lett., **56**, 1984 (1986).]

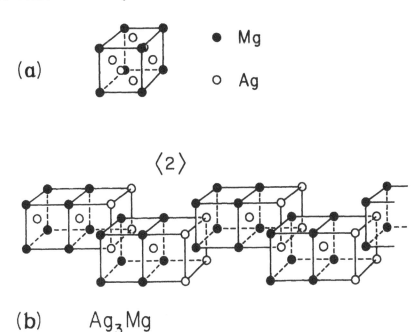

Figure 11. (a) A cubic unit cell illustrating the fcc structure of Ag₃Mg and similar alloys. (b) The "two-up/two-down" or $\langle 2 \rangle$ phase of Ag₃Mg. As in (a) solid circles denote Mg ions, open circles Ag ions; not all the face-centered ions are shown.

is a crystal structure that is seemingly more complicated than it had any need to be; but, as the micrograph in Fig. 12 demonstrates, it is actually found in the laboratory![21] The ordered state of this alloy could have been a perfect fcc structure, and that probably is the ultimate, low temperature ground state; but at moderately high temperatures and appropriate composition range one finds this unexpected $\langle 2 \rangle$ phase—our first, simple example of what we may call *spatially modulated matter* in which there appear extra periodicities, significantly larger than those more-or-less imposed by the atomic or ionic sizes and the gross composition. In this case a wavelength $\Lambda \simeq 16$ Å appears compared to the basic cubic cell edge of approximately 4 Å.

Now where do these larger periodicities come from? And is this simple $\langle 2 \rangle$ phase typical of the possibilities? The answer to this latter question is a resounding "No!" Indeed, in this same silver-magnesium alloy, with very slightly differing composition, one discovers much more complicated patterns: see Fig. 13.[22] The general nature

Figure 12. A photomicrograph of the $\langle 2 \rangle$ phase in Au_3Zn. [After Guymont, Portier and Gratias, Ref. 21.]

of the structures visible in these photomicrographs is illustrated schematically in Fig. 14(a). The patterns consist of two layers of ions displaced "up", then two layers "down", and so on k times but, finally, only *one* layer "up" (or "down"); this whole pattern of $(2k + 1)$ layers then repeats. Figure 14(a) represents the case $k = 4$, while the real examples in Fig. 13 correspond to $k = 3$ and $k = 2$: in short one sees a $\langle 2^3 1 \rangle$ phase with a total repeat distance of $\Lambda \simeq 28$Å and a $\langle 2^2 1 \rangle$ phase with $\Lambda \simeq 40$Å. But this is not all: in Ag_3Mg one also finds the equilibrium phases $\langle 2^4 1 \rangle$, $\langle 2^5 1 \rangle$ and $\langle 2^6 1 \rangle$

---▷

Figure 13. Photomicrographs of Ag_3Mg crystals revealing $\langle 2^3 1 \rangle$ and $\langle 2^2 1 \rangle$ phases (lower and upper parts, respectively). The total repeat distances are $\Lambda \simeq 28$Å and 40Å. [After Kulik, Takeda and de Fontaine, Ref. 22. Reprinted with permission from *Acta Metallurgica* **35**, 1137 (1987), Pergamon Press plc.]

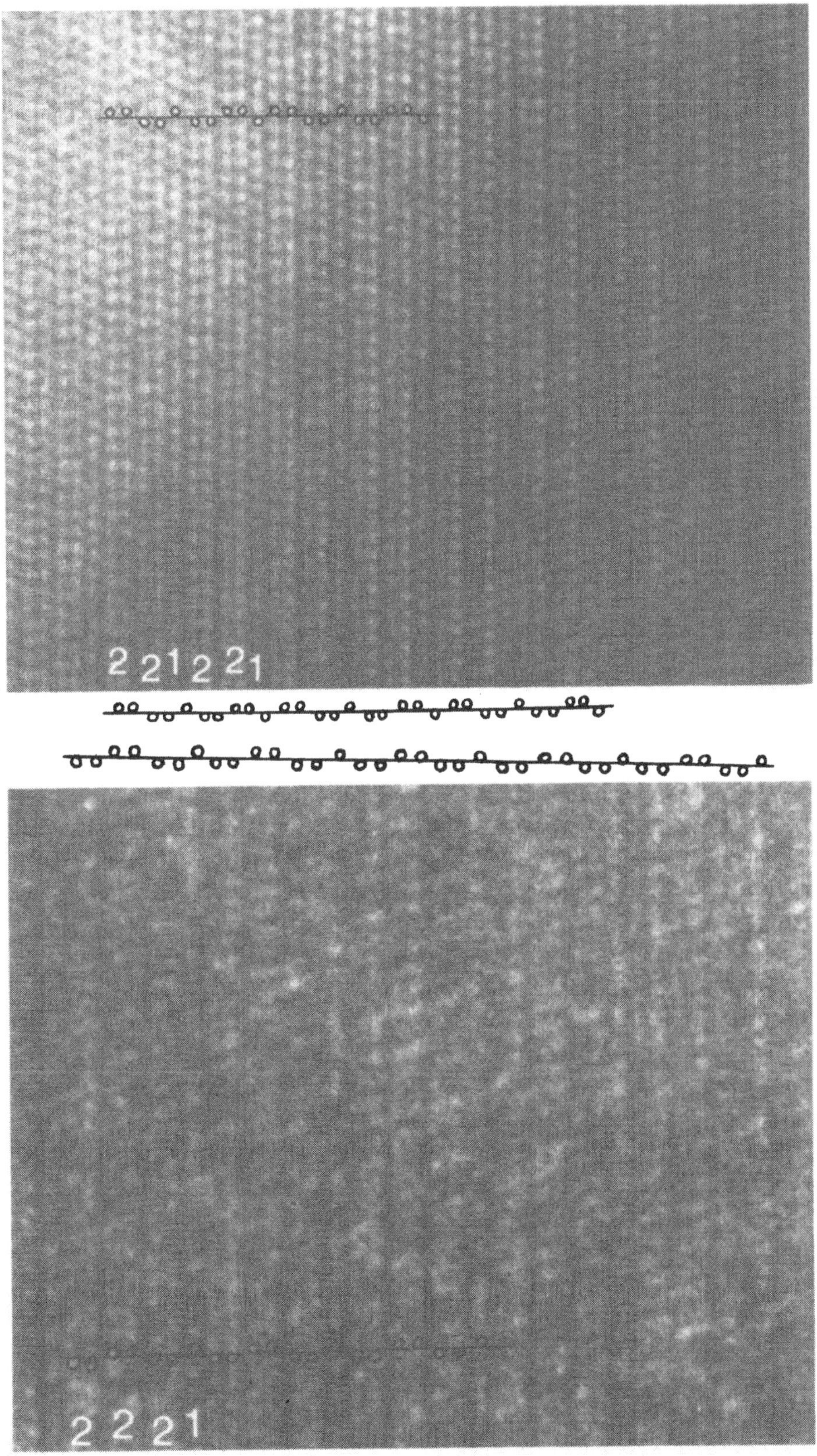
2 2 1 2 2 1
2 2 2 1

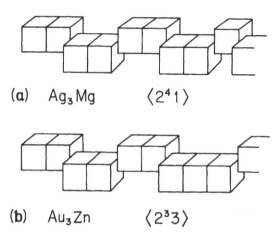

(a) Ag_3Mg $\langle 2^41\rangle$

(b) Au_3Zn $\langle 2^33\rangle$

Figure 14. Schematic depiction of (a) the $\langle 2^41\rangle$ phase, as found in Ag_3Mg, and (b) the $\langle 2^33\rangle$ phase which is observed in Au_3Zn.

while in Au_3Zn one observes, as illustrated schematically in Fig. 14(b), the phase $\langle 2^33\rangle$. There are further variants: for example, one discovers 'mixed' or 'branched' phases[21,22] such as $\langle 2^312^41\rangle$ in which *three* pairs of layers, "two-up/two-down" are followed by a single layer which, in turn, is followed by *four* pairs of layers before the next single layer. The alternate spacing, three-pairs/four-pairs, continues generating an overall period exceeding 120 Å.

How can one understand these large equilibrium periodicities? It is possible that ground-state quantum-mechanical effects associated, for example, with the location of the Fermi surface in the alloys play a role. However, thinking of the ground state alone and attempting to explain only that is *not* adequate; indeed, as mentioned, the ground state might well be just the ordinary fcc lattice. This approach, indeed, illustrates a sinful attitude apt to be taken by 'quantum physicists': it was epitomized earlier in this Symposium when one participant said, no doubt mainly in jest, that he wished he had been the first to say "To understand the hydrogen atom is to understand the whole of physics!" What a revealing and distressing overstatement! Understanding the hydrogen atom, even with its most highly excited states will, alas, not carry one far beyond the ground state of any condensed matter system—and, as alluded to, even leaves one many problems in understanding the quantal states of a simple molecule such as methane, containing only four hydrogen atoms! No, the long-wavelength spatially modulated patterns found at high temperatures, and seen to be sensitive to the temperature,

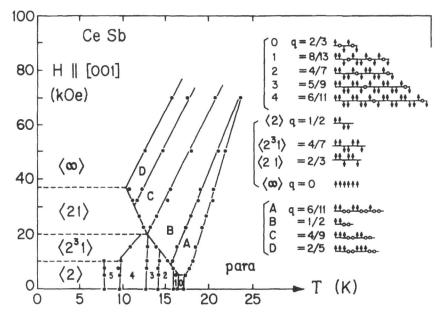

Figure 15. The magnetic phase diagram of cerium antimonide for magnetic field parallel to a [001] axis. The orientation of the spins in successive layers in the various phases is indicated on the right; O denotes a layer of zero average magnetization. (The symbol q denotes the reduced wave number characterising the basic periodicity of a phase.) [After J. Rossat-Mignod et al., Ref. 23.]

depend on entropy and fluctuations—the effects may be subtle but they would have been appreciated by Gibbs, Boltzmann and Maxwell.

In thinking about spatially modulated phases and their interrelations it must be realized that their occurrence is not confined to metallic alloys or represented only by purely structural transformations. Figure 15 shows the phase diagram of the magnetic crystal CeSb, cerium antimonide, in the temperature/magnetic field plane as determined by specific heat measurements made by J. Rossat-Mignod and coworkers.[23] One observes an amazing multiplicity of phases. The nature of the individual phases is revealed by neutron scattering.[23] As indicated in the figure, the majority prove to be long-period, *spatially modulated magnetic phases*. The spins, which are Ising-like in this material, are aligned parallel in ferromagnetic layers but the orientation of successive layers alternate to produce patterns like those found in the alloys. Note, at low T and as a function of H, the sequence of phases $\langle 2^k 1 \rangle$ with $k = 1, 3$ and ∞: it is not unlikely that some higher order phases with $k = 5, 7, \cdots$ have been missed. Similarly, for $H \simeq 0$ when the temperature is reduced below the disordered paramagnetic phase one finds a sequence which may be written $\langle 2^k \bar{3} \rangle$ where $\bar{3}$ denotes a band of three layers, one "up", one "zero" or *unmagnetized* (indicated by $\bigcirc$ in the figure), and one "down", or vice-versa. As T falls, equilibrium phases corresponding to $k = 0, 2, 3, 4$, and ∞ appear.

Now what is going on? Clearly, whatever the underlying quantum mechanical basis, it is different in the alloys than in the spin systems. But the overall similarities of the phenomena suggest that some common principles may be operating. One approach to the issue is to return to the Ising model, more particularly, to a simple three-dimensional Ising model with just one new feature. Take a spin system on a three-dimensional simple cubic lattice with nearest-neighbor couplings of strength J_0 in two of the directions, say, x and z, see Fig. 16, and, J_1 in the third direction, y. If J_0 and J_1 are positive this is just a three-dimensional Ising model, which can have a standard ferromagnetic transition but nothing more. But now, in the third direction put in an *extra* coupling of strength J_2 between second neighbors: see Fig. 16. This yields a three-dimensional model with second-neighbor interactions parallel to one axis which has been dubbed the "ANNNI" model, for "axial next-nearest-neighbor Ising model."

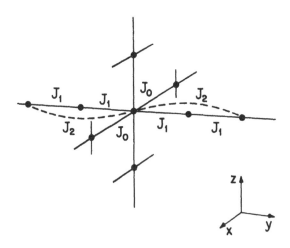

Figure 16. Part of a simple cubic lattice of Ising spins showing the couplings of strength J_0, J_1 and J_2 defining the ANNNI or axial next-nearest-neighbor Ising model.

To explore some of the physics underlying long-period phases one should take the second-neighbor coupling J_2 to be negative or *anti*ferromagnetic so that it *competes* with the nearest-neighbor couplings J_1. Then spins along the y direction get confused: J_1 tends to align them but J_2 tries to make them point antiparallel! If one writes $J_2 = -\kappa J_1$, then the parameter κ measures the degree of competition. It is easy to see that the ground state remains totally ferromagnetic for $\kappa < \frac{1}{2}$; but for $\kappa > \frac{1}{2}$ a $\langle 2 \rangle$-phase configuration provides the lowest energy. Nothing more appears at $T = 0$. However, at nonzero temperatures, where fluctuations and entropy come into play, matters change dramatically. In the first place, at $\kappa = \frac{1}{2}$ the simple modulated phase $\langle 3 \rangle$ consisting of three-up/three-down layers appears; but then as κ increases the competition shifts and yields further modulated phases $\langle 23 \rangle$, $\langle 2^2 3 \rangle$, etc. Indeed, an *infinite* number of new phases arise and all are spatially modulated![24] Figure 17 shows the low-temperature region of the phase diagram in the (κ, T) plane near the so-called *multiphase point* at $\kappa = \frac{1}{2}$, $T = 0$. In addition to the simple periodic phases $\langle 2^k 3 \rangle$ for $k = 0, 1, 2, \cdots$, all the mixed phases $\langle 2^k 3 2^{k+1} 3 \rangle$ also appear.[24] This is not the place to enter further into the details, but it is worth mentioning that the behavior of these, and many other modulated phases can be understood and systematized in terms of effective interactions, mediated largely by statistical fluctuations, between 'domain walls,' 'discommensurations,' 'phase slips,' or 'solitons.'[24,25] In the ANNNI model at low T, the three-layer bands, separated by $2k$ layers of the underlying $\langle 2 \rangle$ phase, constitute the domain walls; pairwise, triplet and higher order effective forces depending on T and κ can be elucidated.[24]

Quasicrystals

To conclude our brief tour of condensed matter physics two very recent and striking discoveries demand attention. Figure 18 records the experimental evidence produced in 1984 by D. Shechtman, I. Blech, D. Gratias and J.W. Cahn:[26] it shows an x-ray scattering pattern of sharp spots which they obtained from a special, quench-cooled aluminum-manganese alloy, roughly of composition Al_6Mn. The crucial feature of this pattern is that it exhibits a ten-fold axis of symmetry. Anyone familiar with x-ray crystallography should become uncomfortable on contemplating this figure! Why? Because

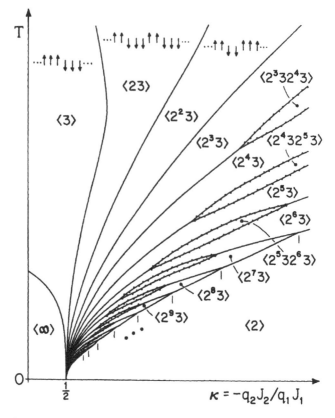

Figure 17. Schematic phase diagram of the ANNNI model in terms of temperature and competition parameter, κ, showing infinitely many modulated phases near the multiphase point $(\kappa, T) = (\frac{1}{2}, 0)$. (The axial coordination numbers, q_1 and q_2, extend the model beyond the simple cubic lattice version described in the text, for which $q_1 = q_2 = 2$.) [After Szpilka and Fisher, Ref. 24.]

a ten-fold or five-fold axis of symmetry is quite incompatible with the translational invariance that is usually taken as defining a crystal. That is a classic result of group theory. This state of matter—whatever it may be—cannot belong to one of the classical crystallographic groups which, it has long been known, exhaust all the possibilities. Now one may and, indeed, should worry lest the ten-fold effect is due merely to the phenomenon of twinning. However, after careful investigation, the answer is unequivocal: the crystals concerned are *not* twins. This new Al_6Mn material was first produced by quenching the liquid alloy by the technique of melt-spinning.[26] The grains of the new material are some microns in diameter. If one

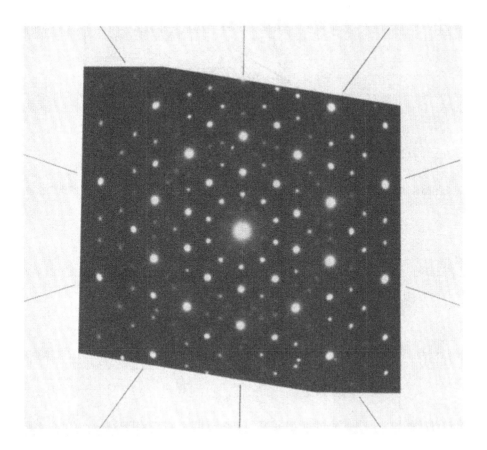

Figure 18. X-ray diffraction pattern from the iscosahedral phase of aluminum manganese. Note ten-fold symmetry axis. [After Schechtman et al., Ref. 26.]

examines microscopically different parts of a grain the same scattering pattern is always seen. The full x-ray scattering displays essentially perfect *icosahedral symmetry*. Figure 19(a) portrays an icosahedron and indicates some of its symmetries: note that it has many "forbidden" five-fold axes. The Al_6Mn icosahedral crystals violate some of the basic, well-established facts of crystallography. Furthermore, the phenomenon is not isolated: precisely similar icosahedral *quasicrystals* have been grown in the alloys $Al_{6+x}M_{1-x}$ with M denoting Cr, Fe, Pd, Pt and Ru.

Now is this a quantum mechanical effect? At one level of description the atoms involved are surely quantum mechanical. But

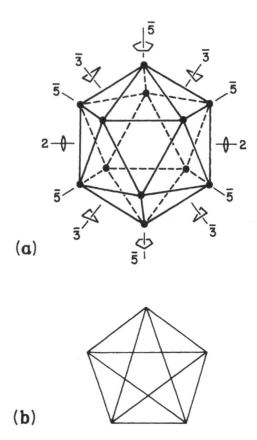

(a)

(b)

Figure 19. (a) An icosahedron and some of its symmetry axes. (b) A pentagram: all line segments are related to one another via powers of the golden ratio, τ.

the way they fit together does not, at first sight, seem to have much to do with that; rather the atomic sizes must matter. Perhaps the fact that aluminum is a light atom and combines with transition-metals atoms in which the electrons act in complex ways will eventually be seen as crucial. But, in the meanwhile, let us at least try to understand the "new" crystallography: after all, one thought one had all of these issues straight, from "day one" of solid state physics! As one can imagine, there has been a lot of fun in understanding the nature of this novel type of material.

Consider, first, two-dimensions and ask about five-fold symmetry. Indeed, pentagonal symmetry, as depicted in Fig. 19(b), poses the same problems: one cannot have a translationally invariant crystal in a plane with such a symmetry. However, Roger Penrose showed

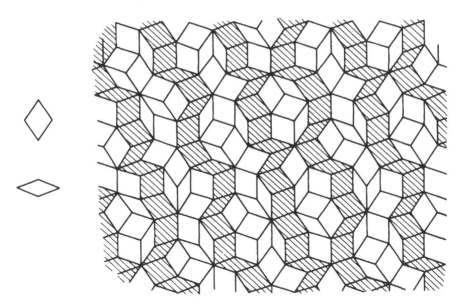

Figure 20. A Penrose tiling of the plane. The acute angles of the two types of tile, shown on the left, are $\pi/5$ and $2\pi/5$, respectively. The shaded tiles indicate a set of quasi-periodic scattering rows. [Adapted from Nelson and Halperin, Ref. 27. © 1985 by the AAAS.]

sometime back,[27] that one can construct two *tiles*, parallelograms of special shapes as illustrated on the left of Fig. 20, that can be fitted together as shown on the left, to build up a pattern that goes on forever, i.e., that fills the plane. The proportions of the tiles entail the so-called *golden ratio*, $\tau = \frac{1}{2}(1 + \sqrt{5})$, known to the ancients for its appealing properties not least of which is its intimate connection with the geometry of a pentagram: indeed, all the line segments in Fig. 19(b) are divided in the ratio $\tau:1$. Clearly this tiling pattern has *long-range bond-orientational order*: all the 'bonds' linking 'sites' at the corners of the tiles are parallel to one of the five pentagonal directions. Now the diffraction pattern of this two-dimensional pentagonal quasicrystal cannot have Bragg spots, because it is not a classical, periodic crystal. Nevertheless, as suggested by the shading in Fig. 20, there are scattering rows in the plane which exhibit a *quasi-periodic* translational order. Consequently the diffraction pattern contains spots distributed with full five-fold symmetry as indicated in Fig. 21. These spots are generated from a set of *five incommensurate* wave-vectors, $q_1, \cdots, q_5$, and are thus actually dense in the plane. However, their intensities fall off with increasing order.

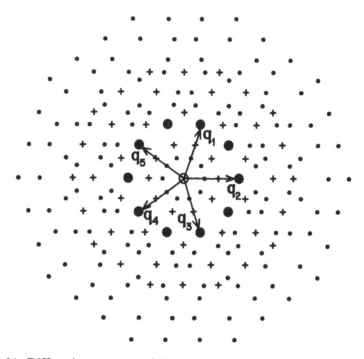

Figure 21. Diffraction pattern arising from the Penrose tiling in Fig. 20. The ten bold spots are generated in first order from the five incommensurate wave-vectors, $q_1, \cdots, q_5$, the crosses in second order, the light dots in third order. [After Nelson and Halperin, Ref. 27.]

Happily, there are three-dimensional generalizations of the Penrose tilings. One can find two distinct rhombohedra, again with proportions determined by the golden ratio. They fit together to fill space and produce an icosahedral quasicrystal. The diffraction pattern is now generated by six incommensurate wave-vectors, $q_1, \cdots, q_6$, in direct analogy to Fig. 21. The rhombohedra may be decorated with two sorts of atoms, representing the Al and Mn atoms, so that a given overall packing corresponds to an array of atoms spaced in such a way as to respect the steric restrictions. The icosahedral symmetry of the resulting diffraction pattern matches that seen in experiments on the real systems—essentially by construction. More significantly, the relative intensities of the diffraction spots correspond reasonably well. The details of the spot intensities should tell us about the locations of the atoms but that information is harder to extract and check unambiguously.

At this point the subject of quasicrystals is well launched and actively "en route". Many research workers, both experimental and theoretical, have been attracted to the area. It is clear that quasi-crystals constitute a new phase of matter—one originally thought to have been ruled out! Basic questions still remain: "Can quasicrystals be genuine ground states?" One does not know. However, it is plausible that if one takes two sorts of atoms of different sizes and treats them quantum-mechanically one will find quasicrystalline ground states if one adjusts the sizes to be just right. It will be most interesting to see. Yet, in the recent past, some of my theoretical colleagues have been trying to prove that all ground states are truly periodic.

Quantum Hall Effect

Now the final topic is a most exciting one, and certainly a case where quantum mechanics matters, namely, the quantum Hall effect, for which Klaus von Klitzing recently received the Nobel prize. This represents two-dimensional physics again, but now two-dimensional *electron* physics. Quantum mechanics always matters for electrons at low temperatures, and in this case temperatures of a few degrees or fractions of a degree are entailed. The inset in Fig. 22 shows some dimensions of the small device, through which one drives a current, I, of some tens of microamperes. The essential feature is that electrons are confined to a two-dimensional layer (of thickness a 100 Å or so). One then measures the Hall voltage, V_H, across the planar strip as a function of the external magnetic field, H, imposed normal to the strip. Nothing unusual happens at small magnetic fields ($H \lesssim 0.5$ T $\equiv 5$ kG), but as H increases one observes well-marked plateaus in V_H: see Fig. 22.* These plateaus get broader and flatter with increasing H and decreasing T. Moreover, they satisfy the following remarkable rule, namely, the Hall resistance $\rho_{xy} \equiv V_H/I$, is given by[28,29]

$$\rho_{xy} = \frac{R_Q}{\nu} \qquad \text{with} \qquad R_Q = \frac{h}{e^2} = 25812.80 \text{ ohms,} \quad (29)$$

---

* One also sees from the righthand scale and lower plot in Fig. 22 that the standard, longitudinal voltage drop, V_x, falls to zero when V_H takes a plateau value. Thus in the quantized regions the ordinary longitudinal resistance ρ_{xx} vanishes! Indeed, a driveless, pseudosupercurrent will persist for some minutes in this regime.

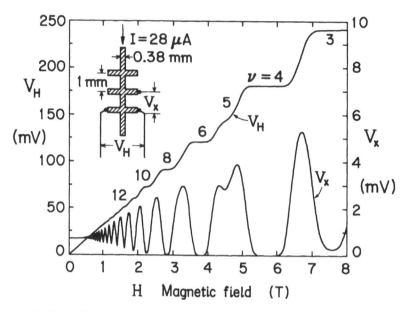

Figure 22. Data illustrating the integral quantum Hall effect. The inset indicates the dimensions of the two-dimensional disordered conducting sample and the voltages measured: the Hall voltage, V_H, on the left-hand scale, exhibits plateaus as a function of the magnetic field, H, at levels accurately given by $V_H/I = h/e^2\nu$ with ν integral; the longitudinal voltage, V_x, on the right hand scale, exhibits sharp dips at the Hall voltage plateaus. [After M.E. Cage et al., Ref. 29. © 1985 IEEE]

where ν is an integer. This holds with a totally unexpected precision of one part in 10^8, and an accuracy of one part in 10^6 or better! The question is, of course, "Why do we see such a precise, universal quantized Hall resistance?" and, from a practical viewpoint "Is this phenomenon going to prove a better way of measuring h and e and the fine structure constant $\alpha = 2\pi e^2/hc$?"

Now the physics of this effect is subtle: but what is even more complicated and surprising is that at higher magnetic fields Tsui, Störmer and Gossard[30] discovered plateaus in which ν is a rational fraction—typically the most strongly-marked one is $\nu = \frac{1}{3}$ but $\frac{2}{3}$ and $\frac{4}{3}$ are frequently observed—and other fractions are also seen, not just any fraction but only certain special rational fractions!

Robert B. Laughlin has been awarded the 1986 Buckley Prize for his contributions to our understanding of both the integral and fractional quantized Hall effect[31]—naturally, quantum mechanics comes in with a vengence![32] To round off the discussion I will sketch some of the main features of the theory.[31,32] (Readers not inclined to follow the somewhat technical details may wish to skip to the 'Concluding Remarks'.)

The first point to make is that, in the presence of a magnetic field, the momentum operator, $\mathbf{p}$, for a quantum-mechanical particle carrying a charge e must be modified by including the vector potential A to get $\mathbf{p} - (e/c)\mathbf{A}$—a subtle and beautiful result. On applying a uniform magnetic field to such a two-dimensional quantal particle one discovers that the usual continuum of energy levels condenses into a set of discrete, highly degenerate 'Landau levels': see Fig. 23(a). The uniform level spacing is $\Delta E = \hbar\omega_c$ where $\omega_c = eH/cm^*$ is the cyclotron frequency, while m^* is the effective mass. Now if one ignores the electron interactions one can just fill these Landau states up to the Fermi level, E_F. Real electrons, of course, interact both with the 'substrate' or ionic medium, which usually will be somewhat disordered, and with one another. These interactions are not easy to deal with theoretically. The first thing one believes is that the Landau levels spread out into bands: see Fig. 23(b). More subtley, one expects the band 'tails', shaded in Fig. 23(b), to consist of localized, nonconducting states. Indeed, some degree of disorder, leading to localized states, seems essential to the full quantum Hall effect.

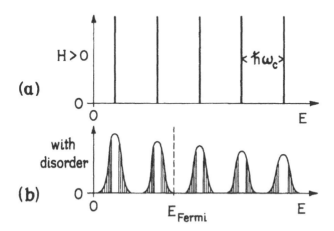

Figure 23. (a) Schematic illustration of the Landau levels for a charged quantal particle moving in two dimensions in a magnetic field. (b) Single-particle density of states derived from the ideal Landau levels in a disordered system: the shading denotes levels with localized wave functions; the states near the centers of the bands are extended. [After R.B. Laughlin, Ref. 31.]

Following Laughlin,[31] let us suppose that the Fermi level lies in a region of localized states between the bands of extended states, as indicated in Fig. 23(b). As the field, H, is changed the cyclotron frequency, ω_c, changes which moves the energy bands through E_F; equivalently, one can think of scanning the Fermi level through a set of fixed levels. Now we will invoke guage invariance: the way that enters quantum mechanics is one of the crucial ingredients. In reality one has some sort of external battery driving current through the planar strip, as shown schematically in Fig. 24(a); but one can, instead, remove the battery and complete the circuit to form a closed strip, as in Fig. 24(b). One may then induce a current in the strip by passing an auxiliary time-varying magnetic flux, $\phi(t)$, through the loop: see Fig. 24(b). Increasing ϕ uniformly drives a steady current around the circuit. If this flux is changed by an integral multiple of the flux quantum

$$\phi_0 = hc/e, \tag{30}$$

the electronic system will, by guage invariance, look just the same

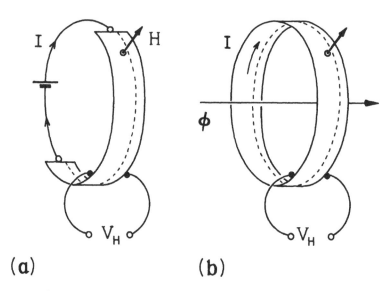

(a)   **(b)**

Figure 24. Illustration of Laughlin's argument explaining the quantum Hall effect. The external battery in (a) which drives a current through the planar sample in a magnetic field H is replaced in (b) by a closed sample threaded by a time-varying auxiliary flux, ϕ, which induces a current I. [After Ref. 31.]

since the Hamiltonian has not changed except for a trivial factor. On the other hand, if ϕ is increased slowly other considerations come into play. Whenever E_F lies in a region of localized states between the extended bands, there is an energy gap and one cannot excite low-energy quasiparticles. Thus a slow process will be adiabatic. On following what happens to the individual two-dimensional electron states one finds that the states that change are those whose wave-functions run completely around the loop: roughly speaking their quantization condition is modulated by ϕ. As ϕ increases adiabatically these extended states are driven sideways across the strip. Thus an electron or a fraction of an electron is effectively moved across the strip. By the time ϕ has changed by one flux quantum, ϕ_0, all the states have been restored to their original form and an integral number of electronic charges have been transported across the width of the strip. Basically one expects one electronic charge transferred for each of the Landau levels lying below E_F: thus if v labels the highest filled level, $n = v$ electrons move across in total.

To complete the argument we appeal to a statistical/thermodynamic relation. If U is the total electronic energy, the current is given by

$$I = c(\partial U / \partial \phi). \tag{31}$$

But for $\Delta \phi = \phi_0 \equiv hc/e$ the change in energy resulting from the transport of n electronic charges is just $\Delta U = neV_H$—by definition of the Hall voltage V_H. Thus one obtains

$$I = \frac{necV_H}{(hc/e)} = \frac{ne^2}{h} V_H, \tag{32}$$

which implies the quantization rule (29) for ρ_{xy}.

One might well object that the argument seems to rest very heavily on a noninteracting or one-electron, picture. That is certainly true and, although great success has been had with one-electron models in the past, one should not, these days, be content with such simplifications; rather one should attempt to understand the electronic interactions in greater depth. Indeed, fascinating progress has been made: on the one hand, it is found that the integer quantum Hall effect reflects profound topological properties of the phase of wave functions—full many-body wave functions, Ψ_N—in the presence of varying magnetic fluxes.[32] On the other hand, it transpires, wonderfully, that replusive interactions between electrons are essential to the *fractional* quantum Hall effect at higher magnetic fields. One now believes that at zero temperature a system of electrons in a fractional Hall regime should be regarded as an *incompressible quantum fluid*—a quite new state of matter.

To attack the problem of describing such states while allowing for the strong Coulombic repulsions between electrons, Laughlin proposed a new type of many-body wave function. It turns out to be a remarkably good guess which embodies the essential ordering features of these new states of matter. To represent Laughlin's wave function it is convenient to take advantage of the two-dimensionality by expressing electronic coordinates as complex numbers via

$$z = x + iy. \tag{33}$$

Then the proposed N-electron ground state wave function is

$$\Psi_0(z_1, z_2, \cdots, z_N) = \prod_{i<j}^{N} (z_i - z_j)^{m_l} \prod_{k=1}^{N} e^{-|z_k|^2/4l^2}. \tag{34}$$

Here $l = (hc/eH)^{1/2} = (\phi_0/H)^{1/2}$ is the cyclotron radius. (Alternatively, one can say that l^2 is the area per state in a given Landau level.) This wave function is not quite as mysterious as appears on first sight. One can, indeed, show that the exponential factor shown *times* a polynomial in the z_i is the form to be expected. More concretely, the prefactors $(z_i - z_j)^{m_l}$ serve to keep different electrons apart so reducing the total repulsive energy. In order to enforce proper Fermi statistics, m_l must be an odd integer. The first case, $m_l = 1$ corresponds simply to a Slater determinant; but the new states with $m_l = 3, 5, \cdots$ turn out to describe the fractional quantized Hall states with index $\nu = \frac{1}{3}, \frac{1}{5}, \frac{1}{7}$. One also discovers that the low lying excitations carry fractional electronic charges, $e^* = e/m_l$! To understand the other fractional states with $\nu = \frac{2}{3}, \frac{2}{5}$, etc. one has to think further: a hierarchy of similar states is required; but the ingenious wave function (34) provides the basis for our understanding.

Amusingly and instructively, one route to gaining an insight into the properties of Laughlin's wave function turns back to classical statistical mechanics. If one squares the wave function to obtain the corresponding many-particle distribution function one sees that it can be written as a Boltzmann weight, i.e., the exponential of a classical Hamiltonian divided by $k_B T$. The Hamiltonian turns out to describe a one-component plasma, in two dimensions. If e_0 is the charge of the classical particles, the plasma parameter $\Gamma \equiv 2e_0^2/k_B T$ is given just by $2m_l$. This classical system had, indeed, been studied previously for its independent interest. So once again we see that quantum-mechanics, if one takes the many-body aspects and the interactions seriously, is so intertwined with statistical mechanics that one has to recognize that classical ideas of fluctuations, ordering and entropy play a significant, if somewhat hidden, role even in this most quantum-mechanical of problems!

Concluding Comments

I was asked by Herman Feshbach to mention major unsolved problems. Certainly, a full understanding of the magnetic properties of solid helium-three[33] and of the microscopic basis of the so-called 'heavy fermion' systems[34] represent outstanding questions. Basically, we have not yet gained a very good grasp of all that can happen when fermions interact strongly. We are lucky that the outcome is frequently that the system generates simpler, more independent entities like atoms, molecules, quasiparticles, Cooper pairs, etc. whose internal structures may be ignored for many purposes. But there are important cases in which the overall behavior is not so readily characterized. Simple models, like the Hubbard model—a lattice description of electrons with short-range repulsions—still defy detailed analysis in many regimes despite manifold attacks including numerical aid from powerful computers. Thus the task of obtaining a reliable and thorough understanding of interacting Fermi particles remains a basic challenge. Indeed, these words, spoken in 1985, have been strengthened by the discovery of high-T_c superconductors[1] fifteen months later!

In conclusion, I hope to have given a picture of the multifaceted character of modern condensed matter physics and the varying degrees to which quantum mechanics is of direct relevance *or* almost total irrelevance. I feel that if Niels Bohr were here with us, in the role in which I would best like to recall him, namely, as a young, active theorist grappling with problems at the blackboard, he would rise to the challenges of condensed matter physics regardless of whether his own earlier work in quantum mechanics had anything to do with the problem or not. Niels Bohr certainly cared for the phenomena seen in the natural world and that, we should agree, is the crucial feature in doing physics.

Acknowledgements

I am grateful to many colleagues and scientific friends, most of whose names do not appear above or in the references, who have educated me through the literature, through correspondence and through patient discussions and arguments. All those who allowed reproduction of data and figures are especially thanked. I am indebted to Professor Peter H. Kleban for preparing, from the taped and transcribed version of the original lecture, a first draft of the manuscript. Professor B.I. Halperin, Professor R.L. Jaffe, Dr. Matthew P.A. Fisher and Professor Daniel S. Fisher kindly commented on and criticized various specific points. Last, the hospitality of the Aspen Center for Physics, where the manuscript was finally written, and the support of the National Science Foundation, primarily through the Condensed Matter Theory program,[35] are gratefully acknowledged.

References

1. J.G. Bednorz and K.A. Müller, Z. Phys. B **64**, 189 (1986); M.K. Wu, J.R. Ashburn, C.J. Torng, P.H. Hor, R.L. Meng, L. Gao, Z.J. Huang, Y.Q. Wang and C.W. Chu, Phys. Rev. Lett. **58**, 908 (1987).
2. F.J. Dyson and A. Lenard, J. Math. Phys. **8**, 423 (1967); F.J. Dyson, J. Math. Phys. **8**, 1538 (1967); A. Lenard and F.J. Dyson, J. Math. Phys. **9**, 698 (1968).
3. See e.g., M.E. Fisher, Arch. Ratl. Mech. Anal. **17**, 377 (1964).
4. E.H. Lieb and J.L. Lebowitz, Advan. Math. **9**, 316 (1972).
5. See the book P.-G. de Gennes, *"Scaling Concepts in Polymer Physics"*, (Cornell Univ. Press, Ithaca, New York, 1979).
6. See the review Y. Oono, Adv. Chem. Phys. **61**, 301 (1985).
7. See e.g., (a) M.E. Fisher in *"Contemporary Physics: Trieste Symposium 1968"*, Vol. I, (Internat. Atomic Energy Agency, Vienna, 1969) pp. 19–46; (b) in *"Essays in Physics"*, Vol. 4 (Academic Press, London, 1972), pp. 43–89; (c) J. Appl. Phys. **38**, 981 (1967).
8. R.B. Griffiths, Phys. Rev. **136** A, 437 (1964).
9. L. Onsager, Phys. Rev. **65**, 117 (1944).
10. P. Nordblad, D.P. Belanger, A.R. King, V. Jaccarino and H. Ikeda, Phys. Rev. B **28**, 278 (1983).
11. H.K. Kim and M.H.W. Chan, Phys. Rev. Lett. **53**, 170 (1984).
12. A.A. Belavin, A.M. Polyakov, and A.B. Zamolodchikov, J. Stat. Phys. **34**, 763 (1984); Nucl. Phys. B **241**, 333 (1984).
13. Vl. S. Dotsenko and V.A. Fateev, Nucl. Phys. B **240**, 312 (1984); D. Friedan, Z. Qiu and S.H. Shenker, Phys. Rev. Lett. **52**, 1575 (1984).

14. J.L. Cardy in *"Phase Transitions and Critical Phenomena,"* Eds. C. Domb and J.L. Lebowitz, Vol. 11 (Academic Press, New York, 1987) p. 55; M.E. Fisher, J. Appl. Phys. **57**, 3265 (1985); J. Mag. Magn. Matls. **54–57**, 646 (1986).
15. B.C. Crooker, E. Hebral, E.N. Smith, Y. Takano and J.D. Reppy, Phys. Rev. Lett. **51**, 666 (1983).
16. (a) M. Rasolt, M.J. Stephen, M.E. Fisher and P.B. Weichman, Phys. Rev. Lett. **53**, 798 (1984); (b) P.B. Weichman, M. Rasolt, M.E. Fisher, and M.J. Stephen, Phys. Rev. B **33**, 4632 (1986).
17. Y. Imry and S.-K. Ma, Phys. Rev. Lett, **35**, 1399 (1975).
18. S. Fishman and A. Aharony, J. Phys. C **12**, L729 (1979).
19. D.P. Belanger, A.R. King and V. Jaccarino, Phys. Rev. B **31**, 4538 (1985); R.J. Birgeneau, Y. Shapira, G. Shirane, R.A. Cowley and H. Yoshizawa, Physica **137B**, 83 (1986).
20. J.Z. Imrie, Phys. Rev. Lett. **53**, 1747 (1984); see also Commun. Math. Phys. **98**, 145 (1985) and D.S. Fisher, J. Fröhlich and T. Spencer, J. Stat. Phys. **34**, 863 (1984); J. Chalker, J. Phys. **16**, 6615 (1983).
21. (a) M. Guymont, R. Portier and D. Gratias, Acta Cryst. A **36**, 792 (1980); (b) R. Portier, D. Gratias, M. Guymont and W.M. Stobbs, Acta Cryst. A **36**, 190 (1980).
22. J. Kulik, S. Takeda and D. de Fontaine, Acta Metall. **35**, 1137 (1987).
23. J. Rossat-Mignod, P. Burlet, H. Bartholin, O. Vogt and R. Lagnier, J. Phys. C **13**, 6381 (1980).
24. A.M. Szpilka and M.E. Fisher, Phys. Rev. Lett. **57**, 1044 (1986).
25. See P. Bak, Repts. Prog. Phys. **45**, 587 (1982).
26. D. Schechtman, I. Blech, D. Gratias and J.W. Cahn, Phys. Rev. Lett. **53**, 1951 (1984).
27. See the review by D.R. Nelson and B.I. Halperin, Science **229**, 233 (1985) and C.L. Henley, Comm. Cond. Mat. Phys. **13**, 59 (1987).
28. K. von Klitzing, G. Dorda and M. Pepper, Phys. Rev. Lett. **45**, 494 (1980).
29. M.E. Cage, R.F. Dziuba and B.F. Field, IEEE Trans. Instr. Meas. **IM–34**, 301 (1985).
30. D.C. Tsui, H.L. Störmer and A.C. Gossard, Phys. Rev. Lett. **48**, 1559 (1982).
31. R.B. Laughlin, Phys. Rev. B **23**, 5632 (1981); Phys. Rev. Lett. **50**, 1395 (1983).
32. See also R.E. Prange and S.M. Girvin, Eds. "The Quantum Hall Effect" (Springer Verlag, Berlin, 1987).
33. M. Cross and D.S. Fisher, Rev. Mod. Phys. **57**, 881 (1985).
34. P.A. Lee, T.M. Rice, J.W. Serene, L.J. Sham and J.W. Wilkins, Comm. Cond. Mat. Phys. **12**, 99 (1986).
35. Under Grants No. DMR 81-17011 and 87-01223.

Nuclear Physics: Comments and Reflections

Herman Feshbach

MASSACHUSETTS INSTITUTE OF TECHNOLOGY

In this paper, three themes which are of general interest to physics and science generally will be emphasized. First is the role of symmetry and conservation rules in nuclear physics. It was here that internal quantum numbers were first introduced into physics and we shall discuss as well the importance of broken symmetry in nuclear physics. The treatment of strong interactions which made its first appearance in nuclear interactions forms the second theme. The third theme refers to the contribution nuclear physics has made to astrophysics not only with regard to energy and element production in the stars but also to the understanding of such stellar objects as pulsars. Finally, we shall conclude with a short discussion of some of the more recent research which emphasizes the issues which originate in the quark-gluon structure of "elementary" particles.

Symmetry and Nuclear Physics

Symmetry principles have played an important role in the understanding of nuclei and nuclear reactions from the very beginning. Modern nuclear physics dates from the early thirties with the discovery of the neutron by Chadwick in 1932 and the formulation by Fermi in 1934 of the neutrino theory of β-decay. These established the neutron as a primary constituent of nuclei resolving thereby the dilemmas presented by data on the spin and statistics of nuclei to the theory that nuclei were made up of electrons and protons. The Fermi theory explained how the emission of electrons by radioactive nuclei could occur in the absence of an explicit presence of electrons in the nucleus. The conservation of linear and angular momenta and of energy and the permutation symmetry required by the identity of the particles in the nucleus were the principles which led to the discovery of these dilemmas and to their resolution by the neutron-proton model of the nucleus.

These symmetry or conservation principles were familiar from studies of the spectra of atoms. But within a few years when low energy neutron-proton scattering and proton-proton scattering were compared, an entirely new symmetry principle was discovered. From these experiments it was surmised by Breit, by Caseen and Condon, and by Wigner that nuclear forces were charge indepen-dent; that is, the neutron-proton interaction for a given state of the neutron-proton system equals the proton-proton interaction when the proton-proton system is in the identical state. This concept was extended to the neutron-neutron interaction, although the supporting evidence was indirect, no data on neutron-neutron scattering being available. Note that charge independence applies to the nuclear forces only. The Coulomb interaction which must be added to the nuclear forces to obtain the total interaction acts only in proton-proton scattering.

Charge independence suggested that the neutron and proton were two different states of a particle called a nucleon. The state of the nucleon is then symbolically given by a two element matrix. The proton state is

$$p = \begin{pmatrix} 1 \\ 0 \end{pmatrix},$$

while the neutron's is

$$n = \begin{pmatrix} 0 \\ 1 \end{pmatrix}.$$

The matrices connecting these two are the isospin operators:

$$\tau_1 = \begin{pmatrix} 0 & 1 \\ 1 & 0 \end{pmatrix} \quad \tau_2 = \begin{pmatrix} 0 & -i \\ i & 0 \end{pmatrix} \quad \tau_3 = \begin{pmatrix} 1 & 0 \\ 0 & -1 \end{pmatrix},$$

as well as the identity

$$1 = \begin{pmatrix} 1 & 0 \\ 0 & 1 \end{pmatrix}.$$

The states p and n are orthogonal axes in a two-dimensional Hilbert space called "isospin space" in which the operators $\tau = (\tau_1, \tau_2, \tau_3)$ act as rotation operators. Charge independence corresponds to invariance against rotation in isospin space. The symmetry implied by these equations is isomorphic with that of ordinary spin and is designated by the symbol $SU(2)$. It is an internal symmetry unconnected with spatial or temporal properties of the interactions or particles. It is a symmetry which is broken, as indicated earlier, by the Coulomb interaction and by the neutron-proton mass difference. The Coulomb interaction is obviously not invariant against a rotation in isospin space.

The consequences of isospin symmetry for the structure of the heavier nuclei are revealed rather spectacularly by the isobar analog resonance illustrated in Fig. 1 obtained in elastic proton scattering. This resonance has a well-defined isospin. It is, in fact, the lowest state with that isospin. If isospin symmetry were not broken, its width would be zero since the isospin conservation implied by charge independence would permit decay only to a state of the same isospin. But isospin symmetry is broken by Coulomb interaction. However, the major effect of the Coulomb interaction is to raise the energy of the state from the value it would have if the projectile were neutral. Its non-diagonal matrix elements connecting states of different isospin are small because of the long range of the Coulomb force. The existence of isobar analogs is thus a direct confirmation of isospin symmetry.

Other examples of the effect of symmetry on nuclear structure are identical to those observed in atoms and molecules. A particular example of great importance is the case of deformed nuclei such as the rare earth nuclei which have been found to have large electric quadrupole moments indicating that the nuclei are not spherical. However, a non-spherical nucleus, say one that has a prolate spheroidal shape violates the principle that there is no preferred direction in space. This symmetry is restored if the axes of the prolate spheroid rotate, and the corresponding wave function is that of the spherical rotator. This leads immediately to a rotational spectrum (see Fig. 2) and the corresponding symmetry is designated by the symbol $O(3)$. The group generators connect the members of the rotational band but at the same time serve as the electromagnetic transition operators. As a consequence the electromagnetic transitions between the rotational levels is greatly enhanced.

The conservation of baryon number provides us with a final example. Breaking this symmetry, that is, by relaxing the condition

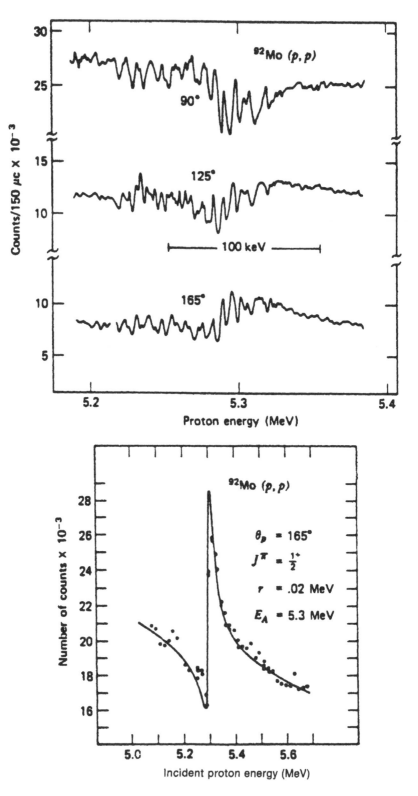

that the number of nucleons is fixed, and then restoring the symmetry by projection, one obtains a band of levels just as in the case of the deformed nuclei. This "superfluid" band consists of the ground states of nuclei where neutron numbers differ by two units. The group in question is $U(1)$. The group generator is also the transition operator for a change in neutron number by two leading to the observed enhancement of two nucleon transfer reactions, *e.g.* (t,p) between the ground states of the initial and final nucleus.

Strong Interactions

The first experiments on proton-proton scattering demonstrated directly the existence of strong and short range nuclear forces. Later experiments showed that these were strongly spin and isospin dependent. Their existence in the many-body nuclear system was most clearly exhibited by the low energy neutron resonances in which Bohr was very interested. Their narrow width was understood on the basis of the strong coupling among the nuclear degrees of freedom which leads to a rapid dissipation of the energy of the incident projectile. It is then trapped within the nucleus for a very long time. Hence the narrow width. The explanation of the properties of nuclei seemed especially daunting as it appeared that the perturbation methods so familiar from atomic physics were inapplicable. How was one to begin to understand the properties of a system of particles interacting via strong short range forces?

The decisive discovery which presented the answer to this question was made in the decade after the end of World War II. Based on the accumulation of data on the binding energies, energies and properties of excited states obtained by observation of electromagnetic transitions and particle exchange reactions $((d,p)$ and $(p,d))$ the shell model of the nucleus was proposed in which it is assumed that a nucleon in the nucleus moves in a *mean field* which is to a very good approximation state independent, that is, independent of the state of the system—at least for the low lying levels. It is hardly a deep remark to say that a nucleon in a nucleus moves in the field

$\longleftarrow$

Figure 1. (a)—Elastic scattering of protons by ^{92}Mo in the neighborhood of the *s*-wave isobar analog resonance at proton energy of 5.3 MeV. (b)—Curve obtained by averaging the data of (a) (from Richard, Moore, Robson, and Fox. Ref. [1]).

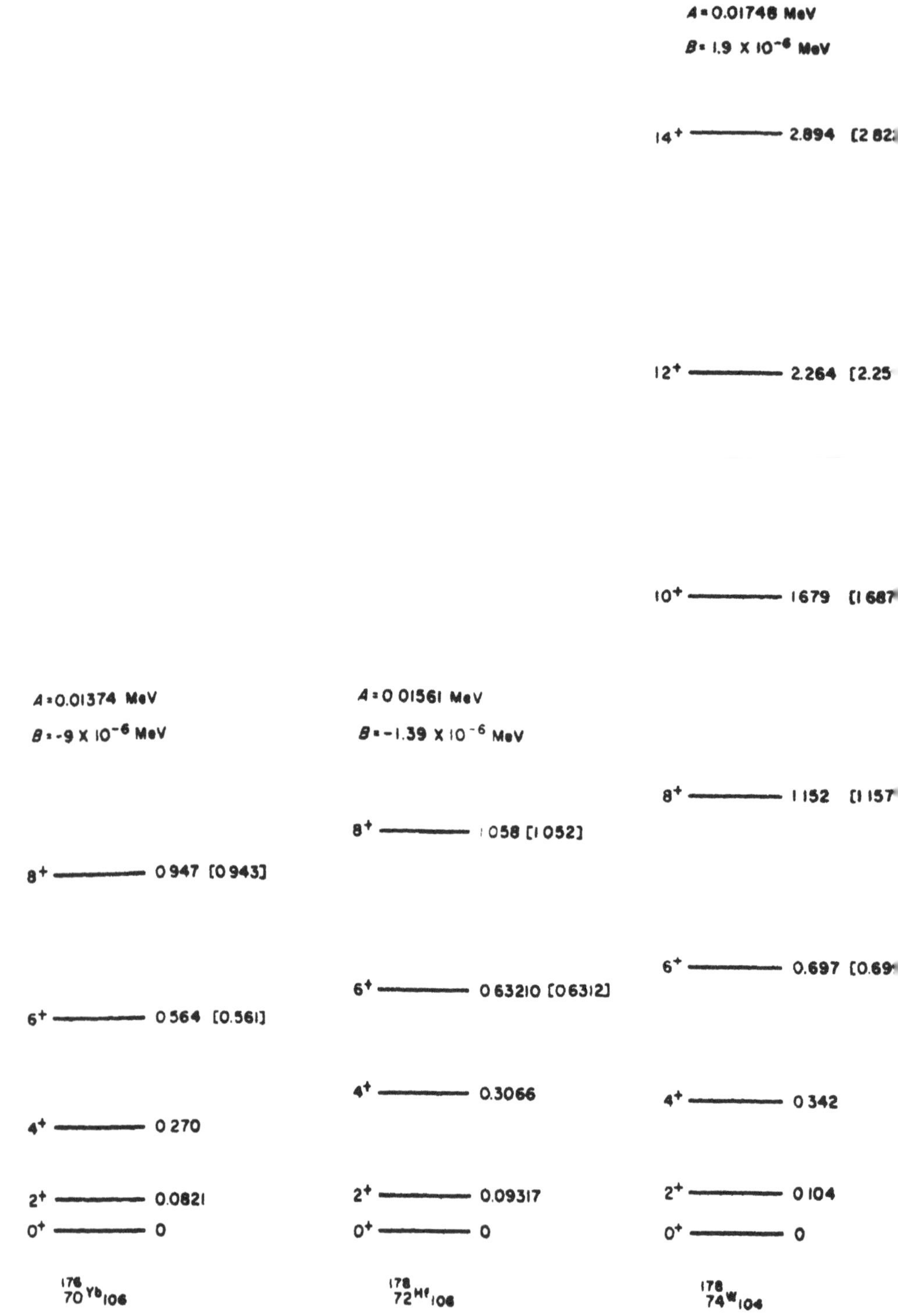

A = 0.01748 MeV
B = 1.9 × 10⁻⁶ MeV

$A = 0.01748$ MeV
$B = 1.9 \times 10^{-6}$ MeV

14⁺ ———— 2.894 [2 82...

12⁺ ———— 2.264 [2.25...

10⁺ ———— 1679 [1 687...

$A = 0.01374$ MeV
$B = -9 \times 10^{-6}$ MeV

$A = 0.01561$ MeV
$B = -1.39 \times 10^{-6}$ MeV

8⁺ ———— 1152 [1 157...

8⁺ ———— 1058 [1 052]

8⁺ ———— 0 947 [0 943]

6⁺ ———— 0.697 [0.69...

6⁺ ———— 0 63210 [0 6312]

6⁺ ———— 0 564 [0.561]

4⁺ ———— 0.3066

4⁺ ———— 0 342

4⁺ ———— 0 270

2⁺ ———— 0.0821

2⁺ ———— 0.09317

2⁺ ———— 0 104

0⁺ ———— 0

0⁺ ———— 0

0⁺ ———— 0

$^{176}_{70}\mathrm{Yb}_{106}$

$^{178}_{72}\mathrm{Hf}_{106}$

$^{178}_{74}\mathrm{W}_{104}$

Figure 2. Excited states of some deformed nuclei. Energies are in MeV. The numbers in brackets give the energies of the expected states as calculated using $AJ(J + 1) + BJ^2(J + 1)^2$ using indicated constants A and B in MeV (from deShalit and Feshbach, Ref. [2]).

122

generated by the other nucleon in the nucleus. Because of the identity of the nucleons, that field is the same for each nucleon. It varies rapidly in both space and time as a consequence of near collisions. The mean field is obtained by time averaging out the sharp fluctuations in the field retaining only those components of the field which vary slowly with time. It immediately follows from the uncertainty principle that the mean field will only describe the low lying excitations. It is not clear from this discussion why the mean field obtained in this way is relatively but not completely independent of the state of the nucleus but it is empirically the case and several derivations which exist which attest to its reasonableness. This isn't the place to go into the details of the calculational methods used to obtain the mean field, but suffice it to say that they represent a major advance in the arsenal of those theorists dealing with the non-relativistic many-body Schrödinger equation with strong interparticle interactions. It is probably unnecessary to emphasize that the nuclear mean field differs sharply from the atomic mean field in both origin and character. The sharp discontinuities in the energies at the closing of a shell are relatively much larger in the atomic case.

But there is another major difference. In the nuclear case the mean field is dynamical. Although the mean field has been "smoothed" with respect to time, slow time-dependence does remain. This is reflected in the existence of collective degrees of freedom in which the nucleus as a whole is involved. We deal here with a tightly coupled self-consistent situation. The motion of the nucleus determines the mean field which determines motion of the nucleons. What one observes are the quantal excited states corresponding to each degree of freedom. Very early, just after World War II, the giant electric dipole ($E1$) resonance was discovered. This was found to correspond to a vibrational degree of freedom in which the center of mass of the neutrons and that of the protons vibrate with respect to each other. This phenomenon is observed widely throughout the periodic table. Since that time, other giant resonances, for example the electric monopole ($E(0)$), electric quadrupole ($E(2)$), and the "Gamow-Teller" type resonances have been observed. It can be shown that there is a symmetry associated with each of these which can be expressed in terms of the operators which connect the ground state with the resonant state. These are of the form $\tau Y(l)$ (where $Y(l)$ is a spherical harmonic of order l), in the giant electric cases and $\sigma \tau Y(l)$ (σ = Pauli spin operator) for the Gamow-Teller. We see the close connection to the Fermi and Gamow-Teller β-decay operators. These symmetries are referred to as *dynamical symmetries* since they have their origin in the dynamical properties of the mean field. These vibrational degrees of freedom can be discussed in what amounts to a quantum many-body theory of small vibrations about the mean field, in an approximation referred to as the RPA. However, such a description cannot be used to describe large amplitude motion as exhibited by fission and heavy ion collisions in which substantial amounts of matter and large amounts of energy are exchanged. One successful attempt makes use of the mean field approach. It is referred to as the time-dependent Hartree-Fock method (TDHF). Briefly, in this method the Hartree-Fock method is extended to the time-dependent many-body Schrödinger equation. The mean field is calculated at each instant in time together with the corresponding single particle wavefunctions. The results of a calculation of heavy ion collision is shown in Fig. 3, where one can follow the progress of the collision. Remarkably, since no correlations in the many-body wavefunction (other than the Pauli principle) have been introduced, the TDHF gives a good account of the phenomena which occur during a heavy ion collision. It can't be the whole story but the system treated in this way is sufficiently rich as

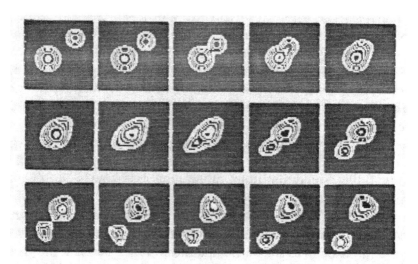

Figure 3a. Contour plots at sequential times of the density in the cm integrated over the normal to the reaction plane for ^{16}C + ^{40}Ca collision at a laboratory energy of 315 MeV. The initial angular momentum $l = 20\,\hbar$ corresponds to a nearly head-on collision. (Taken from Negele, Ref. [3].)

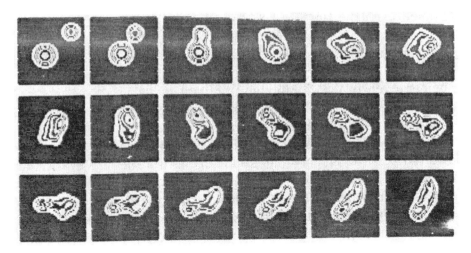

Figure 3b. Contour plots for the same reaction as in Fig. 3a with an initial angular momentum of $l = 60 \, \hbar$.

it provides a useful description of the collision. Another, and surprising, triumph for mean field theory.

The determination of the mean field and the corresponding single particle orbitals suffices to explain the light nuclei. However, for the heavier nuclei, the number of configurations which can be con-

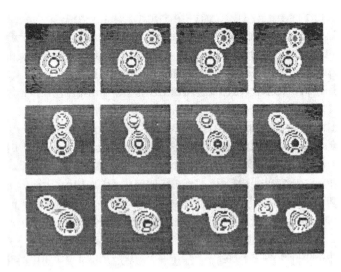

Figure 3c. Contour plots for the same reaction as in Fig. 3a with an initial angular momentum of $l = 80 \, \hbar$.

structed from the single particle states is enormous. For example, in the case of ^{154}Sm there are $\sim 10^{15}$ states possible. Obviously, not all of these are interesting. The Interacting Boson Model (IBM) is an attempt to deal with this problem. It is assumed that the low lying nuclear levels of the vibrational-rotational nuclei are determined by a very small subset of the possible configurations; namely, those which pair to form s and d bosons spin zero and two, respectively. The effective Hamiltonian is assumed to be composed of the Casimir invariants of the chained subgroups of $SU(6)$. This model is quite successful phenomenologically in explaining the spectra of even nuclei in their dependence on N and Z. Of course, establishing the relationship to the underlying shell model is of fundamental importance. When odd nuclei are considered, involving both boson and fermion degrees of freedom, a type of supersymmetry becomes possible and indeed some examples of nuclei which exhibit supersymmetry have been found.

As a last example (many more are possible) we mention the system in which one of the nucleons is excited becoming a Δ. This system is formed when a pion in the pion-nucleon resonance region is absorbed by one of the nucleons of the target nucleus. Understanding the mean field under which the Δ moves is essential to the understanding of the whole range of pion-nucleon interactions.

Let us return to nuclear reactions. We have earlier referred to the slow neutron resonances which were interpreted to be a consequence of the strong interactions between the nucleons within the nucleus. The resonant state is referred to as a state of the compound nucleus. At higher excitation energies one still spoke of the formation of the compound nucleus which in a reaction would be followed by its decay. The Bohr independence hypothesis stated that the branching ratio for various reactions depended only on the properties of the compound nucleus and were independent of the way in which the compound nucleus was found and indeed an "evaporation" description of the decay became possible if the excitation is sufficiently large. Typically, the angular distribution of the decay products is spherical giving no inkling as to the direction of the incident projectile. These results are characteristic of a process which involves a long interaction time. The cross-section for such a process should show rapid fluctuations as a function of the energy. These were observed when the necessary experimental resolution was achieved.

After World War II, an entirely new phenomenon was discovered.

In, for example, pickup (p,d) and stripping (d,p) reactions, the angular distributions were sharply peaked forward. Moreover, the cross-sections varied very slowly with the energy (see Fig. 4). These results were clear indications of processes involving short interaction times totally different from the long interaction times characteristic of the formation of the compound nucleus. The Bohr independence hypothesis fails here as the properties of the reaction products depend importantly on the nature of the incident channel.

The existence of phenomena with relatively rapid reaction time was corroborated by an important experiment in which the elastic scattering of moderate energy neutrons was measured for a wide variety of target nuclei. Regular patterns in the angular distributions which changed systematically with nuclear size were observed. The explanation was a surprise. The observed structure could be explained if it were assumed that the incident neutron moved in a mean field whose extent was determined by the nuclear size. This mean field differed from that discussed earlier in connection with the structure of the nuclei in that the mean field potential is complex indicating the existence of absorption. This complex potential model is now generally referred to as the optical model.

Nuclear physicists were faced with a dilemma. Nuclei reactions with the same target nucleus exhibit both long and short interactions times depending on the reaction under study. The eventual resolution of the dilemma has two facets. First it has been found that the reactions exhibiting a spectrum of interaction times can occur. The compound nucleus reaction is one extreme, the stripping reaction, the other. Second, which of these possibilities one observed depend on the energy resolution of the beam; the poorer the resolution, the shorter the interaction time as one would expect from the uncertainty principle. In principle and in practice, then, by varying the energy resolution one could select specific interaction times and thereby emphasize specific reaction mechanisms. This is illustrated by the complex optical model potential. The absorptive term has two sources. One is a consequence of inelastic processes. But even where the energy is below any of the inelastic thresholds one still finds absorption. This is because the cross-section measures only the prompt flux emerging from the interaction region and not the time delayed emission following the formation of compound nuclear states. In summary, there is a hierarchy of reactions all of which can occur in a given collision. The simplest, one-step reaction, is exemplified by the elastic scattering and stripping; the next more

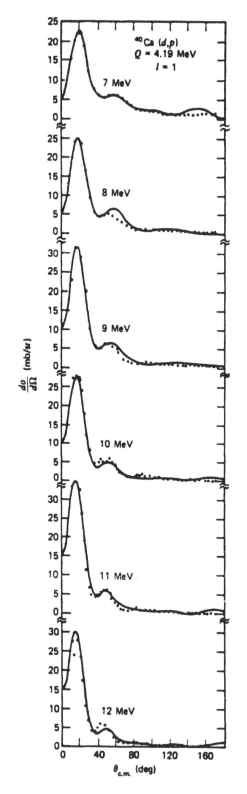

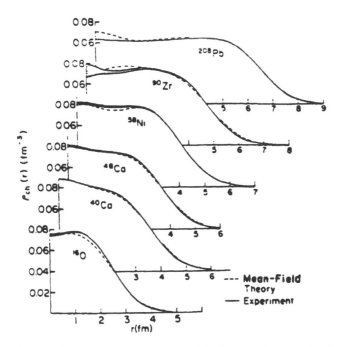

Figure 5. Comparison of DME mean-field theory charge distributions in spherical nuclei (dashed lines) with empirical charge densities. The solid curves and shaded regions represent the error envelope of densities consistent with the measured cross sections and their experimental uncertainties (from Negele, Ref. [3]).

complex is the two-step process which directly suggests the existence of resonances which are called "doorway state" resonances of which the isobar analog resonance described earlier is an example. Multi-step processes, empirically no more than four steps, lead to the formation of the compound nucleus.

There are many successful applications of the mean field in addition to those mentioned above. Rather than listing them all, I shall give a few examples. One can use the mean field to compute the charge density of nuclei and compare it with that density observed using high energy electron scattering. A comparison is shown in Fig. 5. A second example is a super-deformed band in ^{152}Dy, predicted by Bohr and Mottelson. This is illustrated in Fig. 6. Finally, one has

←

Figure 4. The ^{40}Ca (d,p) reaction for bombarding energies in the range 4.19 MeV to 12 MeV exciting a level in ^{41}Ca, the solid lines are theoretical (from L. L. Lee et al., Ref. [4]).

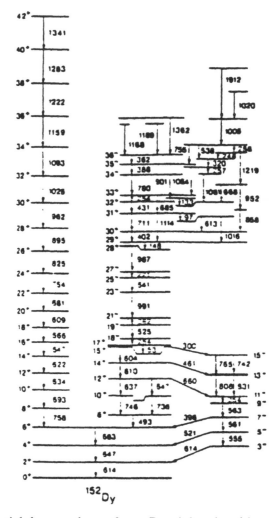

Figure 6a. Partial decay scheme for $_{152}$Dy. A low hand is seen to spin 42^+, the irregular oblate states are seen to near spin 40 (from Nyako et al., Ref. [6]).

recently observed the presence of giant resonances in excited nuclei, observed as these decay electromagnetically.

In concluding this section, we note that the mean field of the shell and optical model were introduced empirically and that a great deal of progress was made in the absence of a direct derivation of the mean field from the empirical nuclear forces. This is not to imply that this problem was not faced. And indeed it was one of the major accomplishments of nuclear theorists that they devised a method

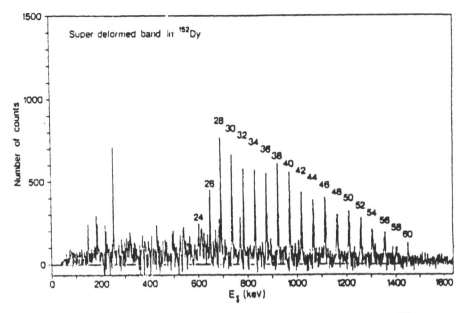

Figure 6b. Gamma-ray spectrum in the superdeformed band in ^{152}Dy following the ^{108}Pd (^{48}Ca, $4n$) ^{152}Dy reaction at 205 MeV (from Twin et al., Ref. [6]).

for deriving the mean field when the underlying forces are not only strong and short ranged but also singular.

Nuclear Astrophysics

Nuclear astrophysics is the scientific discipline with which nuclear physics has the closest intellectual relationship. Nuclear processes play an important role in energy production in stars, in the formation of the elements from the primordial hydrogen and helium, in stellar structure and evolution. Nuclear physics is essential to the understanding of the formation and properties of neutron stars, pulsars, burstars, etc. Not that all problems have been solved.

The solar neutrino puzzle—namely the discrepancy between the flux of solar neutrinos generated by the energy production of the sun arriving at the earth's surface is far fewer than that predicted. Only one experiment has been performed. A second one using Gallium is soon to be done. The resolution of the solar neutrino problem may very well have a significant impact on our understanding of the weak interactions.

We turn to a number of problems now under recent study. In recent years much has been learned theoretically regarding supernova collapse, the nearly constant matter entropy, the role of neutrino interactions and transport and how softening of the equation of state enhances the ejection of the mantle. The spectacular observations of the supernova 1987a, the direct observations of the neutrino bursts and the prospect of observing a pulsar remnant when the debris scatter provide unique data for the theoretical understanding of supernova. Most important for these developments is the equation of state for normal matter densities.

The structure of neutron stars provides even more difficult challenges, involving hadronic matter at densities far higher than nuclear matter. We are interested not only in the equation of state, which governs the equilibrium structure and thus such observables as the mass, radius, and gravitational red shifts, but also in the phases of matter at high density. On the one hand, the equation of state determines the maximum mass of a neutron star. On the other hand, it is very sensitive to the nature of the underlying nuclear forces. Including three body forces makes a substantial difference. Interestingly, calculations of the properties of nuclear matter, saturation energy and density, demonstrate the failure of two-body potentials. At densities at which nucleons and nuclei are the relevant degrees of freedom, we need to understand the superfluid properties of neutron star matter. For the rapidly rotating neutron stars we observe as pulsars, the large angular momentum is realized by a high density of vortices in the superfluid, and thus vortices must migrate out through the surface as the star slows down. The details of the pinning of vortices to nuclei in the crust and the mechanical properties of the crust itself will be crucial in understanding discontinuous jumps in the pulse rate which have been observed in many neutron stars. At higher densities, the question arises as to whether other phases described by hadronic coordinates occur, such as pion or kaon condensates, before the quarks comprising hadrons become completely delocalized and the system becomes quark matter. It is important to note that the non-nucleonic phases have important observable consequences in terms of neutron star cooling. Whereas in the normal neutron phase, the URCA process $n \rightarrow p + e^- + \bar{\nu}_e$, $p + e^- \rightarrow n + \nu_e$ is blocked by the Pauli principle, there is no corresponding hindrance in pion condensed or quark phases, and such stars would cool much more quickly. The role of strangeness also remains to be understood. For example, there is no evidence to rule out the possibility that strange quark matter is in fact the lowest energy state of matter in the universe, in which the core of neutron stars would be comprised entirely of strange matter. However, the presence of "nuggets" of strange matter formed during the early universe are ruled out.

The astrophysical observations richly complement efforts to study matter at high density and energy density in the laboratory using relativistic heavy ion collisions. Together, these frame many of the key questions and challenges for the future.

Particle Structure and Nuclear Physics

The discovery that the particles of nuclear physics, the nucleons, the pions and other bosons have structure has had a profound effect on nuclear physics research and goals. To be sure, it was well-known that nucleons had a structure since the discovery of the Δ and the prediction of exchange currents following from the Yukawa theory of nuclear forces recently verified by electron scattering from few-body nuclei ^{3}He and ^{3}H. But it was still hoped that it was a good first approximation to assume that the nucleons, pions, etc., were point particles. In atomic physics, for example, the structure of the nucleus plays a minor role principally because of its small size. But as it turned out, the nucleon radius is about 0.8 fm, which is very large indeed. Moreover, its structure is quite complicated involving quarks and gluons which are confined so that the nucleons are color singlets.

A number of theoretical problems present themselves. How does one reconcile the by-and-large successful understanding of nuclear structure based upon mean field derived from a nucleon Hamiltonian involving interacting point nucleons with the finite size quark model of the nucleon? The nature of nuclear forces is reopened as the current picture based on the exchange of point bosons by point nucleons must be modified. Yet it must be modified in such a way as to ensure the fact that the long range part of the nuclear forces is determined by pion exchange. And how does one understand three-body forces? It soon became apparent that nuclear theorists would have to be concerned with the structure of the nuclear particles as well as with non-perturbative QCD (quantum chromodynamics) the non-linear non-Abelian theory which is believed to describe the quark-gluon medium. These are formidable problems for which the only rigorous solution appears to be a numerical one based on a discretization of QCD, the lattice gauge theory, although that discretization is not without its problems. It is, for example, not possible to simultaneously have locality, hermiticity and chiral symmetry when fermions on the lattice sites are included in the calculation. Of course, a number of models have been proposed which it is hoped mock up certain solutions of QCD, a situation reminiscent of the early history of nuclear physics.

Happily, there is an experimental program which should help to elucidate some of these issues. A great impetus to the whole field was given by the "EMC" experiment in which the deep inelastic scattering of muons by deuterium (D) and Fe were compared. A similar experiment in which electrons were scattered from various elements was later performed at SLAC. These experiments among other things measure the quark and gluon distributions per nucleon. The difference between these distributions for D and Fe were thought to represent the impact of the nuclear environment on these distributions. These experiments are examples of a class of experiments in which the regime between the scattering of very high energy electrons in which the charged quark distribution plays the dominant role, and the scattering of lower energy electrons in which charged nucleon distribution is observed. What happens between these two limits is obviously of great interest. An accelerator (CEBAF) which could furnish electron beams of sufficient energy (only if the nominal energy is exceeded considerably) is under construction in the U.S.

The collision of relativistic heavy ions provides another test of the quark-gluon description. It is hoped that such a collision at the right energies will lead to the deposition of a large amount of energy per unit volume. If this energy density is large enough the mechanism which confines the quarks will be overcome and the quarks will become deconfined and will have a much larger volume in which they can move. Calculations based on the statistical mechanics of quarks and gluons predict that a quark-gluon plasma consisting of quarks and gluons and essentially hadron free can be formed. The transition from the normal nuclear densities for which the nucleon description, with the quarks are confined to one in which the quarks are not confined is thought to be sharp enough as to be considered a phase transition (see Fig. 7). There are possibly two phase tran-

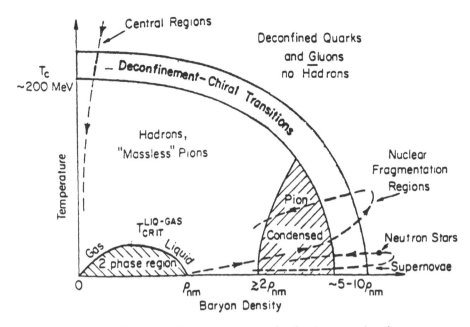

Figure 7. Phase diagram of nuclear matter in the baryon density, temperature plane showing regions of hadronic and deconfined matter. Normal nuclear matter density ρ_{nm} is 0.16 fm^{-3} (from Baym, Ref. [7]).

sitions, one in which chiral symmetry is restored. In a second transition deconfinement takes place. The nuclear-particle physics communities are now engaged in a daring experiment which will test the validity of these ideas. It is thought that the required energy-density deposition will occur when high energy very heavy ions collide. Two such beams are now available. One at the Brookhaven National Laboratory provides beams of mass number up to ^{32}S with an energy 15 Gev/A; the other at CERN provides ^{17}O beams (and later ^{32}S and possibly ^{40}Ca) with an energy of 200 GeV/A. These beams will strike stationary targets so that their effective nucleon-nucleon center of mass energy is 5.6 GeV and 20 GeV for the BNL and CERN experiments, respectively. The projected relativistic heavy ion collider would have 100 GeV/A nuclei colliding with 100 GeV/A nuclei. The nucleon-nucleon center of mass energy is 200 GeV. At this energy the two nuclei will appear to the laboratory observer as two relativistically flattened disks. As the two nuclei collide and pass through each other, each nucleus will have energy deposited in it but in addition there will be energy deposited in the region between the two nuclei. In the former case, the medium will be baryon rich but in the latter it will be baryon-free. It is thought that it is in the baryon free region that a quark-gluon plasma will form. Of course, it is also possible that it will form within the individual nuclei if the energy disposition is large enough. At the lower BNL and CERN energies this will be tested to some extent as the baryon-free region will not be formed. The principal problem with all of these experiments is the identification of a signal indicating the formation of a quark-gluon plasma. If such a plasma is found we shall have a new form of matter, one which prevailed at the very early stage of the history of our universe.

Other New Roles

We shall conclude with a discussion of some classes of experiments which have a great potential but whose major results lie in the future. The availability of low energy anti-proton beam is an example. The collision of these with nuclei and with protons should reveal the nature of the nucleon-antinucleon interaction. Various models give rather different predictions, such as the existence of resonances. One should be readily distinguished amongst these and compare with the prediction of QCD.

A second group of experiments is concerned with hypernuclei. These are nuclei in which one of the baryons is a hyperon that is a Δ or a Σ. Most of the Λ hypernuclei observed so far have been made through a recoiless (K^-, π^-) reaction. This reaction has provided us with a number of states of the hypernuclei for the light nuclei Λ. Hypernuclear states for heavier nuclei seem to be observable in a (π, K) reaction but this possibility is just beginning to be studied experimentally. One of the principal conclusions obtained from these experiments is the weakness of the spin-orbit nucleon-lambda interaction. The discussion of the results of these studies on the basis of the quark model is in progress. Surprisingly, the observation of Σ hypernuclei has been reported, surprising because of the large mass difference between the Λ and the Σ which would imply a large width. But in these experiments the reported result is small, indicating the existence of a narrowing mechanism which may very well reflect a new symmetry principle. This would of course be of great interest.

As a final example, we mention the scattering of polarized electrons by nuclei now being conducted at the MIT Bates Accelerator.

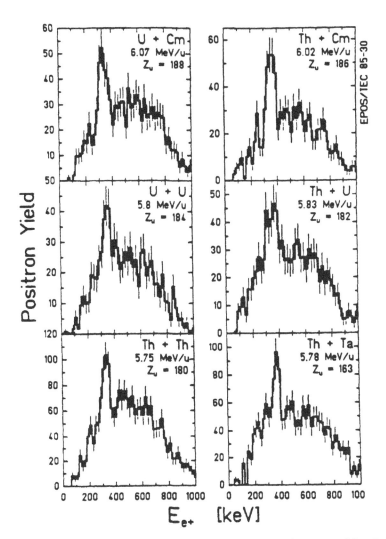

Figure 8. Positron energy for six collision systems with a combined charge $Z_u = Z_{proj.} + Z_{target}$ ranging from $Z_u = 188$ to $Z_u = 163$ and for bombarding energies corresponding to a marginal nuclear contact in head-on collisions. A pronounced peak appears in each spectrum in addition to the continuous atomic and nuclear background. Average peak energies are between 315 and 375 keV, their widths are about 70–80keV, and the c.m. differential cross-sections are similar and of the order 5–10 μb/sr assuming the cross-sections to be constant over the target thickness (≈ 0.15MeV/u) (from Schwalm, Ref. [8]).

These experiments will probe the parity non-conserving components of the electron-nucleon interaction which are predicted by the "standard" theory of the weak and electromagnetic interactions. These experiments will provide important tests of that theory and in particular they will make it possible, using the nuclei as filters, to study the components of the electron-nucleon parity non-conserving interaction.

Narrow Positron Lines Observed in Collisions of Very Heavy Ions

The discussion of nuclear physics issues which are of general interest would not be complete without a mention of the narrow positron lines observed in the collision of heavy ions whose total charge Z_u is very large (see Fig. 8). The bombarding energies are near the Coulomb barrier. It is expected that the large time varying Coulomb interaction would result in positron production, a familiar result which was discussed in the early days of quantum electrodynamics. However, as one can see from Fig. 8, one finds narrow positron lines superimposed on a continuous background for $163 \leq Z_u \leq 188$. The energies of these lines varies slowly if at all with Z_u, the observed energies falling between 250 and 400 keV. The experimental width of the lines is of order of 70 to 80 keV, independent of the heavy ion scattering angles. Positron peaks are found to be correlated with electron peaks in the $U + Th$ collision. The electron energy equals the positron energy and the widths are similar. These facts suggest that a neutral particle of mass 1.6 MeV is formed in these collision decays into a positron-electron pair. However, this interesting hypothesis does not appear to explain all the observations. So as this is being written a puzzle remains.

Acknowledgement

I am indebted to J.W. Negele for his assistance on the section entitled, "Nuclear Astrophysics".

References

1. P. Richard, C.F. Moore, D. Robson and J.D. Fox, *Phys. Rev. Lett.* **13**, 343 (1964).
2. A. deShalit and H. Feshbach, *Theoretical Nuclear Physics, Vol. I: Nuclear Structure* (John Wiley & Sons, Inc., New York, 1974).

3. J.W. Negele, *Rev. Mod. Phys.* **54,** 914 (1982).
4. L.L. Lee, J.P. Schiffer, B. Zeidman, G.R. Satchler, R.M. Drisko and R.H. Bassel, *Phys. Rev.* **136B,** 971 (1964).
5. J.W. Negele, *Rev. Mod. Phys.* **54,** 914 (1982).
6. B.M. Nyako, J. Simpson, R.J. Twin, D. Howe, P.D. Forsyth, and J.F. Sharpey-Schafer, *Phys. Rev. Lett.* **56,** 2690 (1986); P.J. Twin et al., *Phys. Rev. Lett.,* **57,** 811 (1986).
7. G. Baym, in *Proceedings of the 1986 International Nuclear Physics Conference, Harrogate,* J.L. Durell, J.M. Irvine and G.M. Morrison, eds. (Institute of Physics, Bristol, 1987), p. 310.
8. D. Schwalm, in *Proceedings of the 1986 International Nuclear Physics Conference, Harrogate,* J.L. Durell, J.M. Irvine and G.M. Morrison, eds. (Institute of Physics, Bristol, 1987), p. 568.

The New Ether

J.D. Bjorken

FERMI NATIONAL ACCELERATOR LABORATORY

I. Introduction

In this talk, I want to address three topics. The first is to review yet again the standard model of elementary particles and forces. This standard model summarizes a fantastic amount of progress made in the last twenty years in our understanding of the basic building blocks of matter and the basic forces in nature. The second goal of this talk is to connect the standard model with the early universe,[1] the "big bang." The third goal, to which the title refers, is to highlight the emerging importance as time goes on of the vacuum state. By the vacuum state, the "new ether," I mean that quantum state of the world, or of some region of the world, which contains absolutely nothing within, where all the matter has been removed, and where there exists no energy or momentum. It is, in a word, nothing at all. But what can one say about "nothing at all"? At this point I'm tempted to try for the next four minutes and twenty-two seconds to do for particle physics what John Cage did for music. But I doubt that the chairman will let me do that.

Although one probably should not bother talking about nothing at all, nevertheless serious physicists do. The study of the vacuum has become a very sophisticated subject, about which there is a great deal of expertise.[2] There are some experts at this meeting. Of course all these experts are theorists. One of them is Gerard t'Hooft. And I can say with certainty that Gerard understands nothing at all better than I do.

So what is vacuum? Well, experimentalists know what it is; it's what's left in the can after everything is pumped out. But, of course, if one takes a can and pumps everything out, and the temperature of the can is finite, everything isn't pumped out. There are still photons radiated from the walls, hence a gas of photons existing inside the can. In order to get rid of those, one has to go to zero temperature to pump out the photons as well. Now, in thinking about vacuum, that suggests approaching the ideal vacuum from the point of view of the big bang and the early universe. Just take a piece of the universe which at early times has lots of things in it. Then, as the universe expands and cools, fewer and fewer particles per unit volume remain. If we live in an open universe with $\Omega < 1$, maybe in the distant future we'd get something that is as close to real vacuum as a particle physicist imagines it.[3] So in what follows I will try to describe, in the context of big-bang history of the early universe, the raw material of particle physicists and to set the stage for the end of the talk, where we discuss what this vacuum is and why it looks to me more complicated than nineteenth-century ether.

II. Big Bang Overview

The earliest epoch of the big bang we will discuss starts a few femtoseconds after the beginning. This is a long time after the start for contemporary particle theorists; this is a very conservative talk. At this time the contents of the universe are believed to be a hot plasma with a temperature of about 10 TeV, somewhat above what can be reached in terms of energy per particle, in the biggest accelerators. As time goes on, the universe expands and cools. As it goes down through lower temperature scales, it passes through all the energy scales of interest to experimental particle physics and nuclear physics. When the universe is 10 milliseconds old, the temperature is down to 10 MeV or so. Beyond that it's on to the first three minutes, and we need not discuss that.[4] The features we will discuss in detail are shown in Table 1. Needless to say, especially from the nanosecond time scale on back, there is a fair amount of conjecture, because we don't have much in the way of experimental facts beyond the 10–100 GeV mass scale, where the highest energy collisions yet studied in detail have just yielded the intermediate bosons of the weak interactions. In those very early times we use some theoretical hubris; namely that the standard model has been working so well that we think it will, in its gross features, extrapolate quite a way up in the energy scale without serious error.

Table I.

Time	Temperature	Comments
10^{-14} sec	10 TeV	
		Symmetric world
10^{-12} sec	1 TeV	
	400 GeV	second-order electroweak phase transition
10^{-10} sec	100 GeV	W, Z freezeout
10^{-8} sec	10 GeV	top quark freezeout
		bottom ⎫
10^{-6} sec	1 GeV	} freezeout
		charm ⎭
	200 MeV	first-order deconfining phase transition
$\sim 10^{-4}$ sec	100 MeV	baryon asymmetry
$\sim 10^{-2}$ sec	10 MeV	appears

III. Early Times; Highest Energy (10^{-14} sec, 10 TeV)

So let us start at the beginning, when the time is $\sim 10^{14}$ sec and the temperature is ~ 10 TeV. While this is the most speculative epoch we'll be talking about, it's the one which really epitomizes most cleanly our present view of the fundamental building blocks and forces that characterizes the standard model, a view which exhibits a high degree of symmetry. What are the contents of this big bang plasma? Let us build it up by quickly going *up* the temperature scale. At three degrees we have just photons. But as one goes up in the energy scale, the photons start colliding with each other, and they can make other particles. First of all come electrons and positrons. Then they all collide and make still others, such as ordinary hadrons. Going up the temperature scale, one makes all the species that exist, more or less in equal proportion. Table 2 shows the "periodic table" of building blocks.

All of them are there in the plasma at this temperature of 1–10 TeV. There is the ordinary stuff that we're made of: up and down quarks in the three colors, electrons and their neutrinos ν_e. In ad-

Table II. Periodic table of building blocks

	Quarks		Leptons	
4th Generation?	?	?	?	?
3rd Generation	ttt	bbb	τ	ν_τ
2nd Generation	ccc	sss	μ	ν_μ
1st Generation	uuu	ddd	e	ν_e

	Forces	
Sources	Carrier	Strength
Charge	photon	1/137
"weak isospin"	W^+, W^-, Z	1/30
color	8 gluons	1/7

dition there are the two replications of that at higher mass scales; one is the "second family" charm and strange quarks, the muon and its neutrino. The other replication, or "third family", contains top and bottom quarks, the tau lepton and its neutrino. All of these objects are discovered for sure except for the top quark and the tau neutrino. Everybody believes the latter exists, but has never been directly observed. But someday it may be discovered at Fermilab. Now all of these quarks, leptons, and their antiparticles are swimming around in this initial 10 TeV plasma as more or less a free gas. But there do exist interactions between them. The principal forces are the "gauge" forces. The electromagnetic force couples to charged sources; the carrier of the force is the photon. The intrinsic strength of the force is measured by the famous number $\frac{1}{137}$ which is small. (That number should go to $\frac{1}{128}$ or some slightly larger number at this scale, but it's still small) So everything charged, e.g. the top quarks of charge $\frac{2}{3}$, bottom quarks of charge $-\frac{1}{3}$, the leptons with charge ± 1, will interact via the electromagnetic force.

The weak force, mediated by the now famous intermediate bosons $W^\pm$ and Z^0, couple to something called weak isospin. All of the quark and lepton species, at least in part, have weak isospin and couple to the weak force. Its strength is measured by a number which, instead of $\frac{1}{137}$, is about $\frac{1}{30}$. The third "gauge" interaction is the strong force, which acts among the quarks. The quarks, which are "colored", couple to each other through exchange of gluons with a strength of order $\frac{1}{7}$. What makes this picture so satisfying is that there is a great similarity in the way the quarks and the leptons in the various generations behave. Furthermore, all these forces have a great similarity as well. At this 10 TeV scale all the forces are inverse square forces, like the electrical force. All the force carriers, the photon, the $W^{\pm}$ and the Z^0, the gluons, can be considered massless, because the energy scale is much higher than, e.g. the mass scale of the intermediate bosons. All of the force carriers have two polarizations, like photons. The electric vector moves transversely to the direction of motion, either in the horizontal or the vertical. Better, one can circularly polarize the quanta and talk about left and right-handed helicities. All of these forces are based on gauge principles and are described by Maxwell-like equations, although in detail the latter two are more complicated than the electromagnetic. And all the force laws possess lots of symmetry. These underlying symmetries have been vital to the discovery of the force laws and the consistency of the theories that describe them.

The building blocks, the quarks and leptons, have helicity properties similar to the gauge particles. The neutrinos come in only one helicity. Their polarization, if you like, rotates to the left, with their spin angular momentum pointing against the direction of motion. The anti-neutrinos have the opposite helicity. Electrons and positrons, as well as the quarks, possess both left and right helicities for each the particle and the anti-particle. But at this high temperature these are independent degrees of freedom without significant communication between them. And only left-helicity particles or right-helicity antiparticles participate in the weak interaction.

Thus, in summary, at the 10 TeV scale all the quark and lepton degrees of freedom are present in the big bang plasma. There is a high degree of symmetry between different species of quarks and species of leptons. There are no forces between the building blocks which are "strong". Even the strong force mediated by gluons is characterized by a coupling strength $\sim\frac{1}{7}$, only four times as large as that of the "weak" force. The three known forces,—strong, weak and electromagnetic—are at this scale inverse-square forces derived from symmetry principles analogous to electromagnetic gauge invariance. Quarks and leptons can be broken into right and left handed parts; the left-handed parts behave much like neutrinos and participate in the weak interaction. The picture is relatively simple and very symmetrical—and slightly incomplete. We have left some things out. But in this simplest description of the standard model they are usually left out at this stage. When I have to put them back in, I'll return to this temperature scale.

IV. The Electroweak Phase Transition (10^{-12} sec. 400 GeV)

So now we let the clock run, bringing down the temperature of the universe, and watch what happens. Below one TeV we begin to approach the top end of the present and future accelerator ranges of study. At the picosecond time scale, i.e., the several hundred GeV range, there is an occurrence believed to happen by 99 percent of all theorists. This is called the electro-weak phase transition, expected to be of second order. It has an analogy in solid state physics, which is simply the transition from normal metal to superconductor. The solid-state analog of the plasma of quarks and gluons is the gas of electrons in a metal. Below the superconducting transition temperature is formed a kind of Bose condensate of bound electron-hole pairs. The analogous electroweak condensate is very important for elementary-particle phenomenology at low energies. Its presence is believed to be responsible for the fact that three of the gauge bosons, the $W^{\pm}$, W^{-}, and Z^{0}, acquire a big mass of about 100 GeV. This phenomenon also has its analog in solid state physics in the Meissner effect. Long range electromagnetic fields in a charged superfluid or superconducting medium get expelled. Long range electromagnetic fields cannot exist inside of the condensate. Currents are formed at the boundaries of a superconductor, (something which can't happen very easily in vacuum, which doesn't have a boundary) such that the fields fall off exponentially as one goes into the superfluid medium, just like a Yukawa meson field falls off exponentially away from its source. In the solid state analog one can colloquially say that the photon gets a mass. That's loose talk, but roughly speaking the Compton wavelength of the photon is, in the superconducting analog, the London penetration depth. In the particle physics case, the analogous words are that below this critical temperature of several hundred GeV, some sort of condensate is forming which is analogous to the superfluid. By analogy with the Meissner effect, the $W^{\pm}$ and Z obtain a mass. It turns out that the square of the mass is proportional to the weak coupling constant ($\frac{1}{30}$) and to the square of the critical temperature. That means the W

mass comes out to about 80 GeV, eighty times the proton mass. The Z comes out a little heavier.

Now the big question is what this condensate is. Are the degrees of freedom identifiable in terms of the quarks and leptons in the periodic table? While many certainly have tried, I think there is a clear answer of "no." The condensate is not made of the known degrees of freedom, and we must put in something extra. Now the simplest model which does that, the Lagrangian-Higgs model, also has a solid state analog. It is analogous to an obsolete theory of superconductivity called the Ginsburg-Landau theory. One to one the solid-state equations can be mapped over, with some corrections in honor of special relativity.

So if we now add in the minimum addition to the periodic table, we may go back to our original high temperature of 10 TeV and ask what has to be thrown into the plasma to account for this phenomenon. Whatever this turns out to be is dubbed the "Higgs sector." This Higgs sector turns out to be at least two particles along with their anti-particles. One of them is electrically charged and the other is neutral. They possess no helicity or spin and they must have "weak isospin." That is, they couple to the $W^{\pm}$ and the Z. Of course, the charged particle couples to the photon. But none of the four have color and thus they do not couple to gluons. So that what is seen at high temperature in this minimum model, the Lagrangian-Higgs model. At a low temperature below the second order phase transition, a condensate is formed from a combination of the two neutral Higgs particles. The other three are subsumed into the W^+, W^- and the Z^0. Those degrees of freedom are necessary to convert the massless $W^{\pm}$ and the Z^0 with their two transverse polarizations that vibrate in the horizontal and the vertical, into a massive particle which can also have its polarization vibrating in the third, longitudinal direction. There is an extra degree of freedom for the massive spin one particle not present for the massless one, provided by these new degrees of freedom. So 33.3% of the $W^{\pm}$ and Z^0 are made from the Higgs sector. Finally there is a dynamical remnant of the condensate, excitations which have an "energy gap" and correspond to another massive spinless particle, which is called *the* Higgs boson.

Now remember that the minimum model of our world, namely the "first generation" up and down quarks, electron and its neutrino, is far from everything. We don't understand why the rest should be there. And, in the same way, there's no reason why these four extra "Lagrangian-Higgs" particles should be the whole story. Real life may well be much more complicated, with a periodic table for the Higgs degrees of freedom as big as what we have for quarks and leptons. It may even be as big as what is known as the Rosenfeld tables. They are a small book, packed with data on all of the particles (hadrons) that one makes out of quarks, antiquarks, and gluons. The study of the Higgs sector may well be as exploratory as the study of the strong force, from Yukawa to QCD.

V. Freeze-outs (10^{-11}–10^{-6} sec; 400–1 GeV)

As the universe continues to expand and the temperature drops from hundreds of GeV down to, say, 10 GeV, new things happen. The first is what can be called the "freeze-out" of the W's and Z's. As the universe expands and the temperature drops below the W and Z rest energy, they disappear from the plasma. They decay into quarks and leptons. This decay process was present at high temperatures as well, but the W and Z bosons could be replenished by collision processes: they are radiated by quarks and leptons just like photons are radiated by charged particles. At low temperature there is no longer enough collision energy to produce the rest energy of the $W^{\pm}$ and Z.

After the W and Z disappear, quarks and other unstable particles freeze out in a similar way. So far, at these high temperature scales, the rest mass of the quarks and the leptons hasn't entered at all, and in fact has been totally ignored. Where does the rest mass come from? Again it is believed to come from the same mechanism that gives the $W^{\pm}$ and Z their mass. In order to have this happen, the quarks and the leptons must be allowed to couple to the Higgs sector. In particular a Yukawa interaction of Higgs meson emission and absorption is postulated such that say, the left-handed top quark can change to a right-handed one, the one with the opposite helicity, plus a Higgs meson, and vice versa. This is similar to the Yukawa force used for the strong interaction. In fact, the charged Higgs mesons may also be exchanged; consequently the top can go to the charged Higgs plus bottom quark. This interaction between the Higgses and the top quarks suffices to create another phenomenon, which has its analogy in modern BCS superconductivity theory, called the energy gap. In the BCS theory, the electron excitations above the Fermi surface have an energy gap. In this case that is interpreted as the top quark getting a mass, namely, the minimum energy that the top quark can have is a nonvanishing amount above zero. That is simply its rest energy. And the amount of mass the top quark gets is proportional to the amount of coupling that it has to the Higgs bosons. Alas, the mass is not predicted; one gets out no more than what is put in.

Now as the temperature comes down and becomes less than the rest mass of the top quarks, they disappear from the big bang plasma. A loss mechanism is again the weak decay of the top quarks. In addition, they can annihilate with their anti-particles, because they are there in about equal numbers. Again if one wants to get them back into the plasma through inverse processes, it is very hard to do because the rest mass has to be created in a collision. When the temperature is small compared to the rest mass it is very unlikely that there are the particles around to do that. And so the top and antitop drop out of equilibrium ("freeze out") and are lost from the plasma, essentially in the same way the W's and the Z's were lost when the temperature went well below 100 GeV. That is also true for massive Higgs bosons, which have to be there in the plasma at high temperature.

As we go on in time and down in temperature, the rest of the quarks start freezing out by the same mechanism. The bottom quark, with its rest energy of 5 GeV, is next, then the charm quark with its rest energy of 2 GeV, soon followed by the highest mass lepton, the tau lepton, which is the second replication of the electron. It has a mass about the same as the charm quark and will also drop out in the few-GeV temperature regime. Now the time scale is getting almost to a microsecond; the freezouts are the main feature in this epoch following the electroweak phase-transition.

VI. The Confining Phase Transition (10^{-5} sec, 200 MeV)

Downward, below one-GeV temperature and going toward 100 MeV, another very interesting thing happens. It is associated with the change in the content of the big-bang plasma from quarks and gluons to the hadrons, protons, pions, that we're made of, and is usually called the de-confining phase transition. But I will call it the "confining" phase transition: on the way down in temperature it's confining, but looking up in temperature from below (as we are usually obliged to do) it's de-confining.

At temperatures well above about 200 MeV, the important degrees of freedom in the plasma are the quarks and the gluons (and some photons and leptons). At, say, 300 MeV those which haven't frozen out are up, down, and strange quarks, their antiquarks, the gluons, the photons, the muon, the electron, and all the neutrinos, all interacting with each other. Now as the energy scale comes down, the strong force rapidly gets stronger through subtle "vacuum polarization" effects. At about 200 MeV, things come to a head and there is a first order phase transition,[5] with lots of latent heat. Below that first-order phase transition there are no more quarks and gluons; the density and temperature is low enough so that identifiable mesons and baryons can most of the time exist (The available volume per hadron is larger than the hadron size). At the same time something called the QCD non-perturbative vacuum, or condensate, forms. This is a relatively wooly concept, not under as good control, in my opinion, as the idea of the electro-weak condensate. What is this new condensate? It is not quite like a superconductor or a superfluid, but there are some similarities in the implications. To appreciate it, one must understand the main feature of this de-confining phase transition. It is the transition between the strong force law at short distances, which is inverse square, (just like all the other ones) and the force law at long distances, which is a *constant* force, associated with a "string tension." What is supposed to happen is that the color fields surrounding a quark and an anti-quark which are close together look pretty much like the electromagnetic fields associated with a charged dipole. But if one pulls the quark and antiquark apart by more than 10^{-13}cm or so, the field lines somehow are supposed to get squashed together into a flux tube. Colloquially, this can be described as pressure from the vacuum squeezing those flux lines in. One may speak of a gluon condensate or "non-perturbative QCD vacuum" on the outside, excluding the colored flux from it in a manner something like the Meissner-effect. In fact there are nice analogies to magnetic monopoles travelling through type-two superconductors, in which this phenomenon does occur. The field lines of the monopole are crowded together into quantized vortices or flux tubes trailing behind the monopole. But while there are some analogies, the actual mathematical description of this non-perturbative QCD vacuum still remains rather primitive.

At the same time and temperature, something else happens; a superfluid condensate of the quark degrees of freedom forms, again in analogy to BCS superconductivity. This is called the chiral phase transition. It has an order parameter associated with it, which is, roughly speaking, the effective mass of quarks. And that adds to the richness of the vacuum below the confining phase transition.

VII. Later Times and the Emergence of Baryons

As we continue onward on this quick trip through early history, below 100 MeV and down to 10 MeV, we have no quarks and gluons. Strange particles and muons are freezing out as well as the protons, neutrons, and their anti-particles because the temperature is much below their rest mass. Baryons disappear via particle-antiparticle annihilation, but thank goodness they don't all disappear, or we wouldn't be here to talk about it. At temperatures somewhere around 50 MeV or so, the protons and neutrons significantly outnumber the anti-protons and anti-neutrons. The latter very rapidly disappear completely but not the former, because there is a slight asymmetry in the amount of matter relevant to anti-matter in the *early* universe. If we go back again to very early times, say when the temperature was hundreds of MeV or above, less than one part per billion excess of quarks over anti-quarks was needed to account for the amount of baryonic matter in the universe now. The asymmetry was a very tiny effect in most of the history of the early universe. That tiny effect undoubtedly has a very deep physical origin, and probably originated long before the epoch that we have discussed, which started at the first femtosecond or so after the bang. It may well have the same origin as the observed CP violation in the neutral K meson system, a phenomenon found in 1964 by Fitch, Cronin and their collaborators.[6] The neutral K is a down quark bound to an anti-strange quark; it turns out that its decay properties into two pions is slightly different from the decay properties of its antiparticle into two pions. Therefore there is, in the basic laws of the physics that govern these weak decays, a slight asymmetry between properties of matter and anti-matter. The connection between the history of the early universe and this phenomenon goes back to Andrei Sakharov in 1967, in an impressive, visionary, set of ideas.[7] Most theorists now accept this connection as a working hypothesis, and furthermore try to blame both problems on the Higgs sector. (What else is there to blame?)

After this freeze-out, when the baryons are left behind and the anti-baryons are gone, we enter the "classical" period from a temperature of 10 MeV on downward to the present temperature of 10^{-4} electron volts, from the first ten milliseconds to the next 10^{10} years, when a few inconsequential things happened such as the formation of nuclei, atoms, stars, galaxies, Steven Weinberg, and the written history of all of this.[4]

VIII. Comments

Let us now review the implications, experimental or otherwise, of this kind of history. First of all, the scenario is probably wrong in detail, especially at the highest energy scale. The "Higgs mechanism" for the origin of the intermediate boson masses argues for new degrees of freedom and new forces which aren't well understood, and which we are probably describing very inaccurately. Beyond what is "needed" there may well be other new degrees of freedom, new forces, and maybe more phase transitions (or maybe fewer—maybe the idea of the electro-weak phase transition is wrong). In terms of laboratory experiments, this argues for particle searches at all mass scales to see whether something has been left out. There is a lot of room for improvement in that area.

Secondly and more specifically, what really is the nature of that electroweak phase transition? Is it really at 400 GeV, or at some different temperature? Are its properties what we think it is? Is there more than one transition? All this bears upon the delineation of the Higgs sector. The mass scale is quite firmly set as no higher than one TeV, and that is very suggestive that one really needs to experimentally explore up to that mass scale very well. This is why particle physicists are in such forceful unanimity about the need for the superconducting super-collider, or SSC. This twenty TeV machine, with a cost of about three billion dollars, addresses in observational terms the physics of the TeV mass-scale, i.e. the mass-scale of the expected electroweak phase transition.

The next question, a little more modest, is the nature of the confining phase transition.[8] Here there are two areas in which much can be and is being done. On the theory side, one needs to calculate better the equation of state of the plasma as well as its transport properties. Experimentally, the relativistic heavy ion collider RHIC, which is proposed to be built some day at Brookhaven and which I hope is built, may provide means of producing quark-gluon plasma at temperatures above the confining phase transition, much as it was supposed to be during the first microseconds of the big bang.

The origin of the baryon asymmetry is a vital question for us, literally. That invites incisive studies of CP violation anywhere one can find it, not only in the K system, but also possibly in heavy-quark (bottom or charm) meson systems. There is a tremendous technological challenge here. There is also very rapid progress, which will probably evolve for the next couple of decades, in being able to experimentally examine particles containing charm and bottom quarks as well as we do now with K's.

IX. The New Ether

What about the either itself? I think one can't help but be impressed at how complicated the vacuum has become. We see all these phase transitions, with the vacuum ascribed all sorts of nontrivial properties. The particle theorists tend to talk about vacuum in the same language as used by the condensed matter people: condensates, order parameters, etc. It is as if the vacuum really is physical, that there is dynamics associated with it. If so, there is a question that immediately arises, namely whether the vacuum gravitates. If there is dynamics associated with all these phase-transition mechanisms, then there should be energy, at least potential energy, associated with them. If one tries to calculate it, it usually comes out infinite and is subtracted away; it is quite an ambivalent question when approached from the point of view of formal quantum field theory. But were there potential energy associated with those condensation phenomena, it would give an enormous cosmological constant in the equations of general relativity. Experimentally, it is not there to one part in 10^{120} of what might naively be expected. It is a nontrivial theoretical challenge to reduce the discrepancy.

There are a lot of other complications in the vacuum which I haven't even mentioned. In the old days, when I started out in this game, a popular subfield was something called "axiomatic field theory." Before one had a good handle on how to make theories of strong and weak interactions, one started with principles believed to be absolutely safe, and tried to derive general consequences from them. Some of those principles were the properties of the vacuum. These seemed absolutely trivial and self-evident at the time, namely that the vacuum state is unique, that it is Lorentz invariant, has zero energy, and has zero momentum. Other than that last one, all those postulates have nowadays been abandoned.

The vacuum is not unique. In the theory of the strong force (QCD) there are a countably infinite number of vacuua, which differ only by the number of topological knots in pure gauge potentials that are present in vacuum configurations. A pure gauge potential doesn't contribute any energy to the vacuum. But if the topology of the gauge potential is non-trivial, its vacuum is non-trivially different from ones, for example, that don't have any gauge potential at all. Furthermore there are dynamical couplings between all of these different vacuua, associated with the names "tunnelling", "instanton" and t'Hooft. This dynamics is extraordinarily subtle, and is not at all irrelevant. This vacuum degeneracy can potentially lead to observable CP violations in the strong force. This would be a disaster, because Norman Ramsey and his colleagues have measured the absence of the neutron electric dipole moment to very, very high accuracy. This creates no doubt about the fact that there is no substantial CP violation in the strong force. This situation is patched up in the present theory by adding still more Higgs condensates at extremely high mass scales. This is a serious problem, not fully solved, which also has experimental implications. The most exciting one is the search for "invisible axions," which might be a candidate for the dark matter of the universe.

The second of the three postulates, the Lorentz invariance of the vacuum, is abandoned in gauge theories. The description of the vacuum is what are called physical gauges ends up not being Lorentz-invariant.[9] A Lorentz transformation of vacuum is accompanied by a gauge transformation. Of course the physical consequences are Lorentz covariant and gauge-invariant. But the description looks a little clumsy.

As we discussed earlier, the third of the three postulates, zero energy of the vacuum, ought to be true in order to avoid an enormous gravitational cosmological constant. But as we also discussed, we don't really know why. So it would seem at the very least that we are being led into a very complicated description of something which ought to be absolutely simple. And it may be that, just as with the old ether, it is our descriptive structure that is wrong. Sooner or later we may get a description that is as efficient and elegant as special relativity, one that leaves the physics consequences more or less alone and gets rid of a lot of excess conceptual baggage. On the other hand, maybe these complications really imply that there are elements of physical reality somehow involved in this description of the vacuum, ultimately observable. I wouldn't think the observations are just around the corner. If they were to be accessible, I would question at the same time *all* the sacred principles of contemporary physics, including gauge-invariance, Lorentz-covariance, and maybe quantum mechanics itself. If such basic principles were in need of modification, I feel that any replacement would for sure have to be not only subtle but very beautiful.

References

1. See the talk of George Field in these proceedings.
2. There is even a book: J. Rafelski and B. Müller, "The Structured Vacuum," Deutsch (Thun, Switzerland, 1985).
3. A splendid discussion of what might happen-and whether life could survive-is given by Freeman Dyson, Revs. Mod. Phys. **51**, 447(1979).
4. See S. Weinberg, "The First Three Minutes," Basic Books, Inc., (New York, 1977).
5. Probably. The order of the transition is still a matter of theoretical debate. But no matter what, a lot of latent heat is released within a small temperature increment.
6. J. Christenson, J. Cronin, V. Fitch, and R. Turlay, Phys. Rev. Lett. **13**, 138 (1964).
7. A. Sakharov, Sov. Phys. JETP Lett. **5**, 24 (1967).
8. See the talk of Arthur Kerman, these proceedings.
9. "Unphysical" gauges can be Lorentz covariant, but allow unphysical particles in the spectrum, including particles emitted with negative probabilities. It is not clear that, from a physics point of view, this is to be preferred.

Recent Developments in Cosmology

George Field

HARVARD COLLEGE OBSERVATORY

Galaxies

By far the most prominent feature of the universe is galaxies. To get a feel for the distribution of galaxies, a catalogue of galaxies was computerized and the distribution of galaxies put on this photograph (Fig. 1). The catalogue reaches out to about $\frac{1}{20}$ of the radius of the observable universe, which is 10 to 20 billion light years. There are about a million galaxies represented in this photograph.

What is going on in cosmology today? Following Einstein's general relativity of 1915, Friedmann used his theory in 1922 to construct models of the universe, assuming them to be homogeneous and isotropic. The essence of general relativity is that the local curvature of space-time depends upon the matter and energy content. Because the models are homogeneous and isotropic, the space curvature is constant everywhere. It is indicated by the index k, and in Figure 2 we see what that means.

Although Hubble discovered the expansion of the universe in 1929, which is an immediate implication of the Friedmann models, he was unaware that it had been predicted theoretically that the velocity of recession of a galaxy is proportional to its distance. The value of the constant of proportionality at this epoch in the evolution of the universe is called H_0. Hard work since 1929 has succeeded in narrowing down the observational estimate of H_0 to between 50 and 100 kilometers per second for every Mpc of distance. (A megaparsec is a million parsecs or about three million light years.) Although what I have to say depends on the value of H_0, the overall conclusions do not depend very sensitively upon this still-imprecise parameter.

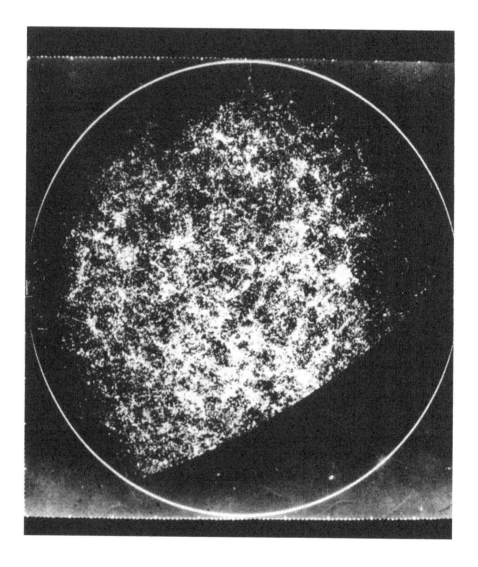

Figure 1. A computer-generated plot showing the distribution of the million brightest galaxies on the sky. The apparent filamentary structure is real, and is due to the bubblelike distribution of galaxies in space (see figure 8).

Figure 2
Three Possible Geometries of Friedmann Models

Type	k	Ω
Open	−1	< 1
Flat	0	= 1
Closed	+1	> 1

There are three possible geometries that Friedmann considered for space-time. An infinite universe with hyperbolic geometry is referred to as the "open universe"; if one could accurately measure huge triangles in such a universe, one would find the sum of the angles less than 180°. There is a "flat universe" to which Euclidean geometry pertains, and there is a "closed universe" in which the sum of the angles is greater than 180°. Each model corresponds to a particular value of k, a free parameter in the theory. The different models can also be described in terms of a dimensionless parameter $\Omega = \rho/\rho_c$, the mean density ρ of the universe divided by a critical value ρ_c of that density. The critical value at this epoch can be calculated in terms of the expansion rate, H_0:

$$\rho_c = \frac{3H_0^2}{8\pi G} = (0.5 - 2) \times 10^{-29} \text{ g cm}^{-3}. \tag{1}$$

Because H_0 is uncertain, so is ρ_c. It follows from Figure 2 that if we could somehow determine Ω, we could also determine k, and therefore the curvature of space. The rest of my talk concerns the value of Ω.

The three different models also correspond to three different histories of the universe, which at present is an expansion: the scale factor R, which measures the distances between galaxies, is increasing with time.

In all of the models, as one goes back in time there is a singularity at a time which is arbitrarily called $t = 0$. The future history of the universe depends upon the model. The open universe ($k = -1$) expands forever, the closed universe ($k = +1$) collapses again, and in the case with $k = 0$, the scale factor reaches infinity with a zero velocity, as with Newtonian two-body motion in which the energy is zero. Corresponding to each curvature index $k = \pm 1$ or 0, then, there is a history of the universe, and a corresponding value of Ω.

In the 1940's Gamow and his collaborators showed that if one goes back early enough, most of the energy in the universe will be radiation, and in that case, there is a unique relationship between the temperature of the universe and its age:

$$T = \frac{1 \text{MeV}}{\sqrt{t}}. \tag{2}$$

Nuclear reactions occur in the first few minutes. Gamow was interested in the production of the light chemical elements that can be observed in astronomical objects. It was known at that time that most of the matter in the universe is either helium or hydrogen, and he wanted to predict the ratio. He and his collaborators, Herman and Alpher, concluded that to get the observed ratio, the current temperature of the universe has to be about 5 K. Steven Weinberg, in his book, *The First Three Minutes*, explores why theorists and experimenters didn't pick up on that prediction, and why little was done on this problem. In 1965, Penzias and Wilson serendipitously discovered the cosmic blackbody radiation, which is believed to be the radiation predicted by Gamow and his collaborators. Contrary to the situation a few years ago, when balloon experiments done at Berkeley seemed to indicate a significant deviation from a blackbody spectrum, recent similar experiments give very good agreement with a blackbody spectrum (Fig. 3). The temperature is now known to within a percent or so, to be $2.74 \pm .04$ K.

Nucleosynthesis in the first few minutes of the universe depends on the entropy, which is dominated by the blackbody radiation. The number of photons in the blackbody radiation is conserved, but each photon lengthens, or redshifts, as the universe expands, so that the gamma rays of the first three minutes are the millimeter-wave photons of today. As the number of baryons has also been conserved, one can define a dimensionless ratio, equal to the number of baryons per photon:

$$\eta = 6 \times 10^{-7} \frac{\Omega_b (H_0/100)^2}{T^3}. \tag{3}$$

The abundances of the various light elements, and the ones that are of greatest interest because they have a reasonably high abundance and therefore might be possible to observe, are H^2, He^3, He^4, and

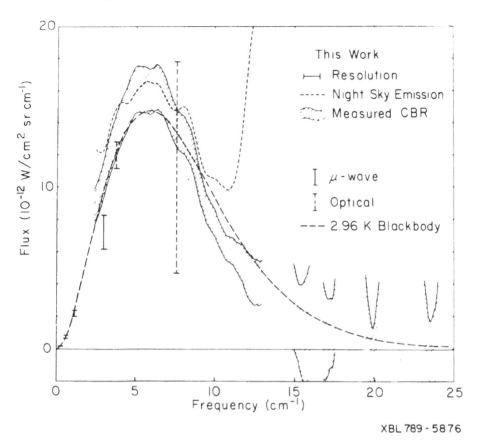

XBL 789 - 5876

Figure 3. Spectrum of the cosmic microwave background. To a first approximation, the spectrum is that of a blackbody at about 3^0 K. Many measurements at wavelengths longward of the peak at about 1 millimeter wavelength, not shown here, confirm this. The measurements shortward of 1 millimeter, made with a balloon- borne telescope above the atmosphere, demonstrate the thermal nature of the spectrum. The best-fit temperature has been revised since this figure was made, to 2.74 ± 0.04^0 K.

Li^7. Those have now all been observed in astronomical sources, and one can fit all of the observed abundances with a value of η between 3 and 10×10^{-10}. Considering the observational inaccuracies involved, I think it is one of the great triumphs of big-bang cosmology that one can fit the data with a narrow range of η's. Once η is known, one can calculate the number of baryons from the number of photons (which we can measure). This determines the contribution to Ω by baryons, Ω_b.

The stuff that the earth, sun, and stars are made of appears to be baryons. Note that the Hubble constant H_0 enters quadradically in Eq. (3), and since we do not know it very well, the uncertainty of the Hubble constant, and the rather uncertain range for the value of η that fits the abundances of the light elements give an order of magnitude uncertainty in Ω_b:

$$\Omega_b = 0.008 - 0.110. \tag{4}$$

Equation (4) indicates that there are not enough baryons in the universe to close it. On the other hand, they fail only by a factor of 10 to 100.

An independent viewpoint originated with Alan Guth, now at MIT, in 1981. He pointed out that the grand unified theory that was then in vogue, SU(5), predicted that there would be a phase transition at a very early time, when the temperature was 10^{15} GeV and the time was 10^{-35} sec. This is obviously extrapolating particle physics to very high temperatures, far beyond the certainty in the theory. One should question whether it is valid to use classical general relativity at all. At times before the Planck time, equal to 10^{-43} sec, general relativity must involve quantum phenomena. The time of the phase transition, 10^{-35} sec, is some eight orders of magnitude later than the Planck time, so it seems reasonable to use quantum field theory in a classical curved space time.

GUTs predicts that as the phase transition took place, the universe "inflated" by a factor of at least 10^{20}. An observer at that time would see a given particle redshift beyond his or her horizon, and as this happened to more and more particles, the observable universe became a tiny fraction of the whole universe. When the inflationary phase ended shortly thereafter, the normal Friedmann expansion began again, and has persisted ever since. During the Friedmann expansion particles have been moving back over our horizon into the observable universe. We see particles appear at an infinite redshift (the horizon). As they are decelerated by the gravitation of the universe, their redshifts decrease to finite values, and they become part of the observable universe. As Heinz Pagels has put it, we should not be saying "hello" to these particles but rather, "hello again." They were once part of the observable universe but they have taken a long ride beyond the horizon before returning to us.

Why is inflation interesting? A number of problems had arisen with the Friedmann models, as discussed by Guth. First of all, there is the so-called "flatness" problem. Recall that Ω_b is roughly 0.01

to 0.1. Certainly a value of Ω which includes all conceivable matter (not just baryons) cannot be much larger than 1, or we would have observed the effects in the redshift-distance relation. Suppose Ω were less than 1 at the moment, so that the universe is open. In the future it would decrease far below 1 as the open universe "coasts" out to infinity. It would be surprising that astronomy would arise just in the early phases of an open model, so that Ω would be less than 1, but not much less. On the other hand, if Ω were somewhat greater than 1, the universe would start to collapse in the future. Again, why would astronomy arise before that happened? The fundamental unit of time which applies here is the Planck time, 10^{-43} sec. Why, then, has the universe lasted 10^{60} Planck times? In short, why is Ω now of order 1? In the inflationary scenario, the actual universe is more than 10^{20} times larger than the observable universe, so that we see only an infinitesimal chunk of the actual universe. If one looks at only a tiny part of a curved space it always appears to be flat, for the same reason that Euclidean geometry applies in everyday life.

What about the isotropy of the universe? The blackbody radiation is known to be quite isotropic. As experiments continue, no one has found deviation from isotropy greater than about 10^{-4}. This is a problem, because if one compares one part of the sky with another when the blackbody radiation decoupled from the matter, the two regions involved were causally disconnected if the areas on the sky are separated by angles greater than a few degrees. It is hard to explain in a Friedmann model that such regions have the same temperatures to one part in 10^4. In the inflationary model all the presently observable matter was within the horizon before the inflation started, and hence in good thermal contact. As matter reappears over the horizon, it carries with it the effect of thermal diffusion that occurred before it inflated over the horizon.

Another problem that arises within the framework of GUTs is that one expects monopoles, but we see very few, if any. Why? Although at one time they were numerous, they have been "inflated" beyond the horizon.

What about deviations from smoothness? As homogeneous as the universe seems to be on a large scale, we do have galaxies—thank God!—and something must account for those deviations from a smooth distribution of matter. Lo! and behold, quantum fluctuations in the GUT phase, which later inflated by a huge factor in size, can account, at least qualitatively, for the fluctuations that we now see in the universe as galaxies. Although the inflationary model originally proposed by Guth ran into some difficulties, a revised version called the "new inflationary universe" does make sense of the universe as we see it.

However, a new problem has arisen. Inflation says that the universe must be flat to within something like 10^{-50}; so that Ω *must* be 1. I've already said that baryons don't provide enough matter. What about other particles envisioned by GUT's? Some are very weakly interacting and so would not have affected the nucleosynthesis calculations. For example, massive neutrinos of about 30 eV, axions of about 10^{-6} eV, or photinos of about 100 GeV could close the universe. The rest of the discussion will consider whether this is plausible.

Astronomers are searching for signs of additional matter through its gravitational interactions. Already in the 1960's observers had discovered that there exists more matter than can be seen. Such matter, called "hidden matter" or "dark matter," seems suggestive in the present context. As Newton's laws are a valid approximation to general relativity within limited regions, one can determine the masses of astronomical objects from the virial theorem. Consider the Milky Way, the galaxy in which we live. We know a lot about it, and we can calculate the mass within the sphere on which the sun is located, about 7000 parsecs from the center. Figure 4 is a photograph of a nearby galaxy, M31, the Andromeda Nebula, which, as far as we can tell, is a twin of our own Galaxy. The sun in this picture would be out at the edge of the visible galaxy. In units of 10^{12} solar masses, the mass of our Galaxy interior to the sun is about 0.15. Figure 5 is a plot of the velocity of rotation of such a galaxy like ours and M31, versus the distance from the center in kiloparsecs. From the light distribution in this galaxy one would expect that the rotation curve would drop off like the inverse square root of the distance, beyond a certain distance, because if most of the mass is concentrated as indicated by the light distribution, objects beyond that distance move in an inverse-square gravitational field, and thus conform to Kepler's laws. Instead, virtually all spiral galaxies have a curve that instead of going down, just continues flat, with velocity independent of distance. The only explanation that seems to make sense is that there is more matter in the outer parts that we don't see, but which keeps the gravitational field from falling off like the inverse square. (To account for flat rotation curves, one needs a density varying like the inverse square of the distance, if it is spher-

Figure 4. The Andromeda Nebula, Messier 31. This galaxy, our nearest neighboring galaxy, is believed to be very similar to the Milky Way galaxy in which the sun is located. It is at a distance of 2 million light years. Like the Milky Way, it is a great flattened disk, rotating in space as shown in Figure 5.

ically distributed.) This leads to the conclusion that spiral galaxies have at least six times as much dark matter as luminous matter.

For elliptical galaxies, it is hard to use the same trick, because they are not in centrifugal equilibrium, but recently Jones and Forman at the Center for Astrophysics discovered a new effect which also indicates the presence of dark matter. Elliptical galaxies have hot coronas around them which can be seen with x-ray telescopes (Fig. 6). Using the hot gas as a tracer of the gravitational field, one comes to the conclusion that the masses of these galaxies are quite high: about 1 in units of 10^{12}. If you look at pairs of galaxies which are orbiting one another, again one gets about 1 times 10^{12}. If one looks at the x-rays from a cluster of galaxies like the cluster in Virgo (Fig. 7), which has a large number of galaxies in a small region, one finds hot gas apparently distributed between the galaxies. Analyzing the distribution of the hot gas, one can estimate the total virial mass of the cluster. Dividing by number of galaxies, one gets an estimate

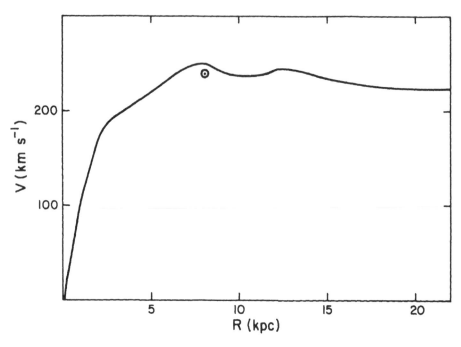

Figure 5. A composite "rotation curve" for the Milky Way and M31 galaxies, showing how the velocity of rotation, V, varies as one moves away from the center of the galaxy. The unit of radius R is 1 kiloparsec, about 3000 light years. The concentration of light toward the center of M31 leads to the expectation that V will decrease like $R^{1/2}$, owing to inverse-square variation of gravitational attraction. Nothing like that is seen, leading to the conclusion that there must be large amounts of matter far from the center of these galaxies. As the matter is not observed (Figure 7), it is referred to as "hidden mass."

for the mass of a typical bright galaxy. Again, the result is about 1 times 10^{12} solar masses. (See Fig. 7.)

All of this I think is consistent with an estimate of about 10^{12} solar masses for a typical galaxy. (Technically, I am speaking of a galaxy with a fixed luminosity, called L^*. We know how to correct for the fact that not all galaxies have the same luminosity, so don't worry about that.) Now take the estimated mass of a galaxy, multiply by the number of galaxies out to a certain distance, and divide by the corresponding volume; one finds Ω for all the material in galaxies, *including* the dark matter. One finds that this Ω is about 0.1, and almost certainly less than 1. I should stress that this estimate of Ω is independent of the value of the Hubble constant, unlike that one

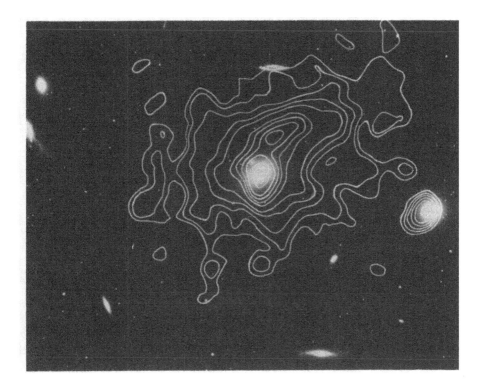

Figure 6. X-ray contours of the elliptical galaxies M84 (center) and M86 (right), superposed on optical images of the two galaxies. The x-ray data reveal that these galaxies contain very hot gas, at several million degrees, invisible in the optical images. The distribution of gas indicates that the gravitational attraction holding the gas to the parent galaxy is at least 10 times that which can be accounted for by the visible matter in the optical image.

gets from the theory of nucleosynthesis. Recall that within the errors, $\Omega_b = 0.01$ to 0.1. It is not excluded, therefore, that all the matter in galaxies, including the hidden matter, could be baryonic, for example, very faint stars, whose masses are so low that they emit too little radiation to observe.

How then, can one make an inflationary universe? It seems very likely that the matter required by inflation is non-baryonic, perhaps of one of the types that I mentioned. Most of this matter has to be outside of galaxies, because when we look at galaxies we don't see its gravitational effects. There is a hint of where the non-baryonic matter is. Radio astronomers have been observing "gravitational

Figure 7

Masses of L* Galaxies
unit: $10^{12} M_\odot$

Milky Way galaxy: Matter inside Sun's orbit	0.15
Other spiral galaxies: larger radii	>0.6
Elliptical galaxies (using hot gas as tracer)	~1
Binaries (Pairs of galaxies in orbit around each other)	~1
Clusters of galaxies	0.8

lenses." Sometimes when one observes a quasar, which is a very distant point-like radio source, one sees not one but two images that appear to be exactly the same, and very close together on the sky. The members of the pair are observed to have exactly the same redshift, the same spectrum, and brightness. It seems that one is dealing not with two objects, but two images of the same object, or in some cases three images of the same object, which can be explained if there is gravitational lensing by a foreground object. The required masses for the lenses are of the order of 10^{12} solar masses, just like the galaxies we've been talking about. But in many cases astronomers have failed to observe a foreground galaxy of the mass required. It is as if there is a "shadow galaxy" which contains mass but which emits little light. Could this be explained if the non-baryonic matter making up most of the matter of the universe were condensed into "shadow galaxies?" Perhaps, and one might speculate further that some of hidden matter in normal galaxies is also non-baryonic.

The picture that emerges is of a population of shadow galaxies composed of non-baryonic matter, of which only a fraction, say 10%, contains baryons. Because we see only this fraction, we mistakenly conclude that $\Omega \approx 0.1$, not 1. This scenario requires a method for keeping the baryons out of 90% of the shadow galaxies. A candidate for this is hydrodynamic motions (which would not affect weakly-interacting particles) such as shock waves induced by supernova or

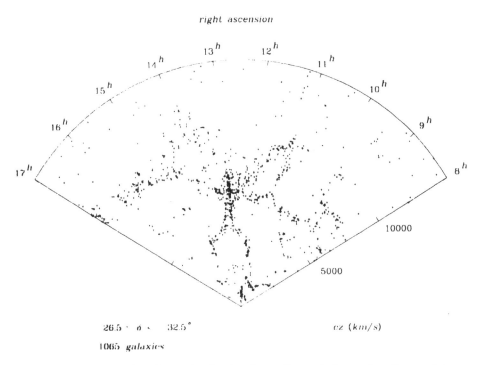

Figure 8. The distribution of galaxies in a slice of sky extending out to a redshift of 15,000 km/sec. The conspicuous group at the center is the Coma Cluster, a well known cluster of galaxies. The other galaxies are at distances indicated by their redshifts; they appear to lie on the edges of giant bubbles.

quasar explosions. Perhaps such shock waves clear the baryons out of most shadow galaxies, leaving only a small fraction to become visible galaxies.

There is another hint in the work at the Center for Astrophysics by de Lapparent, Geller and Huchra. When I first saw Figure 8 about a month ago, I felt that it was a great discovery in astronomy. De Lapparent, et al., have plotted the positions in space, as opposed to the positions on the sky, of individual galaxies. This is done by measuring the redshift of each galaxy, which according to the Hubble law, is an indication of its distance. Figure 8 is a two-dimensional map of the galaxies with right ascensions between 8 hours and 17 hours at a particular declination. The distribution of galaxies shown here, which was presaged by work indicating that there are voids in the distribution of galaxies, shows that most of the galaxies lie in "sheets" or "shells" of some kind on the surfaces of voids. This was totally unexpected. Galaxies are not simply clustered in certain regions; they also seem to avoid other regions.

How can one rationalize this with the concept of "dark matter" or "shadow galaxies?" According to the shadow galaxy scenario I discussed, the voids in the galaxy distribution may not be empty, but filled with shadow galaxies, which reveal themselves from time to time as gravitational lenses. If this is correct, the shock waves which remove baryons from some galaxies somehow deposit them on shells, where they fall into shadow galaxies and form stars. Ostriker, Cowie and Bertschinger have proposed (in another context) that shock waves in an expanding universe will cool and form dense shells, and numerical simulations support that idea. Now the question is whether we can find explosions powerful enough to do the job.

To conclude, there is about ten times as much dark matter as visible matter in ordinary galaxies. The number of baryons deduced from Big Bang nucleosynthesis (Ω_b up to 0.1) may be sufficient to account for this. One needs $\Omega = 1$ for inflationary models to work, and there is not enough matter, either visible or dark, in ordinary galaxies to account for this. Non-baryonic matter is required; it might be condensed into "shadow galaxies" from which baryons have been removed by shock waves.

Black Holes and the Foundations of Quantum Mechanics

Gerard 't Hooft

UNIVERSITY OF UTRECHT

Abstract

A new relativity principle in quantum mechanics may have observable consequences in the Hawking radiation of a Schwarzschild black hole. The original formulation of this phenomenon is questioned.

1. Introduction

Understanding the quantum mechanical properties of black holes may well be crucial if one wishes to devise a credible model of elementary particles in the region of the Planck length scale. Concerning a black hole with Schwarzschild radius much greater than the Planck length one expects to be able to learn much from considerations of ordinary quantum field theory in the modestly curved background of its metric. Indeed until shortly it seemed that one result is well established [1]: due to vacuum fluctuations around the Schwarzschild horizon all heavy black holes must emit particles of all sorts, and this emission has a thermal spectrum corresponding to a temperature $T = 1/(8\pi M)$, where M is the mass of the black hole, in Planck units. The importance of this result is that it suggests well-determined decay properties of all "fundamental" particles heavier than the Planck mass.

Unfortunately, two difficulties were standing in the way of further progress. One is that this new piece of information is clearly only statistical in nature. Purely thermal radiation can only be expected from sufficiently large systems so it is reasonable to assume that only very large black holes behave this way on the average. For not too large black holes we would like to have the quantummechanical amplitudes, not only the probabilities. Can we also say something about these amplitudes by applying quantum field theory in a curved space-time? The standard picture appears to deny such a possibility.

Secondly, in an attempt to find an improved formalism, this author recently proposed an alternative model [2] for the radiating black hole. Although not yet all consequences of this model have been elaborated, it has one striking feature: black holes radiate, as before, but here the temperature is $1/(4\pi M)$, not $1/(8\pi M)$. For this reason, most investigators dismissed this model. Hawkings value for the temperature, $1/(8\pi M)$, seemed to be an inescapable consequence of known laws of physics. So how can one possibly believe a model that gives a conflicting value? Now this would have been the end of just another untenable theory if the author had not further analysed the cause of the discrepancy. The flaw, he now claims, is not in the logic of the new theory but in the logic of the old one, in as far as inescapability of Hawkings result is claimed. In fact, the new model suggests a new alley to produce a completely quantum mechanical, not just statistical, description of large black holes.

We claim that the very foundation of quantum mechanics and relativity theory are touched upon.

Since the time the author first proposed the alternative model several years have past and by now he is convinced that it is presumably as naive or worse than the standard picture. So we do not desire to defend the factor 2 at all cost, but there is this curious possibility that the probability interpretation of wave functions needs to be reconsidered with extreme care when a black hole horizon is present.

In the next section we present the most essential steps in the derivation of Hawkings familiar result, and then in section 3 describe briefly our alternative. In section 4 we explain our attitude concerning the difference between the old and new models as a new "relativistic" effect. We then give in section 5 an extensive discussion of the objections against our views that are most commonly heard.

2. The Conventional Theory

The original Schwarzschild metric is defined in a region $r \geqslant 2M$ and is independent of its time coordinate t. As is well known there is a natural extension of this metric showing that $r = 2M$ is not a singular point. One writes, for instance [3],

$$uv = \left(1 - \frac{r}{2M}\right) e^{r/2M}, \tag{2.1}$$

$$\frac{v}{u} = -e^{t/2M}, \tag{2.2}$$

the so-called Kruskal coordinates. Defining

$$\rho = v - u; \tau = v + u, \tag{2.3}$$

we find that the Schwarzschild region, I,

$$\rho \geqslant |\tau|, \tag{2.4}$$

is solidly connected to another region, II;

$$\rho \leqslant -|\tau| \tag{2.5}$$

via regions III ($\tau > |\rho|$) and IV ($\tau < -|\rho|$). The metric in II is the mirror image of region I, a statement that remains true also for the Kerr, Reissner-Nordstrom and Kerr-Newman solutions. Regions II and IV do not exist or carry a different metric when the imploding object that give birth to the black hole is taken into account. If we nevertheless use these regions of the metric (2.1), (2.2) then they should be considered as nothing but convenient analytic extensions of the metric, to which we can also extrapolate the fields of the outgoing particles.

Consider the generator H of a translation in the Kruskal time coordinate τ. Since this is not a symmetry of the metric this hamiltonian is not exactly conserved but as long as the curvature of space-time is not too strong it does describe the evolution of a quantum field in this metric. We can write

$$H = \int \mathcal{H}(\mathbf{x})d\mathbf{x} \geqslant o \tag{2.6}$$

and a single ground state $|o\rangle_k$ with

$$H|o\rangle_k = o \tag{2.7}$$

is to be expected (after normal ordering of $\mathcal{H}$). Due to curvature this vacuum is not exactly but approximately conserved.

We wish however to describe the time evolution of the

Schwarzschild system as seen by an outside observer. The generator of a translation in t turns out to be

$$h = H_I - H_{II}, \qquad (2.8)$$

with

$$H_I = \frac{1}{2M} \int_{\rho=o}^{\infty} \rho \, \mathcal{H} d^3x; \; H_{II} = \frac{1}{2M} \int_{\rho=-\infty}^{o} |\rho| \mathcal{H} d^3x, \quad (2.9)$$

and

$$[H_I, H_{II}] = o. \qquad (2.10)$$

We can write the eigenstates of H_I and H_{II} as $|n,m>$ with

$$H_I \, | \, n,m\rangle = n \, | \, n,m\rangle; \; H_{II} \, | \, n,m\rangle = m \, | \, n,m\rangle. \qquad (2.11)$$

A straightforward calculation now shows that the "Kruskal vacuum" $| \, o\rangle_k$, does not coincide with the "Schwarzschild vacuum" $| \, o,o\rangle$, but instead we have

$$| \, o\rangle_k = C \sum_n | \, n,n\rangle e^{-4\pi Mn}, \qquad (2.12)$$

where C is a normalization factor. Note that we do have

$$h \, | \, o\rangle_k = o, \qquad (2.13)$$

which is due to t-invariance of $| \, o\rangle_k$.

In the view of the establishment the way to proceed is now as follows. Consider any observable operator O, built from the field operators ϕ in space I. Necessarily we have

$$[O, H_{II}] = o; \langle n',m'| \, O \, |n,m\rangle = O_{nn'}\delta_{mm'}. \qquad (2.14)$$

We find

$$O_k = {}_k\langle o \, | \, O \, | \, o\rangle_k$$

$$= C^2 \sum_{n,n'} e^{-4\pi M(n+n')} \langle n',n' \, | \, O \, | \, n,n\rangle \qquad (2.15)$$

$$= C^2 \sum_n e^{-8\pi Mn} \, O_{nn}.$$

We recognize a Boltzmann factor $e^{-\beta n}$ with $\beta = 8\pi M$, corresponding to a temperature

$$T = 1/(8\pi M). \qquad (2.16)$$

Hence, the Kruskal vacuum $| \, o\rangle_k$ corresponds to a mixed state as seen by observers in I with temperature as given by (2.16); the black hole radiates. The reason why we have a mixed state is clear: all particles in space II are fundamentally unobservable and any quantum mechanical superposition of states in I and II corresponds to mixing of states in I only.

The above constitutes only the main lines of the argument. It can be strengthened by confirming that indeed $| o \rangle_k$ is the state that will emerge when matter collapses, and by checking that eq. (2.15) is quite generally valid also for non-scalar fields.

Thus, the logic that was used to obtain this result seems to be impeccable and unavoidable. Yet it could be wrong. In any case, the conclusion that collapsing matter in a pure state would produce a black hole in a mixed state would imply that abundant "mixture" of quantum mechanical states would occur at the Planck scale of length, time and energy. It seems an impossible task then to recover any kind of unified particle theory at that scale such that ordinary "pure" quantum mechanics would emerge at our presently available energies. More importantly, this conclusion seems to contradict the pure field theoretic arguments that *were* used as a starting point for the calculations.

3. The Alternative Model and Its Temperature

In our alternative model the information that went into space II is not entirely lost. We use the same Kruskal metric as before. Again we assume that at equilibrium the particle content of the black hole is described by the Kruskal vacuum $| o \rangle_k$ (or else a state that contains only a negligible amount of particles as seen by the Kruskal observer). All we do is give a different interpretation of the basis $| n,m \rangle$ in terms of the states in the Hilbert space for an outside observer. We noticed that the time evolution is

$$| n,m \rangle_t = e^{-int + imt} | n,m \rangle_o, \qquad (3.1)$$

which is exactly that of a density matrix ρ_{nm}, or equivalently the product $| n \rangle \langle m |$ of bras and kets both defined *only* in the Hilbert space spanned by observables in space I. The fact that the metric in space II is the exact mirror image of space I (in black holes where all matter sources have been removed) implies that the Hamiltonian H_I has the same spectrum as H_{II}. This, as stated before, remains true for rotating and charged black holes. In our alternative theory we use this fact to postulate that states $| o \rangle_k$ or $| n,m \rangle$ in Kruskal space described not states but density matrix elements for an outside observer in Schwarzschild space.

That this is a profound departure from the previously described picture becomes apparent when the temperature of a radiating black hole is computed. Again our starting point is the Kruskal vacuum $| o \rangle_k$. We now rewrite eq. (2.12) as

$$| o \rangle_k = C \sum_n | n \rangle \langle n | e^{-4\pi n} \tag{3.2}$$

Clearly this is a density matrix. It is diagonal and therefore stationary in the time t. We recognize the Boltzmann factor

$$e^{-4\pi n} = e^{-\beta n}; \ T = 1/\beta = 1/(4\pi M). \tag{3.3}$$

The temperature is twice the usual value, eq. (2.17).

This different result should be easily detectable experimentally whenever a light black hole (light in astronomical sense) would be found.

Non-diagonal density matrices where the black holes in the bras and kets have *different* masses can also be constructed but then the intertial observer sees matter close to the horizon [2,4]. The back-reaction of this matter causes the mass-difference.

In the conventional model the inclusion of the back-reaction of Hawking's radiation required to describe a shrinking black hole is problematic, though interesting attempts are being made [5].

We note that the density matrix implicit in the standard theory (see eq. 2.15) is the square of ours (eq. 3.2). One may therefore summarize our proposal by emphasizing that we postulate a *linear* mapping between states $| \psi \rangle_k$ and density matrices ρ in Schwarzschild space I. In a quadratic mapping all states in II would be averaged over, which is not what we want.

4. A Reconciliation: Relativity Revisited

Should we dismiss the above model as being impossible? As long as we restrict ourselves to those systems to which the postulate can be applied (a black hole emitting and absorbing radiation close to its stationary equilibrium point) there seems to be no contradiction whatsoever. So we asked the question whether a principle can be formulated that is ubiquitously valid so that the theory can be made complete. To explain the principle which we found, we first remind the reader of the fundamental notions of relativity.

Relativity tells us that different observers might interpret their measurements of certain phenomena quite differently. When one observer sees an electric field, another might see a superposition of an electric and a magnetic field. When one observer states that the chronological order of two events A and B was (A,B), another might get the conflicting result (B,A). In short, rather than requiring physical observations to be invariant under Lorentz transformations, it might be necessary to require only *co*variance. What this really means, in our view, is the following:—there is no such thing as Lorentz-invariance of physically observable notions such as electromagnetic fields, or the chronological order of events. After all, the whole universe *does* have a preferred coordinate frame (moving with the same velocity as the average of the stars). What relativity means is that the laws of physics as experienced by observers who move with a certain velocity *can be derived* from the laws of physics as seen by a stationary one. But in the derivation transformations of all sorts may be necessary. This is called covariance, as opposed to invariance. All transformation rules are in principle acceptable. In fact, the further we let our phantasy go, the more likely we are to discover new forms of covariance, and with that new laws of nature.

General relativity fits well in this picture. We can derive the laws of physics as seen by an observer in a gravitational field if we know what is observed by a freely falling observer. But we should not necessarily require the observations to be identical. A transformation may be necessary. What we require in the case at hand is a further transformation in Hilbert space which is of a quite unusual type: pure states for an observer in free fall are linearly mapped onto elements of the density matrix for an accelerated observer. Wherever $\partial t/\partial \tau$ is negative, for certain values of $\mathbf{x}$, we have that an observable $\phi(\mathbf{x},\tau)$ for the freely falling observer acts on ket-space, but for the accelerated one on bra-space.

All this implies that if the density matrix is used to compute probabilities, as it should, then one observer may experience probabilities differently from others. As before, not the physics itself is invariant under general coordinate transformation, but by means of a transformation we can deduce the laws of physics in one frame from those in another.

One can verify that this covariance principle is sufficient to ensure conventional general invariance in the classical (i.e. non-quantum mechanical) limit and conventional quantum mechanics remains true locally (as long as horizons stay far away). In short we claim that there is no contradiction with the known laws of physics.

The situation in our theory could be viewed as being not fundamentally different from that in the famous Stern-Gerlach experiment. There the question which event actually occurred (which slit was passed by the electron) cannot be answered unambiguously. Here, the question with which probability an event occurred depends on the observer. One might of course add that, anyway, the infalling observer will not be able to communicate his observations to those who stay outside. Worse even, he and his entire laboratory are absorbed and will form part of the Hilbert space studied by the outside observer!

The conclusion of this section is that the difference in the predictions of the two theories lies in the fact that we require different transformation laws relating the observed results from the different observers. The difference only occurs when a horizon is present and nowhere else. The crucial factor is time reversal. It could be that time-reversal has more far-reaching effects on the laws of physics than previously taken for granted.

5. Objections Raised to the Theory, and the Answers

Anyone is invited to shoot down our alternative theory, and as stated in the Introduction, many attempts were made. We here list the objections most often heard together with the replies one could give. Although the latter are still far from perfect one might conclude that the objections are not insuperable.

OBJECTION: One might object against the use of the regions II and IV as described by the Kruskal metric: one should instead consider a black hole that has been formed by collapse of matter and in that case there is no mirror image space II at all. The bulk of matter that went in, no matter how long ago, removes it.

REPLY: Our alternative model, as formulated, is only suitable for describing a black hole close to its stationary equilibrium. During collapse it was not at all close to equilibrium. The model still has to be extended to describe large amounts of matter going in and/or out. Progress has already been made [2,4] but the problem is not yet completely understood. Note however that in any case we are dealing with a quantum theory here, in which some form of PCT invariance is to be expected. It would be unreasonable to treat region IV ($\tau < -|\rho|$) and region III ($\tau > |\rho|$) entirely differently. The standard picture, used in the objections, assumes that the Penrose diagram of the complete history of a blackhole can always be given. We however assume that this is only so for black holes in certain mixed states. A "pure" black hole always evolves. Note that the black hole is not stable, it decays, so that if one selects a pure state or a differently mixed state at a given time, one might inevitably obtain a *superposition* of metrics both in the far past and in the far future. Only one mixed configuration is stationary in time and we use the full Kruskal metric to describe it.

OBJECTION: Consider now a dust cloud as large as our galaxy. Take as an initial condition that all particles move slowly inward. An observer will still feel quite comfortable while he moves through the horizon. One can "compute" what will be seen by the observer who stays outside using nothing but standard physics. One gets $T = 1/(8\pi M)$.

REPLY: That is a low value indeed. But let's assume the outside observer can measure it. Even though standard physics was used in the calculation, one assumption was made whether one likes it or not. It is the assumption that the outside observer has the same Hilbert space to his disposal as the infalling one and that the probabilities he measures are the same as those of the other. But his Hilbert space is smaller because there are things he cannot see and that are irrelevant to him. Thus a further transformation law might be necessary. We may have covariance, rather than invariant probabilities. The infalling observer cannot communicate his results to the outside one and therefore there are no contradictions in our alternative model.

OBJECTION: Both space I and space II were used to describe the Hilbert space seen by the outsider. That corresponds to identifying points (u,v) with $(-u,-v)$ of the Kruskal coordinates. If we continue to Euclidean space then such an identification would indeed divide the periodicity β by two and the temperature doubles. But, such an identification would lead to a conical singularity at the origin of Kruskal space, and an extremely singular curvature there, whereas no appreciable amount of matter could be present there that could be responsible for such an enormous curvature.

REPLY: In the *classical* (i.e. non-quantum mechanical) *limit* points (u,v) and $(-u,-v)$ might indeed be considered identified. It doesn't really matter much because, remember, regions II and IV were analytic continuations. And in the classical limit no one can check whether matter was or wasn't present at $u = v = o$. But in our description of the quantum mechanics where we used the state $|o\rangle_k$, those points are definitely not identified. This state has no conical singularity at the origin so no singular amount of matter is needed there. As for Euclidean space, this should always be regarded as a mathematical tool, not as a physical reality.

OBJECTION: In the more detailed descriptions of the alternative models in refs [2] it is claimed that gravitational interactions between infalling matter and outgoing Hawking radiation diverges and that this is a reason to reconsider the complete formulation. Yet the standard arguments show that, in Kruskal coordinates, the metric tensor $T_{\mu\nu}$ is basically regular at the horizon so fear of large gravitational effects there is totally unjustified.

REPLY: This is only true in the description of the black hole in its mixed state, as generated by $|o\rangle_k$. If we try to describe a *pure state*, either by producing a density matrix with only one column, or with eigenvalues 1 and 0 only, then the corresponding state in Kruskal space would have infinite amounts of matter at the horizon, leading indeed to uncontrollable gravitational forces at the origin. We assume that in our model transitions between pure and mixed states do not take place and therefore "pure" black holes exist also.

OBJECTION: The alternative model suggests a complete Hamiltonian with a discrete spectrum for a black hole. Where is it?

REPLY: It is being worked on but no promises are made. We do not pretend to solve the world.

OBJECTION: In eq. (3.2) the right hand side is not properly normalized if the left hand side is (normalization is left unaffected in the standard theory). Indeed any definition of an inner product between "bra" states in space II and "het" states in space I seems to be unnatural or ad hoc as seen by a Kruskal observer.

REPLY: This may perhaps be the most severe objection to which we have no completely satisfactory answer. Of course it could be that a simple adaption of the normalization constant when a mapping is performed is the logical answer, and we note that on both sides of the equation the norm does not vary with time. But it is indeed strange that the "Identity matrix" **1** for a Schwarzschild observer corresponds to a highly excited state in Kruskal space. In another variant of our model it is this identity matrix which is mapped onto the vacuum state $|o>_k$. In that case the temperature of a black hole may appear to be infinite!

6. Conclusion

There is no way to reject a priori a fundamental requirement to further transform the data from the infalling observer before they are interpreted by an outside observer. We have made no attempt (yet) to *prove* that such a requirement *must* be made. The temperature of a black hole could be twice the usual value, and as a consequence the final explosion of a mini-black-hole would have a little more than twice the energy as computed in the conventional way [6].

As indicated in the title of this paper the arguments presented here may affect the interpretative basis of quantum mechanics itself. Whether or not one accepts the views presented here may depend on the way in which one adheres to the dogmas of this theory.

Acknowledgment

The author acknowledges a fruitful correspondence with R.M. Wald.

Note Added in Proof

More recent attempts by this author and others to construct coherently quantum mechanical models for black holes usually favor the conventional picture, not our alternative one, so that Hawking's original value for the temperature emerges. Region II may refer to some other area *outside* the black hole horizon. But the matter is not settled, and we would still like to emphasize the possibility of a philosophy as sketched in this paper.

References

[1] S.W. Hawking, "Particle creation by black holes", Comm. Math. Phys. **43** (1975) 199.
J.B. Hartle and S.W. Hawking, "Path integral derivation of black hole radiance", Phys. Rev. **D13** (1976) 2188.
W.G. Unruh, "Notes on black hole evaporation", Phys. Rev. **D14** (1976) 870.
R.M. Wald, "On particle creation by black holes", Comm. Math. Phys. **45** (1975) 9.
R.M. Wald, "Entropy and black hole theromodynamics", Phys. Rev. **D20** (1979) 1271.
[2] G. 't Hooft, "Ambiguity of the equivalence principle and Hawking's temperature". J. Geometry and Phys. **1** (1984) 45.
G. 't Hooft, "On the quantum structure of a black hole", Nucl. Phys. **B256** (1985) 727.
[3] S.W. Hawking and G.F.R. Ellis, "The large scale structure of space-time", Cambridge Univ. Press (1973).
[4] T. Dray and G. 't Hooft, "The effect of spherical shells of matter on the Schwarzschild black hole", Commun Math. Phys. **99** (1985) 613.
[5] R. Balbinot, "Hawking radiation and the back reaction—a first approach", Class. Quantum Gravity **1** (1984) 573.
[6] D.N. Page and S.W. Hawking, "Gamma rays from primordial black holes", Astrophys. Journal **206** (1976) 1.

Do Black Holes Emit Black-Body Radiation?

T.D. Lee

COLUMBIA UNIVERSITY

I. EPR Phenomena and Inclusive Experiment

Outside a black hole, or in a constant accelerating frame, one deals with space-time manifolds that possess horizons. It is well-known [1,2] that observers bounded within such a manifold can detect radiation whose characteristics bear a remarkable resemblance to that from a black body. However, before discussing this subject, let me briefly mention two basic points about quantum mechanics:

1. It is well known that a pure quantum-mechanical state can carry coherent phase correlations between dynamical variables that are localized in different regions outside each other's light cones. For example, in the decay

$$\pi^0 \to \gamma(1) + \gamma(2), \qquad (1.1)$$

the electric field of $\gamma(1)$ is parallel to the magnetic field of $\gamma(2)$, no matter how far apart these two photons are. Such correlations have to be expressed in terms of the quantum-mechanical amplitude, not the classical probability. This point was questioned by the celebrated Einstein-Podolsky-Rosen article [3]; the subsequent papers by John Bell [4] and others have completely clarified this non-classical aspect of quantum mechanics.

A quantum-mechanical state vector $|S\rangle$ is defined globally, depending on a space-like surface S which spreads to infinity. Therefore, its coherence can extend to local variables that are far away from one another, outside each others' light cones. In the literature most of the discussions on the Einstein-Podolsky-Rosen phenomena

This research was supported in part by the U.S. Department of Energy.

are for microscopic systems like π^0 or positronium. As we shall see, they are also applicable to macroscopic systems, such as the black-hole radiation.

2. Quite often in physics, even for a pure state it is difficult to measure all commuting variables in the system. For example, we may measure only the single pion distribution in a high energy e^+e^- collision:

$$e^+ + e^- \rightarrow \pi(1) + \pi(2) + \cdots + K(1) + K(2) + \cdots . \quad (1.2)$$

These are usually referred to as inclusive measurements, in which only part of the final system is included in the detection. These inclusive measurements account for most of the experiments in physics. Of course, in such measurements certain information concerning the final state is lost. But there is a difference between this type of loss of information and the entropy concept used in statistical mechanics.

In statistical mechanics the system is described *initially* as an ensemble of incoherent states, whereas in the example of the e^+e^- reaction the system is in a pure coherent state. In the inclusive (say, single pion) measurement, the loss of information is not intrinsic to the e^+e^- collision, but only to the particular choice made by the observer. Insofar as these choices are arbitrary, such loss of information is clearly different from the entropy in statistical mechanics. While entropy does imply a certain degree of ignorance, not all ignorance can be usefully expressed as entropy. For example, if I close my eyes and refuse to observe anything, this clearly makes me ignorant; however, it does not mean that the entropy of the objective world has increased accordingly.

Both in statistical mechanics and in an inclusive experiment, the measurement of any physical observable O is given by

$$\frac{\text{trace } \rho O}{\text{trace } \rho} \quad (1.3)$$

where ρ is the (un-normalized) density matrix. In statistical mechanics, for a canonical ensemble, the density matrix is

$$\rho_{st} = e^{-H/kT} \quad (1.4)$$

where H is the Hamiltonian, k the Boltzmann constant, and T the absolute temperature.

For an inclusive measurement, let q and q' denote the complete set of commuting variables, with q referring to the set (of inclusive variables) that are to be measured and q' the remaining ones (which are not to be measured). To each state vector $\langle q,q' \mid \rangle$, the matrix elements of the density matrix are given by

$$\langle q_f \mid \rho \mid q_{\text{in}} \rangle \equiv \sum_{q'} \langle q_f,q' \mid X \mid q_{\text{in}},q' \rangle. \qquad (1.5)$$

Any observable in an inclusive measurement is, by definition, represented by an operator O which depends only on q and their conjugate momenta $\partial/\partial q$; the expectation value $\langle \mid O \mid \rangle$ is then equal to (1.3). The crucial question is whether for certain state vectors $\mid \rangle$ there could exist a $H(q,\partial/\partial q)$ such that

$$\rho = \rho_{\text{incl}} = e^{-H/kT}? \qquad (1.6)$$

If affirmative, then these two cases, (1.4) and (1.6), are identical so far as all physical measurements (within the stated restrictions) are concerned; if negative, then they are not.

These seemingly obvious remarks will play a central role in resolving the difference between a black-hole radiation and a black-body radiation.

II. A New Derivation

For clarity, we shall first discuss the Rindler case. Consider the case of a scalar field ϕ with an arbitrary interaction. To make the problem even simpler, we assume a flat two-dimensional Minkowski space of coordinates X and T. (The theorem proved below is valid in any dimension and also holds when there are fermions.) The Lagrangian $\mathscr{L}$ and Hamiltonian $\mathscr{H}$ are given by (in units $\hbar = c = 1$)

$$\mathscr{L} = \int_{-\infty}^{\infty} \left[\frac{1}{2} \left(\frac{\partial\phi}{\partial T} \right)^2 - \frac{1}{2} \left(\frac{\partial\phi}{\partial X} \right)^2 - V(\phi) \right] dX \qquad (2.1)$$

and

$$\mathscr{H} = \int_{-\infty}^{\infty} \left[\frac{1}{2}\mathscr{P}^2 + \frac{1}{2} \left(\frac{\partial\phi}{\partial X} \right)^2 + V(\phi) \right] dX \qquad (2.2)$$

where $\mathscr{P} = \partial\phi/\partial T$ and V can be any function of ϕ. At equal time T, we have the usual commutation relation

$$[\mathscr{P}(X,T), \phi(X',T)] = -i\delta(X - X'). \qquad (2.3)$$

The Minkowski coordinate system X and T will be referred to as Σ. The equation of motion in Σ is

$$\frac{\partial^2\phi}{\partial X^2} - \frac{\partial^2\phi}{\partial T^2} - V'(\phi) = 0, \qquad (2.4)$$

where $V'(\phi) \equiv dV/d\phi$.

Next, we introduce the familiar coordinates x and t in a constant accelerating frame Σ_{acc} (or, the Rindler frame):

$$X = \frac{1}{g} e^{gx} \cosh gt \qquad \text{and} \qquad T = \frac{1}{g} e^{gx} \sinh gt \quad (2.5)$$

where g denotes the constant acceleration. In Σ_{acc}, the entire domain $-\infty \leqslant x \leqslant \infty$ and $-\infty \leqslant t \leqslant \infty$ covers only the quarter

$$X \geqslant |T| \geqslant 0, \qquad (2.6)$$

called I in Σ. The horizon is defined by $X = |T|$, and I refers to its inside. In I and in terms of x and t, the field equation (2.4) becomes

$$\frac{\partial^2\phi}{\partial x^2} - \frac{\partial^2\phi}{\partial t^2} - e^{2gx} V'(\phi) = 0. \qquad (2.7)$$

The Lagrangian and Hamiltonian in the accelerating frame Σ_{acc} are

$$L = \int_{-\infty}^{\infty} \left[\frac{1}{2}\left(\frac{\partial\phi}{\partial t}\right)^2 - \frac{1}{2}\left(\frac{\partial\phi}{\partial x}\right)^2 - e^{2gx} V(\phi) \right] dx \qquad (2.8)$$

and

$$H = \int_{-\infty}^{\infty} \left[\frac{1}{2}P^2 + \frac{1}{2}\left(\frac{\partial\phi}{\partial x}\right) + e^{2gx} V(\phi) \right] dx \qquad (2.9)$$

so that (2.7) can also be derived by Lagrange's or Hamilton's equations. In (2.9), $P = \partial\phi/\partial t$ satisfies in Σ_{acc}, similar to (2.3),

$$[P(x,t), \phi(x',t)] = -i\delta(x - x'). \qquad (2.10)$$

In the Minkowski frame $\Sigma(X,T)$, we may regard the field operator $\phi(X,T)$ at $T = 0$ as the generalized coordinates. Define

$$Q(X) \equiv \phi(X,0). \qquad (2.11)$$

The conjugate momentum $\mathcal{P}(X,T) = \partial\phi/\partial T$ at $T = 0$ becomes

$$\mathcal{P}(X,0) = -i\frac{\delta}{\delta Q(X)}, \qquad (2.12)$$

so that (2.3) is satisfied. The Hamiltonian (2.2) can then be expressed in terms of $Q(X)$ and $\delta/\delta Q(X)$:

$$\mathcal{H} = \mathcal{H}\left(Q, \frac{\delta}{\delta Q}\right) \tag{2.13}$$

where

$$Q \equiv \text{set } \{Q(X)\} \tag{2.14}$$

and

$$\frac{\delta}{\delta Q} \equiv \text{set } \left\{\frac{\delta}{\delta Q(X)}\right\}. \tag{2.15}$$

Since Q describes a complete set of coordinates, any state vector $|\rangle$ in the Q representation can be written as

$$\langle Q \,|\rangle. \tag{2.16}$$

A similar coordinate representation can be constructed in the accelerating frame. Define

$$q(x) \equiv \phi(x,0). \tag{2.17}$$

Since the $X > 0$ portion of the surface $T = 0$ coincides with the entire $t = 0$ surface in Σ_{acc}, we have

$$q(x) = Q(X) \qquad \text{when} \qquad X = \frac{1}{g}\, e^{gx} > 0. \tag{2.18}$$

In this q-coordinate representation, the conjugate momentum $P(x,t) = \partial\phi/\partial t$ at $t = 0$ assumes the form

$$P(x,0) = -i\, \frac{\delta}{\delta q(x)} \tag{2.19}$$

and the accelerating frame Hamiltonian (2.9) can be written as

$$H = H\left(q, \frac{\delta}{\delta q}\right) \tag{2.20}$$

where

$$q \equiv \text{set}\{q(x)\} \tag{2.21}$$

and

$$\frac{\delta}{\delta q} \equiv \text{set } \left\{\frac{\delta}{\delta q(x)}\right\}. \tag{2.22}$$

Viewed in Σ, q does not describe a complete set of coordinates. To supplement (2.18), we introduce

$$q'(x') \equiv Q(X) \qquad \text{when} \qquad X = -\frac{1}{g} e^{gx'} < 0. \quad (2.23)$$

Like x, the range of x' is also from $-\infty$ to ∞; together, x and x' cover the entire X-region. Let

$$q' \equiv \text{set}\{q'(x')\}. \tag{2.24}$$

By definition, the union of these two sets q and q' is Q. Hence, (2.16) can also be written as

$$\langle q,q' \,| \rangle, \tag{2.25}$$

in which the set q denotes the field coordinates in Σ_{acc}, and q' those outside Σ_{acc}.

Within the horizon in Σ_{acc} (i.e., in the region $X \geq |T|$), the field operator $\phi(X,T) = \phi(x,t)$ satisfies the field equation (2.7); in addition, there is the boundary condition that ϕ and $\partial\phi/\partial t$ at $t = 0$ satisfy (2.17) and (2.19). Consequently, by using the Green's function of (2.7), at any point (x,t) in the entire frame Σ_{acc}, $\phi(x,t)$ can be expressed as a function of the operators q and $\delta/\delta q$. An observer in the accelerating frame Σ_{acc} can only measure functions of the field $\phi(x,t)$ within his horizon; hence, his observables can always be written as

$$O = O\left(q, \frac{\delta}{\delta q}\right). \tag{2.26}$$

Conversely, *any* observable of the form (2.26) can, in principle, be measured by the observer in the accelerating frame. (By taking this general approach, we do not have to enter into any detailed discussions of measuring apparatus used in the accelerating frame.) On the other hand, the expectation value of a measurement in Σ_{acc} is given by $\langle| 0 |\rangle$ which, as shown by (2.25), depends on $\langle q,q' |\rangle$ with the coordinates q' lying outside Σ_{acc} and therefore undetected. Quite often, we are interested in a problem in which the state vector $|\rangle$ refers to some low-lying states of the total Hamiltonian $\mathcal{H}$. In that case, mathematically is there any way to construct $|\rangle$ in the larger frame Σ by using only operators in the smaller frame Σ_{acc}? The answer is yes, as will be shown by Theorems 1 and 2 in this and the next sections.

Let $| VAC \rangle$ be the ground state of the total Hamiltonian $\mathcal{H}$ given by (2.2) in the Minkowski frame Σ. The following theorem [5–7] relates the coordinate-representation $\langle q,q' | VAC \rangle$ of this ground state to a matrix element of the accelerating frame Hamiltonian H, given by (2.20); it also gives the density matrix in Σ_{acc} as $\exp(-2\pi H/g)$ when the state vector in Σ is $| VAC \rangle$.

Theorem 1

$$\langle VAC | q,q' \rangle = \frac{\langle q' | e^{-\pi H/g} | q \rangle}{(\text{trace } e^{-2\pi H/g})^{1/2}} \qquad (2.27)$$

and

$$\langle VAC | O | VAC \rangle = \frac{\text{trace } O\rho_V}{\text{trace } \rho_V} \qquad (2.28)$$

where O can be any observable (2.26) in Σ_{acc} and ρ_V is the corresponding (un-normalized) density matrix given by

$$\rho_V = e^{-2\pi H/g}. \qquad (2.29)$$

Note that in (2.27), on the left-hand side we have $| q,q' \rangle$, denoting the eigenstates of the operators q_{op} and q'_{op}, defined previously by (2.17)–(2.18) and (2.23), with the subscript op added here to emphasize the operator nature of these symbols

$$q_{\mathrm{op}} | q,q' \rangle = q | q,q' \rangle \qquad (2.30)$$

and

$$q'_{\mathrm{op}} | q,q' \rangle = q' | q,q' \rangle \qquad (2.31)$$

where, as before, q'_{op} lies outside Σ_{acc}, and q,q' refer to the corresponding ($c.$ number) eigenvalues. However, on the right-hand side of (2.27), both the bra and ket vectors $\langle q' |$ and $| q \rangle$ are eigenstates of q_{op}, independent of q'_{op}; they satisfy

$$q_{\mathrm{op}} | q \rangle = q | q \rangle \qquad (2.32)$$

and

$$\langle q' \mid q_{\text{op}} = \langle q' \mid q'. \tag{2.33}$$

The c. number eigenvalues q' in (2.31) and (2.33) happen to be identical.

Proof. Introduce the euclidean coordinates by setting in (2.5)

$$T = -iY \qquad \text{and} \qquad t = -i\theta/g, \tag{2.34}$$

and let Y and θ be real. Thus,

$$X = r \cos \theta \qquad \text{and} \qquad Y = r \sin \theta \tag{2.35}$$

where

$$r = \frac{1}{g} e^{gx}$$

which, unlike (2.5) in the Minkowski space, is valid everywhere in $\Sigma(X,Y)$. Express the matrix elements $\langle Q' \mid e^{-\beta\mathcal{H}} \mid Q \rangle$ and $\langle q' \mid e^{-\theta H/g} \mid q \rangle$ in terms of Feynman's path integrals in the euclidean space

$$\langle Q' \mid e^{-\beta\mathcal{H}} \mid Q \rangle = \int [d\phi] \exp[-A(\phi)] \tag{2.36}$$

and

$$\langle q' \mid e^{-\theta H/g} \mid q \rangle = \int [d\phi] \exp[-A'(\phi)] \tag{2.37}$$

where in the first expression the action is

$$A(\phi) \equiv \int_0^\beta dY \int_{-\infty}^\infty dX \left[\frac{1}{2} \left(\frac{\partial\phi}{\partial X} \right)^2 + \frac{1}{2} \left(\frac{\partial\phi}{\partial Y} \right)^2 + V(\phi) \right] \tag{2.38}$$

with the boundary conditions that at $Y = 0$, the initial field configuration is given by $Q(X)$ through (2.11) and (2.14), and at $Y = \beta$ the final $\phi(X,\beta)$ is given by $Q'(X)$; in the second expression (2.37), the action is

$$A'(\phi) \equiv \int_0^\theta d\theta \int_0^\infty r \, dr \left[\frac{1}{2} \left(\frac{\partial\phi}{\partial r} \right)^2 + \frac{1}{2r^2} \left(\frac{\partial\phi}{\partial\theta} \right)^2 + V(\phi) \right] \tag{2.39}$$

with the boundary condition that at $\theta = 0$ the initial field configuration ϕ is given by q through (2.17)–(2.18) and at the final θ, q is replaced by q'. For definiteness, we take the boundary conditions at ∞ to be

$$\phi = 0 \qquad \text{at} \qquad X = \pm\infty \qquad \text{in} \quad A(\phi) \qquad (2.40)$$

and

$$\phi = 0 \qquad \text{at} \qquad r = \infty \qquad \text{in} \quad A'(\phi). \qquad (2.41)$$

The precise definition of the differential $[d\phi]$ is most conveniently given in terms of a lattice. Introduce any lattice L in which the average distance between lattice sites in the (X,Y) plane is ϵ. (If one wishes, one may take L to be simply a square lattice in the (X,Y) plane; of course, viewed in the (x,θ) plane, L is no longer a square lattice.) Let $A_L(\phi)$ and $A'_L(\phi)$ be the integrals (2.38) and (2.39) evaluated on the lattice L; i.e., the appropriate lattice approximations of (2.38) and (2.39), so that when the lattice spacing $\epsilon \to 0$, $A_L(\phi)$ and $A'_L(\phi)$ become the continuum expressions $A(\phi)$ and $A'(\phi)$. We note that, except for the different boundaries assumed in (2.38) and (2.39), the integrands of $A(\phi)$ and $A'(\phi)$ would be identical; the same holds for $A_L(\phi)$ and $A'_L(\phi)$. [See Reference 8 for the construction of a lattice action for an arbitrary lattice L.] Equations (2.36) and (2.37) can then be written as

$$\langle Q' \mid e^{-\beta\mathcal{H}} \mid Q \rangle = \lim_{\epsilon\to 0} \int J \prod_i d\phi_i \, \exp[-A_L(\phi)],$$

$$\langle q' \mid e^{-\theta H/g} \mid q \rangle = \lim_{\epsilon\to 0} \int J' \prod_i d\phi_i \, \exp[-A'_L(\phi)],$$

where ϕ_i denotes the field amplitude on the ith lattice site, J and J' denote the appropriate Jacobians. These Jacobians depend on ϵ and the lattice structure in the (X,Y) and (x,θ) planes; however, they are independent of the field amplitudes ϕ_i. It is convenient to eliminate them by considering the ratios

$$\frac{\langle Q' \mid e^{-\beta\mathcal{H}} \mid Q \rangle}{\langle 0 \mid e^{-\beta\mathcal{H}} \mid 0 \rangle} = C \lim_{\epsilon\to 0} \int \prod_i d\phi_i \, \exp[-A_L(\phi)] \qquad (2.42)$$

and

$$\frac{\langle q' \mid e^{-\theta H/g} \mid q \rangle}{\langle 0 \mid e^{-\theta H/g} \mid 0 \rangle} = C' \lim_{\epsilon\to 0} \int \prod_i d\phi_i \, \exp[-A'_L(\phi)], \qquad (2.43)$$

where C and C' are constants independent of Q, Q' and q, q'; $\mid 0 \rangle$

refers to $|Q\rangle$ or $|q\rangle$ when $Q = 0$ or $q = 0$; i.e., all the corresponding set of fields is zero. Next, consider the limit

$$\beta = \infty \qquad (2.44)$$

and set

$$\theta = \pi, \qquad Q = Q(q,q'), \qquad \text{and} \qquad Q' = 0 \quad (2.45)$$

(i.e., Q is related to q and q' by (2.18) and (2.23), and all $Q'(X) = 0$). When $\theta = \pi$ and $Q = Q(q,q')$, the boundary condition of ϕ described by Q at $Y = 0$ is identical to that described by q at $\theta = 0$ and q' at $\theta = \pi$. Also, $Q' = 0$ plus the boundary condition (2.40) become equivalent to the boundary condition (2.41); therefore, under (2.44) and (2.45), for any given function ϕ on the lattice L, we have

$$A_L(\phi) = A_L'(\phi);$$

correspondingly, (2.42) is proportional to (2.43). The ratio C/C' is unity, because the left-hand sides of (2.42) and (2.43) are both unity if $Q = Q' = 0$ and $q = q' = 0$; in the present case, $Q' = 0$ on account of (2.45), and when $q = q' = 0$ we must also have $Q = Q(q,q') = 0$, in accordance with (2.18) and (2.23). Hence,

$$\frac{\langle q' \,|\, e^{-\pi H/g} \,|\, q\rangle}{\langle 0 \,|\, e^{-\pi H/g} \,|\, 0\rangle} = \underset{\beta \to \infty}{\text{Lim}} \frac{\langle 0 \,|\, e^{-\beta \mathcal{H}} \,|\, Q\rangle}{\langle 0 \,|\, e^{-\beta \mathcal{H}} \,|\, 0\rangle}. \qquad (2.46)$$

As $\beta \to \infty$,

$$\langle Q' \,|\, e^{-\beta \mathcal{H}} \,|\, Q\rangle \to e^{-E_0\beta} \langle Q' \,|\, VAC\rangle \langle VAC \,|\, Q\rangle$$

where E_0 is the vacuum energy (eigenvalue of $\mathcal{H}$ in the ground state $|VAC\rangle$). Combining these formulas, we find

$$\langle q' \,|\, e^{-\pi H/g} \,|\, q\rangle = \text{constant} \langle VAC \,|\, Q\rangle.$$

Since $Q = Q(q,q')$, we derive (2.27), with the denominator on its right-hand side to insure $\langle VAC \,|\, VAC\rangle = 1$.

Since H is Hermitian, the complex conjugation of (2.27) gives

$$\langle q,q' \,|\, VAC\rangle = \langle q \,|\, e^{-\pi H/g} \,|\, q'\rangle/(\text{trace } e^{-2\pi H/g})^{1/2}. \qquad (2.47)$$

From (2.26), we see that any observable O in the accelerating frame Σ_{acc} depends only on the operators $q(x)$ and $\delta/\delta q(x)$. In the q-representation, O takes on the matrix form $\langle q_f \,|\, O \,|\, q_{\text{in}}\rangle$, where q_f and q_{in} refer to the final and initial values of q_{op}. Hence

$$\langle VAC \mid O \mid VAC \rangle$$

$$= \sum_{q'} \sum_{q_f, q_{in}} \langle VAC \mid q_f, q' \rangle \langle q_f \mid O \mid q_{in} \rangle \langle q_{in}, q' \mid VAC \rangle.$$

Substituting (2.27) and (2.47) into the above expression and noting that O is independent of q', we have, after first carrying out the sum over q', and then those over q_f and q_{in},

$$\langle VAC \mid O \mid VAC \rangle = \frac{\text{trace}[e^{-2\pi H/g} O]}{\text{trace } e^{-2\pi H/g}} \qquad (2.48)$$

which establishes (2.28)–(2.29) and completes the proof of Theorem 1.

From this proof, it is clear that the theorem can be readily extended to fields of *arbitrary spin in any space-dimension with arbitrary interactions and in any given metric* (provided the assumption of euclidean extension and the symmetry between $\theta = 0$ and π are valid); this includes interacting fields in a Rindler or Schwarzschild space.

Remarks

1. For an eternal black hole of mass M, we replace g by the gravitational acceleration of the Schwarzschild radius $R = 2GM$:

$$g = \frac{GM}{R^2} = \frac{1}{4GM}. \qquad (2.49)$$

Theorem 1 can then be applied to any observable $O(p,q)$ that depends only on field variables and their conjugate momenta p operators outside the black hole (i.e., $r > 2GM$):

$$\langle VAC \mid O(p,q) \mid VAC \rangle = \frac{\text{trace}[e^{-8\pi GMH} O(p,q)]}{\text{trace } e^{-8\pi GMH}}, \qquad (2.50)$$

where the state vector $\mid VAC \rangle$ is the ground state of $\mathcal{H}$ in the Kruskal frame, H is the Hamiltonian outside the black hole, which depends also only on q and p, and $(8\pi GM)^{-1}$ is the Hawking temperature times the Boltzmann constant. Hence, in this case, a black hole is a black body.

2. For a free field, and for the observable O = occupation number n_k of momentum k (restricted to the space bounded by the horizon), theorem 1 gives the Planck formula

$$\langle VAC \mid n_k \mid VAC \rangle = (e^{2\pi\omega/g} - 1)^{-1} \tag{2.51}$$

which is the Hawking formula. Our proof generalizes the result to include also fields with interactions, for which the simple Planck formula is no longer applicable.

3. For the subsequent discussions, we wish to ask three questions:

(i) Can any $\langle q,q' \mid \rangle$ in the larger frame Σ be written as a functional of q and $\delta/\delta q$ in the smaller frame Σ_{acc}?

In other words, can we extend Theorem 1 to include other eigenstates of $\mathcal{H}$, or an arbitrary superposition of these states in Σ?

(ii) If the relevant state vector in Σ is not $\mid VAC \rangle$, then the corresponding density matrix ρ may not be $e^{-2\pi H/g}$. How much can an observer restricted in Σ_{acc} (through ρ) infer about the nature of $\mid \rangle$ in Σ (which extends beyond Σ_{acc})?

(iii) If an observer restricted in Σ_{acc} refuses to believe in any physical reality beyond his horizon, can he always maintain a hardline black body radiation interpretation?

As we shall see, the answer to question (i) is "yes", that to question (ii) is "a lot" and that to question (iii) is "not always, and sometimes can be very, very difficult". Hence, depending on $\mid \rangle$, a black hole may or may not be a black body.

III. Equivalence Theorem

The ground state $\mid VAC \rangle$ of the total Hamiltonian $\mathcal{H}$ is a coherent state correlating the operators q_{op} in Σ_{acc} with q'_{op} outside. Because of this, we can prove a number of general identities.

Let I be an arbitrary function of q_{op} and its variational derivative $\delta/\delta q_{\text{op}}$; for each such function, we can define two operators,

$$I_{\text{op}} \equiv I(q_{\text{op}}, \delta/\delta q_{\text{op}}) \tag{3.1}$$

and

$$I'_{\text{op}} \equiv I(q'_{\text{op}}, \delta/\delta q'_{\text{op}}), \tag{3.2}$$

so that I_{op} lies inside the accelerating frame Σ_{acc}, and I'_{op} outside. As in (2.30)–(2.31), $q_{\text{op}}(x)$ and $q'_{\text{op}}(x')$ refer to the coordinate operators defined by (2.18) and (2.23); i.e., they are related to the field $\phi(X,T)$ at $T = 0$ by

$$\frac{1}{g}\,e^{gx} > 0$$

$$\phi(X,0) = \begin{cases} q_{op}(x) \\ q'_{op}(x') \end{cases} \qquad \text{when} \qquad X = \qquad (3.3)$$

$$-\frac{1}{g}\,e^{gx'} < 0.$$

Theorem 2

To any operator I'_{op} (outside Σ_{acc}), of the form (3.2), there exists an operator J_{op} (inside Σ_{acc}) such that

$$I'_{op}\,|\,VAC\rangle = J_{op}\,|\,VAC\rangle, \qquad (3.4)$$

where J_{op} is related to the transpose of (3.1) by

$$J_{op} = e^{-\pi H/g}\,\tilde{I}_{op}\,e^{\pi H/g}. \qquad (3.5)$$

Proof

As before, let $|\,q,q'\rangle$ be an eigenstate of q_{op} and q'_{op} with eigenvalues q and q'. In the matrix notation

$$\langle q,q'\,|\,I'_{op}\,|\,VAC\rangle = \langle q'\,|\,I'_{op}\,|\,q'_m\rangle\,\langle q,q'_m\,|\,VAC\rangle$$

and

$$\langle q,q'\,|\,J_{op}\,|\,VAC\rangle = \langle q\,|\,J_{op}\,|\,q_m\rangle\,\langle q_m,q'\,|\,VAC\rangle \qquad (3.6)$$

where the repeated variables (in the middle) q'_m and q_m are to be summed over. By using (2.47), we can convert the equality (3.4) to be proved into

$$\langle q'\,|\,I'_{op}\,|\,q'_m\rangle\langle q\,|\,e^{-\pi H/g}\,|\,q'_m\rangle = \langle q\,|\,J_{op}\,|\,q_m\rangle\langle q_m\,|\,e^{-\pi H/g}\,|\,q'\rangle. \qquad (3.7)$$

From (3.1)–(3.2) we see that

$$\langle q'\,|\,I'_{op}\,|\,q'_m\rangle = \langle q'\,|\,I_{op}\,|\,q'_m\rangle \qquad (3.8)$$

in which $\langle q'\,|$ and $|\,q'_m\rangle$ on the left refer to eigenvectors of q'_{op}, while the ones on the right denote those of q_{op}, but with identical eigenvalues q' and q'_m. Furthermore, (3.8) equals $\langle q'_m\,|\,\tilde{I}_{op}\,|\,q'\rangle$ which, when we substitute it into (3.7), leads to

$$e^{-\pi H/g}\tilde{I}_{op} = J_{op}e^{-\pi H/g}.$$

This gives (3.4) and completes the proof. This theorem, called the equivalence theorem, states that the effect of any operator I'_{op} outside Σ_{acc} when operating on $|\,VAC\rangle$ becomes completely equivalent to $J_{op}|\,VAC\rangle$ when the operator J_{op} lines completely within Σ_{acc}.

At first sight, one may be surprised by the implication of the equivalence theorem: I'_{op} can be any operator outside Σ_{acc}, while J_{op} is always inside Σ_{acc}. Suppose that the state vector under consideration is not $|\,VAC\rangle$, but

$$|\rangle = I'_{op}|\,VAC\rangle. \tag{3.9}$$

To an observer in the accelerating frame, I'_{op} is, by definition, beyond his reach; nevertheless, because of the coherence contained in $|\,VAC\rangle$, the same state $|\rangle$ can also be written as

$$|\rangle = J_{op}|\,VAC\rangle. \tag{3.10}$$

Since J_{op} is within his horizon, he may detect deviation of $|\rangle$ from $|\,VAC\rangle$. It turns out that this is possible only when I'_{op} is not unitary, as will be discussed in Section 4. (See also Remark 1 below.)

The physical content of Theorem 2 becomes more explicit if we express it in terms of the annihilation and creation operators. Take *any* complete orthonormal set of real functions $\{F_K(X)\}$ that satisfy

$$\int_{-\infty}^{\infty} F_K(X)\, F_{K'}(X)\, dX = \delta_{KK'} \tag{3.11}$$

with K and K' as the running labels, and $\delta_{KK'}$ the Kronecker symbol. Expand the field $\phi(X,T)$ and its conjugate momentum $\mathcal{P}(X,T)$ at a fixed T, say $T = 0$:

$$\phi(X,0) = \sum_K (2\Omega_K)^{-1/2}(A_K + A_K^\dagger)\, F_K(X)$$

and

$$\mathcal{P}(X,0) = \sum_K i(\Omega_K/2)^{1/2}(-A_K + A_K^\dagger)\, F_K(X), \tag{3.12}$$

where $\{\Omega_K\}$ can be any set of nonzero constants. The inverse of (3.12) expresses A_K and its hermitian conjugate $A_K^\dagger$ in terms of ϕ and $\mathcal{P}$. By using (2.3), we find that A_K and $A_K^\dagger$ satisfy the commutation relations of annihilation and creation operators:

$$[A_K, A_{K'}^\dagger] = \delta_{KK'}$$

and

$$[A_K, A_{K'}] = 0. \tag{3.13}$$

These formulas are valid for any interacting scalar field ϕ. So far, we have not specified the precise form of F_K nor of Ω_K. A convenient choice is to use the plane waves for F_K with $\Omega_K = (K^2 + m^2)^{1/2}$ and m = physical mass of ϕ.

Likewise, in the accelerating frame $\Sigma_{acc}(x,t)$ we may take any complete set of real functions $\{f_k(x)\}$, with k as the running label; the corresponding orthonormal relations are

$$\int_{-\infty}^{\infty} f_k(x) f_{k'}(x) \, dx = \delta_{kk'}. \tag{3.14}$$

Expand, at $t = 0$,

$$\phi(x,0) = \sum_k (2\omega_k)^{-1/2}(a_k + a_k^\dagger) f_k(x)$$

and its conjugate momentum $P = \partial\phi/\partial t$

$$P(x,0) = \sum_k i(\omega_k/2)^{1/2}(-a_k + a_k^\dagger) f_k(x) \tag{3.15}$$

where $\{\omega_k\}$ can, again, be any set of nonzero constants. The inverse of (3.15) is

$$a_k = \int_{-\infty}^{\infty} (\tfrac{1}{2}\omega_k)^{1/2} f_k(x) \left[\phi(x,0) + \frac{i}{\omega_k} P(x,0) \right] dx$$

and

$$a_k^\dagger = \int_{-\infty}^{\infty} (\tfrac{1}{2}\omega_k)^{1/2} f_k(x) \left[\phi(x,0) - \frac{i}{\omega_k} P(x,0) \right] dx. \tag{3.16}$$

Thus, on account of (2.10), we find

$$[a_k, a_{k'}^\dagger] = \delta_{kk'}$$

and

$$[a_k, a_{k'}] = 0. \tag{3.17}$$

Substituting (2.17) and (2.19) into (3.16), we see that a_k and $a_k^\dagger$ can be expressed as functions of q and $\delta/\delta q$

$$a_k = I\left(q, \frac{\delta}{\delta q}\right) \qquad \text{and} \qquad a_k^\dagger = I^*\left(q, \frac{\delta}{\delta q}\right). \tag{3.18}$$

where

$$I\left(q,\frac{\delta}{\delta q}\right) \equiv \int_{-\infty}^{\infty} (\tfrac{1}{2}\omega_k)^{1/2} f_k(x) \left[q(x) + \frac{1}{\omega_k}\frac{\delta}{\delta q(x)}\right] dx$$

and

$$I^{\dagger}\left(q,\frac{\delta}{\delta q}\right) \equiv \int_{-\infty}^{\infty} (\tfrac{1}{2}\omega_k)^{1/2} f_k(x) \left[q(x) - \frac{1}{\omega_k}\frac{\delta}{\delta q(x)}\right] dx. \quad (3.19)$$

In the following, we shall adopt the representation that a_k and $a_k^{\dagger}$ are real (and therefore, $\tilde{a}_k = a_k^{\dagger}$). By using (3.4)–(3.5), proved in Theorem 2, we establish

Cor. 1

$$a_k' \mid VAC\rangle = e^{-\pi H/g}\, a_k^{\dagger}\, e^{\pi H/g} \mid VAC\rangle$$

and

$$a_k'^{\dagger} \mid VAC\rangle = e^{-\pi H/g}\, a_k\, e^{\pi H/g} \mid VAC\rangle, \quad (3.20)$$

where a_k' and $a_k'^{\dagger}$ are related to the functions I and $I^{\dagger}$ of (3.19) by

$$a_k' = I\left(q',\frac{\delta}{\delta q'}\right) \quad \text{and} \quad a_k'^{\dagger} = I^{\dagger}\left(q',\frac{\delta}{\delta q'}\right) \quad (3.21)$$

and q' is defined by (2.23).

If one wishes, one may introduce another accelerating system, $\Sigma_{acc}'(x',t)$, which lies outside the original Σ_{acc}; when viewed in Σ, the new system Σ_{acc}' has an acceleration of the same magnitude as g of Σ_{acc} but in the opposite direction. Similarly to (2.5), the new coordinates (x',t) are given by

$$X \equiv -\frac{1}{g}\, e^{gx'} \cosh gt$$

and

$$T \equiv -\frac{1}{g}\, e^{gx'} \sinh gt. \quad (3.22)$$

While $\Sigma_{acc}(x,t)$ covers the quarter $X \geqslant \mid T\mid$ in Σ, $\Sigma_{acc}'(x',t)$ covers a different quarter $X \leqslant -\mid T\mid$. The a_k and $a_k^{\dagger}$ are the annihilation and creation operators in Σ_{acc}, whereas a_k' and $a_k'^{\dagger}$ are those in Σ_{acc}'.

By identifying $e^{-\pi H/g}\, a_k^\dagger\, e^{\pi H/g}$ and $e^{-\pi H/g}\, a_k\, e^{\pi H/g}$ as I_{op} in (3.1) and assuming H real, like a_k, we derive the converse of Cor. 1:

Cor.2

$$a_k\,|\,VAC\rangle \;=\; e^{-\pi H'/g}\, a_k'^{\,\dagger}\, e^{\pi H'/g}\,|\,VAC\rangle$$

and

$$a_k^\dagger\,|\,VAC\rangle \;=\; e^{-\pi H'/g}\, a_k'\, e^{\pi H'/g}\,|\,VAC\rangle, \tag{3.23}$$

where H' is the Hamiltonian in Σ'_{acc}. As before, these formulas are valid for *any interacting* scalar *field* ϕ, and the choice $\{f_k(x)\}$ is as yet *quite arbitrary.*

For a massive free field, the field equation (2.7) in $\Sigma_{acc}(x,t)$ takes the form

$$-\frac{\partial^2\phi}{\partial x^2} + \frac{\partial^2\phi}{\partial t^2} + e^{2gx}m^2\phi = 0. \tag{3.24}$$

A convenient choice of $f_k(x)$ is the c. number eigensolution of

$$\left(-\frac{d^2}{dx^2} + m^2 e^{2gx}\right) f_k = \omega_k^2 f_k. \tag{3.25}$$

Correspondingly, the Hamiltonian in Σ_{acc} is

$$H = \sum_k \omega_k(a_k^\dagger a_k + \tfrac{1}{2}). \tag{3.26}$$

Thus, we have

Cor. 3

$$a_k'\,|\,VAC\rangle = e^{-\pi\omega_k/g}\, a_k^\dagger\,|\,VAC\rangle,$$

$$a_k'^{\,\dagger}\,|\,VAC\rangle = e^{\pi\omega_k/g}\, a_k\,|\,VAC\rangle, \tag{3.27}$$

and therefore

$$|\,VAC\rangle = \text{constant }[\exp(\sum_k e^{-\pi\omega_k/g} a_k^\dagger\, a_k'^{\,\dagger})]|\,0\rangle, \tag{3.28}$$

where $|\,0\rangle$ is the ground state of H in Σ_{acc} and H' in Σ'_{acc}, satisfying

$$a_k\,|\,0\rangle = a_k'\,|\,0\rangle = 0. \tag{3.29}$$

Remarks

1. From (3.28) we see that $| \, VAC \rangle$ is a coherent state with a B.C.S. type pair correlation [9] between the $a_k^\dagger$-quantum in Σ_{acc} and the $a_k'^\dagger$-quantum in Σ_{acc}'. Therefore, when applied to $| \, VAC \rangle$, the creation of an additional $a_k'^\dagger$-quantum in Σ_{acc}' is equivalent to the annihilation of an $a_k^\dagger$-quantum in Σ_{acc}. Likewise, the annihilation of an $a_k'^\dagger$-quantum in Σ_{acc}' is equivalent to the creation of an $a_k^\dagger$-quantum in Σ_{acc}. This is the underlying reason for the identity (3.27).

2. When $m = 0$, the transformation from (X,T) to (x,t) is conformal. A most convenient choice of $f_k(x)$ is e^{ikx}. (There is a trivial modification in (3.14)–(3.16), because f_k is now complex.) Correspondingly, $\omega_k = | \, k \, |$, and a_k, $a_k^\dagger$ refer to the annihilation and creation operators of a quantum of momentum k in Σ_{acc}, whereas a_k' and $a_k'^\dagger$ refer to those in Σ_{acc}'. Viewed in the Minkowski frame Σ, the quantum $a_k^\dagger \, | \, 0 \rangle$ moves in the same k-direction, but $a_k'^\dagger \, | \, 0 \rangle$ goes in the opposite direction. Thus, the ground state $| \, VAC \rangle$ correlates each $a_k^\dagger$-quantum in Σ_{acc} with an opposite moving $a_k'^\dagger$-quantum in Σ_{acc}'.

3. Theorem 1 states that $| \, VAC \rangle$, the ground state of $\mathcal{H}$ in the Minkowski frame Σ, can be expressed as a matrix element of the operator, $e^{-2\pi H/g}$, in Σ_{acc}. An excited state of $\mathcal{H}$ can be written as a linear superposition of $| \, VAC \rangle$, $A_K^\dagger \, | \, VAC \rangle$ and $A_K^\dagger A_{K'}^\dagger \, | \, VAC \rangle$, etc. Now, $A_K^\dagger$ is a linear function of a_k, $a_k^\dagger$ and a_k', $a_k'^\dagger$; when operating on $| \, VAC \rangle$, its dependence on a_k', $a_k'^\dagger$ can be converted into a_k, $a_k^\dagger$, on account of Cor. 1. Thus, *any excited state* of $\mathcal{H}$ can be cast in the form $u(a_k,a_k^\dagger) \, | \, VAC \rangle$, and because of Theorem 1, *can also be constructed only out of the operators available in* Σ_{acc}. This gives, then, an affirmative answer to the first question asked in Remark 3 of Section 2; in the following, we turn to the other two questions.

IV. An Example

Let us assume that the state vector $| \rangle$ is not $| \, VAC \rangle$; instead, it is

$$| \rangle = \left[C_0 + \int \frac{dk}{2\pi} C_1(k) \, a_k'^\dagger \right] \, \Big| \, VAC \Big\rangle. \qquad (4.1)$$

Note that the "primed" quantum generated by $a_k'^\dagger$ is beyond the horizon of Σ_{acc} and therefore outside the direct reach of an observer in Σ_{acc}. In Figure 1, Σ_{acc} is restricted to region I. Through $C_1(k)$,

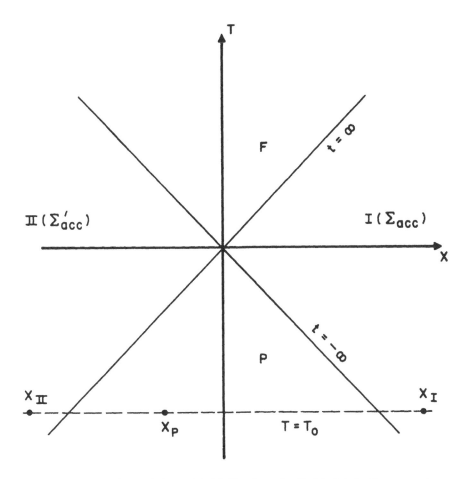

Figure 1. The entire Minkowski (X,T) plane is divided into four quarters: I, II, F and P. For a state vector prepared at $T = T_0 \ll 0$ (except for the part due to $\phi(X_{\mathrm{II}})$), its dependence on $\phi(X_P)$ and $\phi(X_I)$ can be detected by an observer in I. Since the effect of $\phi(X_P)$ propagates to II as well as I, that portion of the information about II can also be deduced by the observer in I.

the "primed" quantum may be viewed as a wave packet confined to the region II in Figure 1. Nevertheless, the observer in Σ_{acc} can infer its existence through an observation within the region I by measuring, say,

$$\langle | a_k | \rangle = \tfrac{1}{2}C_1(-k)^* \, C_0 \operatorname{csch} \pi\omega/g. \tag{4.2}$$

If there is a genuine thermal bath of black-body radiation, then the same expectation value would necessarily be zero. A nonzero value of (4.1) can tell the observer in Σ_{acc} that the state is a coherent mixture of $|\, VAC\rangle$ plus an additional amplitude of a quantum beyond the horizon; in addition, from (4.1) he can infer the B.C.S. nature of $|\, VAC\rangle$, which pair-correlates the $a_k^\dagger$-quanta in Σ_{acc} with the $a'^\dagger_{-k}$-quanta in Σ'_{acc}.

This example illustrates the possibility of inference about $|\rangle$ in the larger frame Σ by doing experiments in the smaller frame Σ_{acc} (without violation of causality, just as in any Einstein-Podolsky-Rosen experiment). We shall now give a more complete analysis.

5. Density Matrix

An observer in the accelerating frame Σ_{acc} can only deal with fields at points (X,T) located in the quarter $X \geqslant |\,T\,|$, labeled I as before. Hence, his oservables O are restricted to

$$O\left(q,\frac{\delta}{\delta q}\right) \tag{5.1}$$

of (2.26). On the other hand, as given by (2.25), the state vector

$$\langle q,q'\,|\rangle \tag{5.2}$$

depends also on q' not being detected by the observer in Σ_{acc}. Thus, given any state $|\rangle$, the expectation value of O is best expressed in terms of a density matrix; we have

$$\langle\!\langle\, O\,|\rangle = \frac{\mathrm{trace}\ O\rho}{\mathrm{trace}\ \rho} \tag{5.3}$$

for all observables O, where

$$\rho = \rho\left(q,\frac{\delta}{\delta q}\right) \tag{5.4}$$

is the (un-normalized) density matrix. Apart from the normalization constant, ρ is completely determined by the state $|\rangle$; its matrix elements are given by (1.5). When $|\rangle = |\, VAC\rangle$, ρ is given by

$$\rho_V = e^{-2\pi H/g}. \tag{5.5}$$

In this section, we investigate the density matrix ρ when the state in question is not $|\, VAC\rangle$. Let us restate the remaining two questions asked in Remark 3 of Section 2 as follows:

1. Assume that the observer in Σ_{acc} is aware of the larger frame Σ and the existence of $|\rangle$ in Σ. From ρ and its possible deviation from ρ_V, what can the observer infer about the corresponding deviation of $|\rangle$ from $|VAC\rangle$?

2. Since the result of any measurement in the accelerating frame is always describable by a density matrix ρ, an observer in Σ_{acc} may well deny the concept of a coherent state $|\rangle$ in the larger frame Σ, or even refuse to believe the very existence of Σ. If the state is $|VAC\rangle$, then because of (5.5) it is perfectly reasonable for him to take such a view, and to think that his system is really in thermal equilibrium with a black body at temperature $(2\pi)^{-1}$ g/Boltzmann constant. However, if the state is not $|VAC\rangle$, can he always maintain a similar hard-line black-body interpretation?

Throughout this section we adopt the Heisenberg picture so that, e.g., the field $\phi(X,T)$ satisfies the field equation (2.4) in the entire Minkowski frame; when (X,T) is restricted to the points (x,t) in the accelerating frame $\Sigma_{acc}(x,t)$, by equating $\phi(X,T) = \phi(x,t)$, we have the corresponding field equation (2.7) in Σ_{acc}. The complete X, T plane in the Minkowski frame $\Sigma(X,T)$ can be decomposed into four quarters (as shown in Fig. 1):

$$\text{I: } \quad X \geq |T| \geq 0, \qquad\qquad \text{II: } \quad X \leq -|T| \leq 0,$$

$$\text{P: } \quad T \leq -|X| \leq 0 \quad \text{and} \quad \text{F: } \quad T \geq |X| \geq 0. \qquad (5.6)$$

The accelerating frame $\Sigma_{acc}(x,t)$ covers only the quarter I, whereas $\Sigma'_{acc}(x',t)$ covers the quarter II, in accordance with (2.6) and (3.22).

Any normalized state vector in $\Sigma(X,T)$ can be written as (in terms of the annihilation and creation operators a_k, $a_k^{\dagger}$ in Σ_{acc} and a'_k, $a_k'^{\dagger}$ in Σ'_{acc} introduced in Sect. 3):

$$|\rangle = U(a_k, a_k^{\dagger}, a'_k, a_k'^{\dagger}) | VAC\rangle \qquad (5.7)$$

where U is unitary. Using (3.20) and (3.23), we may eliminate the dependence in the above expression on either a'_k, $a_k'^{\dagger}$ or a_k, $a_k^{\dagger}$ in U; the same equation can be written as

$$|\rangle = u \, | VAC\rangle = v' \, | VAC\rangle \qquad (5.8)$$

where

$$u = u(a_k, a_k^{\dagger}) \qquad (5.9)$$

and

$$v' = v(a'_k, a_k'^{\dagger}). \qquad (5.10)$$

However, u and v' may or may not be unitary. By using (5.3), (5.7)–(5.10), and Theorem 1, we find

$$\rho = u e^{-2\pi H/g} u^\dagger = e^{-\pi H/g}(v^\dagger v)^* e^{-\pi H/g} \qquad (5.11)$$

where * denotes complex conjugation, † hermitian conjugation and

$$v = v(a_k, a_k^\dagger) \qquad (5.12)$$

can be obtained by replacing a_k', $a_k'^\dagger$ by a_k, $a_k^\dagger$ in (5.10).

The definition of $|\rangle$ depends on a complete set of commuting observables. The usual method is to take the field variables on a space-like surface S. For definiteness, let us assume S to be

$$T = T_0 \ll 0, \qquad (5.13)$$

i.e., the state vector $|\rangle$ is prepared at some distant past: $T = T_0$. The surface $T = T_0$ is indicated by a dashed line in Figure 1; it cuts into three of the four quarters in (5.6). Let X_I, X_P and X_II be the points on this surface that lie within each of these three quarters, I, P and II. Represent the field and its conjugate momentum collectively at these points by $\phi(X_\mathrm{I})$, $\phi(X_P)$ and $\phi(X_\mathrm{II})$. The unitary matrix U in (5.7) is a functional of $\phi(X_\mathrm{I})$, $\phi(X_P)$ and $\phi(X_\mathrm{II})$.

(a) If U depends only on fields in II, i.e.,

$$U = U(\phi(X_\mathrm{II})) = U(a_k', a_k'^\dagger), \qquad (5.14)$$

then since in this case $U = v(a_k', a_k'^\dagger)$ in (5.10), we have $v^\dagger v = 1$ and therefore, according to (5.11), the density matrix remains $\rho = \rho_V = e^{-2\pi H/g}$. This is obvious, because X_II lies outside the light cone of any point in I; hence, there cannot be any observable effect* to the viewer in Σ_acc.

(b) If the unitary matrix U depends only on fields in I; i.e.,

$$U = U(\phi(X_\mathrm{I})) = U(a_k, a_k^\dagger), \qquad (5.15)$$

then since we now have $U = u(a_k, a_k^\dagger)$ in (5.8)–(5.9), the density matrix is given by

$$\rho = u\, e^{-2\pi H/g}\, u^\dagger \qquad (5.16)$$

---

* This is to be contrasted with the situation, say, when the state $|\rangle = a_k'^\dagger$ $|VAC\rangle$; then, through the equality (3.20) the effect of $a_k'^\dagger$, acting on the coherent state $|VAC\rangle$, would be detectable by an observer restricted in Σ_acc, even though $a_k'^\dagger$ is an operator outside Σ_acc. This is because $a_k'^\dagger$ is not unitary. The same state $a_k' |VAC\rangle$ cast in the form (5.7), would require a unitary matrix U that depends on a_k, $a_k^\dagger$ as well as a_k', $a_k'^\dagger$.

with

$$u^+u = 1. \tag{5.17}$$

In this case, one can interpret the result still as a black-body radiation, but with the labeling of states altered by the unitary transformation u. (Hence, e.g., the entropy of ρ is the same as that of ρ_V.) The effect of u, of course, can be detected by the observer in Σ_{acc}.

(c) A more complicated case arises when U is not of type (a) or (b); e.g., U depends also on points X_P. Then, while U is unitary, the matrices u and v in (5.11) do not have to be unitary, and that makes it difficult to maintain the black-body interpretation, as we shall see.

We may summarize our conclusion concerning the first question we asked. Assuming that the state (5.7) is prepared at some remote time T_0 in the past, the density matrix ρ is invariant under the change, as in (a),

$$U \to W(\phi(X_{\mathrm{II}}))U, \tag{5.18}$$

where $W(\phi(X_{\mathrm{II}}))$ is unitary, depending only on $\phi(X_{\mathrm{II}})$ at the far left corner of the dashed line $(T = T_0)$ in Figure 1. Apart from the ambiguity of such a multiplicative factor, the rest of the U-dependence on both $\phi(X_{\mathrm{I}})$ and $\phi(X_P)$ is, in principle, determinable by an observer in Σ_{acc} through his density matrix. Since the effect of $\phi(X_P)$ propagates to II as well as I; that portion of the physical information in II can also be inferred by the observer in I. (All of this is hardly surprising, since it is exactly what one should expect from an elementary reasoning based on causality.)

Take the familiar scenario of setting up, say, an incoming state in a typical scattering problem: one assumes then that the state vector is prepared at $T_0 = -\infty$. (Correspondingly, X_{I} has to be at ∞, and X_{II} at $-\infty$.) Hence, U depends essentially on $\phi(X_P)$. In that case, an observer restricted in Σ_{acc} can deduce, through his measurements, a great deal concerning the state vector $|\,\rangle$, which is of a global character and extends beyond Σ_{acc}. Yet, suppose that the observer refuses to accept any interpretation regarding the physical reality beyond his horizon. Suppose that to him his density matrix is everything. (Viewed from the larger frame Σ, such a description may be incomplete, but clearly not inconsistent.) Nevertheless, there remains the question whether such a density matrix can always be cast into a natural one in statistical mechanics, like $\rho_V = \exp(-2\pi H/g)$ when the state vector is $|\,VAC\rangle$?

This brings us back to our second question: Can the observer in Σ_{acc} always maintain a hard-line black-body interpretation of his density matrix ρ, even when the state $|\rangle$ is not $|VAC\rangle$?

As one may expect, the answer depends on the state in question. In the above cases (a) and (b), the answer is affirmative, as discussed before. For case (c), the answer would be mostly negative; an exception [9] is when

$$|\rangle = e^{iL} | VAC \rangle \qquad (5.19)$$

where

$$L = L^\dagger = \text{linear function of } a_k,\ a_k^\dagger,\ a_k'\ \text{and}\ a_k'^\dagger. \qquad (5.20)$$

Physically, this corresponds to expanding the quantum field ϕ around a classical background field. Therefore, even though $U = e^{iL}$ depends on $\phi(X_P)$ as well as $\phi(X_I)$ and $\phi(X_{II})$, the problem can be reduced to case (b): Write

$$L = l + l'$$

where

$$l = l^\dagger = \text{linear function of } a^k \text{ and } a^k \qquad (5.21)$$

and

$$l' = l'^\dagger = \text{linear function of } a_k' \text{ and } a_k'^\dagger. \qquad (5.22)$$

It is straightforward to show that the density matrix ρ in Σ_{acc} is

$$\rho = u\, e^{-2\pi H/g}\, u^\dagger \qquad (5.23)$$

where

$$u = e^{il}. \qquad (5.24)$$

Since $u^\dagger u = 1$, we can interpret the result as a black-body radiation, but in the background of a classical field.

Except in this example, for a general case (c) with U depending on points X_P, a hard-line black-body approach may lead one into serious problems, as we shall see.

Consider the physical situation when the state $|\rangle$ is known to be of low excitation, as observed in the larger frame Σ. Such a state $|\rangle$ can be well approximated by an amplitude consisting of only a small number of excited quanta:

$$|\rangle = [C_0 + \sum_k C_1(K)A_K^\dagger + \sum_{k,k'} C_2(K,k')A_K^\dagger A_{K'}^\dagger + \cdots] | VAC \rangle.$$

$$(5.25)$$

Table I.

	State vector	Density matrix $\rho(a_k, a_k^\dagger)$	Hard-line black-body interpretation
(i)	$\lvert VAC\rangle$	$e^{-2\pi H/g}$	black-body (b.b.) radiation
(ii)	$a_k'^\dagger \lvert VAC\rangle$ or $a_k \lvert VAC\rangle$	$a_k e^{-2\pi H/g} a_k^\dagger$	(unusual) b.b. radiation in which momentum k has a two-fold degeneracy, while all other momenta are nondegenerate (for explanations of (ii)–(v) see comments in the text).
(iii)	$a_k^\dagger \lvert VAC\rangle$ or $a_k' \lvert VAC\rangle$	$a_k^\dagger e^{-2\pi H/g} a_k$	an "ensemble" mixture of (i) and (ii), with a relative probability -1 for (i), and $(1 - e^{-2\pi\omega k/g})^{-1}$ for (ii).
(iv)	$(a_k'^\dagger)^m \lvert VAC\rangle$ or $a_k^m \lvert VAC\rangle$	$a_k^m e^{-2\pi H/g} (a_k^\dagger)^m$	(unusual) b.b. radiation in which momentum k has an $(m + 1)$-fold degeneracy, while all other momenta are non-degenerate.
(v)	$(a_k^\dagger)^m \lvert VAC\rangle$ or $a_k'^m \lvert VAC\rangle$	$(a_k^\dagger)^m e^{-2\pi H/g} a_k^m$	ensemble mixture of (i)–(iv), some with negative probability.
(vi)	$e^{iL} \lvert VAC\rangle$ where $L = L^\dagger = $ linear function of $a_k, a_k^\dagger, a_k', a_k'^\dagger$	$u e^{-2\pi H/g} u^\dagger$ where $u = e^{il}$ is unitary, and $l = l^\dagger = $ linear function of a_k and $a_k^\dagger$ only.	b.b. radiation plus a classical background field.
(vii)	$e^{iQ} \lvert VAC\rangle$ where $Q = Q^\dagger$ is quadratic in $a_k, a_k^\dagger, a_k', a_k'^\dagger$	$u e^{-2\pi H/g} u^\dagger$ but $u(a_k, a_k^\dagger)$ is, in general, not unitary	difficult

Table I. (*continued*)

State vector	Density matrix $\rho\,(a_k, a_k^\dagger)$	Hard-line black-body interpretation
example: $Q = \theta(a_k^\dagger a_k'^\dagger - a_k a_k)$	$e^{-2\pi \overline{H}/g}$ where $$\overline{H} = \sum_{p \neq k} \omega_p a_p^\dagger a_p + \kappa \omega_k a_k^\dagger a_k$$ with $$\kappa = \frac{\ln \tanh(\theta + \theta_0)}{\ln \tanh \theta_0}$$ and $$e^{-\pi\omega k/g} = \tanh \theta_0$$	another unusual b.b. radiation, in which the momentum k has a temperature different from that of other momenta by a factor $1/\kappa$. Since the field equation remains $$\left(\frac{\partial^2}{\partial s^2} - \frac{\partial^2}{\partial t^2}\right)\phi = 0$$ in $\sum_{acc}(x, t)$, the frequency associated with momentum k is still $\omega_k = |k|$.

Take the case of a free field. Since $A_K^\dagger$ is a linear function of a_k, $a_k^\dagger$, a_k' and $a_k'^\dagger$, by using (3.27) we can rewrite the factor in the square bracket as a finite polynomial in a_k, $a_k^\dagger$, or in a_k', $a_k'^\dagger$, as in (5.8)–(5.10). Typically, they can be written as linear superpositions of $|\,VAC\rangle$,

$$a_k'^\dagger\,|\,VAC\rangle \qquad \text{or} \qquad a_k\,|\,VAC\rangle,$$

$$a_k'\,|\,VAC\rangle \qquad \text{or} \qquad a_k^\dagger\,|\,VAC\rangle,$$

$$(a_{k\dagger}')^2\,|\,VAC\rangle \qquad \text{or} \qquad a_k^2\,|\,VAC\rangle, \text{ etc.}$$

Our concern is the properties of the corresponding density matrix ρ. For an observer in $\sum_{acc}$ who has no idea about the larger frame $\sum$, what kind of statistical interpretation can he give to these density matrices? Examples are tabulated in Table 1 under the heading "hard-line black-body interpretation". (For simplicity, we set the mass of the particle to be zero, so that k denotes its momentum in the accelerating frame.)

Comments

1. In a pure black-body radiation, the partition function for each mode k is

$$p(\zeta) \equiv \text{trace } \zeta^{a^\dagger a} e^{-\beta a^\dagger a} = (1 - \zeta e^{-\beta})^{-1}$$

where a denotes a_k, ζ is the fugacity and $\beta = 2\pi\omega_k/g$. In case (iv) of Table 1, the corresponding partition function is

$$\text{trace } \zeta^{a^\dagger a} a^m e^{-\beta a^\dagger a} (a^\dagger)^m = \sum_n \frac{(n+m)!}{n!} \zeta^n e^{-(n+m)\beta}$$

$$= m! e^{-\beta m} (1 - \zeta e^{-\beta})^{-(m+1)} \quad (5.26)$$

which, apart from a ζ-independent factor $m! e^{-\beta m}$, is exactly $p(\zeta)^{m+1}$; thus, it gives a distribution identical to that of $(m+1)$ independent modes with the same frequency ω_k. For case (ii), $m = 1$.

2. For case (iii), its partition function can be written as a sum

$$P(\zeta) \equiv \text{trace } \zeta^{a^\dagger a} a^\dagger e^{-\beta a^\dagger a} a = e^{\beta}[p(\zeta)^2 - p(\zeta)]. \quad (5.27)$$

Thus, e.g., the average of $n^l \equiv (a^\dagger a)^l$ is given by

$$\langle n^l(P)\rangle = \frac{1}{P(1)}\left[\frac{\partial^l}{(\partial \ln \zeta)^l} P(\zeta)\right]_{\zeta=1} = \frac{e^\beta p(1)}{P(1)} \times [p(1)\langle n^l(p^2)\rangle - \langle n^l(p)\rangle]$$

$$(5.28)$$

where

$$\langle n^l(p^2)\rangle = \frac{1}{p(1)^2}\left[\frac{\partial^l}{(\partial \ln \zeta)^l} p(\zeta)^2\right]_{\zeta=1} \quad (5.29)$$

and

$$\langle n^l(p)\rangle = \frac{1}{p(1)}\left[\frac{\partial^l}{(\partial \ln \zeta)^l} p(\zeta)\right]_{\zeta=1}, \quad (5.30)$$

as if we had a mixture of cases (i) and (ii), with relative probability $p(1) = (1 - e^{-\beta})^{-1}$ when the distribution is $p(\zeta)^2$, and relative probability -1 when the distribution is $p(\zeta)^2$. (Of course, ρ itself is positive definite.)

3. Even in statistical mechanics, deviations from a canonical ensemble distribution may arise. Take the naive example of "a fly plus a black-body radiation", in which we assume that there is an additional fly, external to the radiation. Neglecting the interaction between the fly and the radiation, we see that the physical description of each state has changed from n quanta to "a fly plus n quanta". Hence, a simple relabeling of the states can restore the distribution

back to the standard black-body distribution; i.e., the density matrix differs from $\rho_V \equiv \exp(-2\pi H/g)$ by a unitary transformation

$$u\rho_V u^\dagger$$

with $uu^\dagger = 1$, as in (5.16) and (5.23). Consequently, the entropy of the system remains the same. A more general situation could be an incoherent mixture of such changes, each occurring with a positive probability. In that case, the entropy would always increase.

For the problem that we are considering, there is no genuine ensemble average. It is not difficult to construct coherent states $|\rangle$ in the larger frame Σ, so that the entropy contained in the corresponding density matrices ρ can be quite different from that of ρ_V, either smaller or larger; this could make a persistent hard-line black-body interpretation rather difficult.

VI. World of a Dedicated Manhattanite

In Figure 2 the space-like surface

$$S = S_M + S'$$

consists of a finite part S_M which denotes Manhattan Island, and an infinite S' representing the remaining part. Imagine that light beacons are erected all along the edges of the island; the light cones

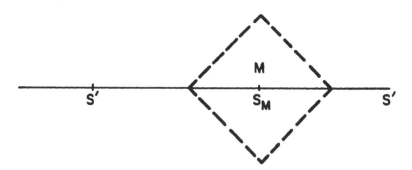

Figure 2. The space-like region occupied by Manhattan is s_M. The dashed line denotes light cones which enclose a space-time region M. A dedicated Manhattanite believes M to be the entire universe. This view, though narrow, is logically possible if the Manhattanite understands only classical physics, but not quantum mechanics.

emanating from the beacons serve as the boundary of a space-time domain called M, which contains S_M. A Manhattanite is one who lives in Manhattan. Naturally, almost all Manhattanites believe Manhattan (M) to be the center of the world. But for a dedicated Manhattanite, M is the entire world. Nothing else exists. Classically, by using the retarded (or advanced) Greens function, the dedicated Manhattanite can completely ignore the region outside M; to him any effect related to the region outside M can and should be cast in terms of suitable boundary conditions within M. Needless to say, these boundary conditions may be rather convoluted. This is why he has to subscribe to the New York Times. To the dedicated Manhattanite this does not at all mean that the news about the world outside M has anything to do with reality; it is merely a convenient method through which this voluminous daily paper can provide him with the needed boundary conditions.

Classically such an attitude is logically possible and totally consistent with the principles of physics. However, quantum mechanically there are difficulties. The state vector

$$| \rangle = | S \rangle$$

depends on an entire space-like surface S which goes beyond M. The preparation of $| S \rangle$ is outside the control of the Manhattanite. Of course, the observables O available to him are restricted to a function of the field inside M, and the observation

$$\langle | O | \rangle = \text{trace } \rho O / \text{trace } \rho$$

can always be expressed in terms of a density matrix ρ within M. However, the interpretation of such a density matrix can be quite different than the ensemble averages in statistical mechanics. This is amply demonstrated by the examples given in Table 1. A sensible and sophisticated Manhattanite (one who understands quantum mechanics) would perhaps be less dedicated.

Acknowledgment

I wish to thank Sidney Coleman for raising pertinent questions after my talk at the Niels Bohr Centennial Symposium in Cambridge, Massachusetts. Table 1 in Section 5 was compiled with the help of R. Friedberg, and was added as a result of Coleman's questions. (Case (vi) in that Table was analyzed by S.C.)

References

1. S.W. Hawking, Nature **248,** 30 (1974); Comm. Math. Phys. **43,** 199 (1975); Phys. Rev. **D14,** 2460 (1976).
2. S. Fulling, Phys. Rev. **D7,** 2850 (1973); W. G. Unruh, Phys. Rev. **D14,** 870 (1976).
3. A. Einstein, B. Podolsky and N. Rosen, Phys. Rev. **47,** 777 (1935).
4. J.S. Bell, Physics **1,** 195 (1964).
5. T.D. Lee, Nucl. Phys. **B264,** 437 (1986).
6. T.D. Lee, Progr. Theor. Phys. Suppl. No. 85 (*The Jubilee of Meson Theory,* ed. M. Bando, R. Kawabe and N. Nakanishi, 1986), p. 271.
7. R. Friedberg, T.D. Lee and Y. Pang, Nucl. Phys. **B276,** 549 (1986).
8. T.D. Lee, J. Stat. Phys. *46,* 843 (1987).
9. S. Coleman (private communication).

Quaternionic Quantum Field Theory

Stephen L. Adler

THE INSTITUTE FOR ADVANCED STUDY

What I want to talk about is whether the standard quantum mechanics, which was invented in Copenhagen in the late 1920's, is the only type of quantum mechanics, or whether there are more general ones; and in particular, I want to talk about the possibilities of a quaternionic quantum field theory.

Let me begin by asking why try to make a new kind of quantum mechanics? I can give two reasons: one is a mathematical reason, which is that we get a better understanding of a system of postulates if we have more than one realization; and specifically we'll get a better understanding of complex quantum mechanics, if we understand which features of the usual complex quantum mechanics are more general than others. The second reason is a physical motivation: the current state of particle theory can be represented pretty much like this:

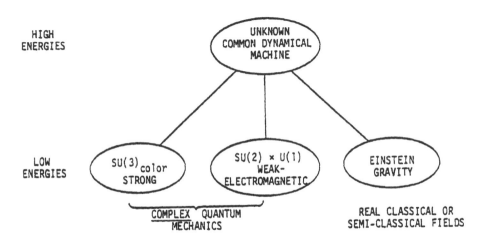

At low energies we know well what's going on. There's an effective
theory, the $SU(3)$ color theory of the strong interactions and the
$SU(2) \times U(1)$ weak-electromagnetic theory, which are both described by complex quantum mechanics, and Einstein gravity, which
is described by real classical or semi-classical fields. At high energies, what we believe is that there's some common dynamical machine which gives rise to all of these three low-energy theories as
effective field theories. And what we don't know, the big question
of high energy physics, is to figure out what is this common dynamical machine. Now attempts over the last fifteen years to construct
the common dynamical machine, using local complex quantum mechanics, have run into difficulties. At least, there isn't a natural,
obvious, simple unifying model. So that raises the question whether
something new is needed, and I think most of the focus of physics
for the last year or two has been on trying new things. For instance,
the superstring generalizes from local to non-local field theory. Here
I'm going to discuss a generalization in a different direction, which
is a generalization from complex numbers to quaternions.

Let me begin by reviewing what it is that distinguishes a quantum
system from a classical one. I'm going to follow very closely here
a very nice discussion in Feynman's famous Reviews of Modern
Physics article, where he develops the path integral, because in fact
essentially everything I do is an attempt to take what Feynman does
and to generalize it from complex numbers to quaternions.

Let B be a set of attributes which completely specify a state in
the quantum mechanical sense, and let P_{ab} be the probability that
if a measurement of the set of attributes A gives the result a, then
the measurement of B gives b; and let P_{bc}, similarly, be the probability that if the measurement of B gives b, the measurement of C
gives c. Classically, the probability that if a measurement of A gives
a, then the measurement of C will give c, is simply given by the law
of conditional probability. We sum over all intermediate outcomes
and just multiply the probabilities. Thus classically, P_{ac} is the sum
over b of P_{ab} times P_{bc}:

$$P_{ac} = \sum_b P_{ab}P_{bc}.$$

This would be the rule you'd use in flipping dice or calculating the
outcomes in a card game. (And this, incidentally, is also what's used
in the derivation of Bell's theorem that we'll hear about later on.)

However, this law of superposition of probabilities is not generally true of quantum mechanics. Instead, in quantum mechanics, we're guaranteed the existence of a probability amplitude, ϕ_{ab}. (In Dirac notation it's the transition matrix element $\langle b \mid a \rangle$.) And we're told that the probabilities are the squares in magnitude of the ϕ's: P_{ab} is the magnitude $\mid \phi_{ab} \mid^2$, P_{bc} is the magnitude $\mid \phi_{bc} \mid^2$, and so forth. And then we're told that ϕ_{ac}, which is the probability amplitude to go from a to c, is equal to the sum over intermediate states b of ϕ_{bc} times ϕ_{ab},

$$\phi_{ac} = \sum_b \phi_{bc}\phi_{ab} = \sum_b \langle c \mid b \rangle \langle b \mid a \rangle.$$

Now let's notice, for later reference, that when you write things in the Dirac notation, there's a natural factor ordering that comes in also: if you simply write the bra's and ket's in the natural order, from left to right, there is an implicit factor ordering. And that's what's going to allow for the possibility of going from a commutative to a non-commutative number system, while still preserving the law of superposition of probability amplitudes. So to summarize, then, in classical mechanics, probability is superimposed; in quantum mechanics, probability amplitudes—and not probability—are superimposed.

Let's now ask what kinds of number systems can we use for the ϕ's? There's a theorem, stemming from the work of Birkhoff and von Neumann which established an axiomatic basis for quantum mechanics, that states that the ϕ's can be real, in which case you have real quantum mechanics; complex, which gives the standard quantum mechanics that we have in all our textbooks; and quaternion, which gives quaternionic quantum mechanics. Let me review briefly what we mean by complex and quaternion numbers. We all know what real numbers are. A complex number, z, has real components

$$z = x + yi,$$

where x and y are real, and i^2 is -1. A quaternion number is a generalization of a complex number, with three imaginary units. So a quaternion q has the form,

$$q = q_0 + q_1e_1 + q_2e_2 + q_3e_3,$$

where the components q_0, q_1, q_2, and q_3 are all real, and e_1, e_2, and

e_3 are quaternion units—in the older literature they're often called $i, j,$ and k—and they satisfy an algebra:

$$e_a e_b = -\delta_{ab} + \varepsilon_{abc} e_c.$$

Writing this out, it says that

$$e_1^2 = e_2^2 = e_3^2 = -1,$$

so $e_{1,2,3}$ are all quantities like imaginary units. However, they don't commute;

$$e_1 e_2 = e_3, \quad \text{but} \quad e_2 e_1 = -e_3,$$

and similarly for the cyclic permutations, so they obey a non-commutative multiplication law. There is a simple complex matrix realization of the quaternions. You write:

$$e_a = -i\tau_a$$

where the τ_a are the standard Pauli matrices, $\tau_1, \tau_2,$ and τ_3, and this complex matrix realization satisfies the abstract algebra. However, what I will say will not make specific use of the complex realization. I'm going to talk about a quantum mechanics where the e's are regarded as fundamental abstract quantities, just as in doing complex number theory we don't normally make use of the fact that i can be given a real matrix representation [i is represented by the matrix

$$\begin{pmatrix} 0 & 1 \\ -1 & 0 \end{pmatrix},$$

because the square of this matrix is -1].

Now what is it that the real, complex and quaternion number systems have in common? Let me follow a discussion of Pontryagin and introduce the concept of a number field. It's a number system which has two operations, an addition and a multiplication. The addition and the multiplication are associative, so,

$$a + (b + c) = (a + b) + c$$

$$a(bc) = (ab)c,$$

and the multiplication is distributive over the addition,

$$a(b + c) = ab + ac.$$

The addition is commutative,

$$a + b = b + a,$$

but the multiplication is not necessarily commutative, and generally

$$ab \neq ba.$$

Finally, there are additive and multiplicative inverses: for every a, there's a $-a$, such that

$$a + (-a) = 0,$$

and for every non-zero a, there's an a inverse, such that

$$aa^{-1} = a^{-1}a = 1.$$

Let me also introduce the idea of topological structure. A number system is said to be topological, roughly speaking, if continuous variations are possible. Then there's a theorem in Pontryagin which states that the only arcwise-connected, topological number fields are the real, complex and quaternion numbers. Now, the properties assumed in this theorem are clearly ones that we would like our probability amplitudes in quantum mechanics to have. Note that sacrificing the commutative law of multiplication is not necessarily a problem, because as I showed, if you simply write out the intermediate state sum in Dirac notation, there's already a natural factor ordering. Therefore this permits a natural generalization of quantum mechanics to the case where the matrix elements individually are elements of a non-commutative number system.

Real quantum mechanics was investigated a long time ago by Stueckelberg. What he showed is that to describe physics as we know it, real quantum mechanics must reduce back to complex, or quaternion, quantum mechanics. Basically, what he found is that it must contain an operator, J, such that

$$J^T = -J, \qquad J^2 = -1$$

which commutes with all observables, but that's, of course, just a real representation for i. The reason you have to have this (it's not Stueckelberg's argument, but a simpler one) is that if you want

$$\frac{d}{dt} \int |\Psi|^2 = 0,$$

i.e., if you want probability to be conserved, but Ψ to have a non-trivial dynamics, you have to have an imaginary unit present; it's the conjugation from i to $-i$ that makes the Hamiltonian cancel out giving conservation of the wave function normalization. If you don't have the i, then Ψ can't have a non-trivial dynamics while having a conserved total norm.

So, real quantum mechanics is not interesting; you have to have an imaginary unit. It can be either the imaginary unit of the complex numbers or it can be one of the quaternion imaginary units. So the interesting cases are the standard one, complex quantum mechanics, which describes the observable world, or possibly quaternion quantum mechanics. Now quaternionic quantum mechanics is potentially interesting because Yang-Mills structures start to appear naturally. This was observed first by Yang in a comment at the 1957 Rochester conference, that served as the motivation for Finkelstein and collaborators to start looking at the quaternionic case. There followed some early investigations by Finkelstein, Jauch, Speiser and others, which led to very interesting mathematical structural theorems about quaternion quantum mechanics. But there were stumbling blocks that prevented them from going on to a complete theory. The stumbling blocks were the following:

First, the question of what you use for dynamics. If you simply write

$$\hat{e}\,\frac{d\Psi}{dt} = H\Psi$$

as a Schrödinger equation, where H is a self-adjoint Hamiltonian, and $\hat{e}$ is some quaternion unit, this is simply a rewriting of the complex case, with a different name for the imaginary unit. And although the language of the Finkelstein paper suggests something more general, when they started writing down wave equations, this is essentially what they did. So what they did, in specifics, was really complex quantum mechanics, giving something very close to the Georgi-Glashow model, in fact, written in a disguised notation.

The second problem is in dealing with composite systems. If Ψ_a and Ψ_b are quaternion-valued wave functions, then clearly

$$\Psi_a\Psi_b \neq \Psi_b\Psi_a.$$

Therefore, we don't have a commutative tensor product. Having a commutative tensor product is the basis for the whole second quantization treatment of field theory. So the second quantization approach to treating many degrees of freedom breaks down in the quaternionic case.

In the rest of the talk, what I will do is to describe, in outline, a new approach which I've been looking at for the last year, to make a quaternionic quantum mechanics using functional integral methods. It turns out to be possible to proceed fairly far in analogy with the complex case. What I'll do first is to review how, from Dirac's observation in the complex case, one gets the functional integral for a transition amplitude

$$\langle b, t \mid a, 0 \rangle.$$

Then I'll show how this generalizes to quaternions, and sketch briefly how by using the functional integral plus a Gaussian integral formula you can derive the Schrödinger equation—this generalizes what's in Feynman's famous paper; and conversely the generalization of what's in all the standard field theory textbooks—how by starting from the Schrödinger equation, or the equivalent transformation theory, plus the same Gaussian integral formula, you get back to the functional integral. Thus, all of this standard complex quantum mechanics does generalize to the quaternionic case. However, the analogy is only partial. At the end of the talk, I'll state a long list of complex quantum theory results that fail in the quaternionic case. Only a subset of complex quantum theory generalizes to the quaternion number system.

The fundamental technical tool on which the whole analysis is based is a Gaussian integral formula. Gaussian integral formulas have a long history in physics, because the theory of wave optics, which is what quantum mechanics originated from, is based on Gaussian integrals. This is where my investigation began. I came across some articles by Dyson showing how to construct quaternion determinants, and reasoned that if there are quaternion determinants then there must be quaternion Gaussian integrals.

Let me begin by reviewing the Gaussian integral formulas in the complex case. This is usually section one in the field theory textbooks: before you develop the path integral, you first state the Gaussian integral formula, because that's the essential technical tool in developing the standard Feynman path integral. So let me first state it for bosons. Let z and u be column vectors containing N complex numbers, and let A be an $N \times N$ complex matrix. Then the standard Gaussian integral formula for complex numbers says that

$$\left(\prod_1^N \int \frac{d\bar{z}\, dz}{2\pi i} \right) e^{-\bar{z}Az + \bar{u}z + \bar{z}u} = (\det A)^{-1} e^{\bar{u}A^{-1}u}.$$

This is very simply derived; you do a shift of the integration variable z by something proportional to A^{-1}, which eliminates the linear terms, so you've completed the square. The $\exp(\bar{u}A^{-1}u)$ piece factors out, and then you simply make a unitary transformation to diagonalize A, and integrate mode by mode, leading to a product of eigenvalues in the denominator which gives determinant A. That's a sketch of the proof; you can of course give a neater derivation.

We'll also need a similar formula for fermions. Now in field theory, to handle fermions one has to introduce peculiar objects called Grassman numbers. They are anti-commuting numbers. In other words,

$$\psi_a \psi_b + \psi_b \psi_a = 0,$$

for any a and b. That means that if a and b are equal, we have

$$\psi_a^2 = 0.$$

So the square of a Grassman number is 0, and any two Grassman numbers anti-commute. Then you define integration of a Grassman variable by the formulas

$$\int d\psi_a \psi_b = \delta_{ab}$$

$$\int d\psi_a (\text{any other monomial}) = 0.$$

This means that $\int d\psi_a$ is effectively simply a derivative with respect to the fermionic variable ψ_a at $\psi_a = 0$. But these derivatives anti-commute, because the Grassman numbers themselves anti-commute.

Now let ψ, ξ be column vectors of N complex Grassman numbers. The analog of the bosonic formula—and this is also in the first section of any standard field theory textbook—says:

$$\left(\prod_1^N \int d\bar\psi \, d\psi \right) e^{-\bar\psi A \psi + \bar\xi \psi + \bar\psi \xi} = \det A e^{\bar\xi A^{-1} \xi}.$$

A is, again, a general $N \times N$ complex matrix. In the complex case, I don't have to restrict A to be Hermitian or anti-Hermitian, but in the quaternionic case I will have to say something more specific about what A is. The reason $\det A$ appears upstairs in the fermionic formula is the anti-symmetry of the Grassman multiplication: you simply Taylor-expand the formula, and it terminates at the nth order term, because there you've exhausted all the $\int d\bar\psi$'s and $\int d\psi$'s, and you just get the expansion for the determinant as a sum over products of one element in each row and column with an alternating sign.

Let's try now and make a quaternionic analog of these Gaussian integrals. I'll replace my complex variable, z, by a quaternion ϕ, which is,

$$\phi = \phi_0 + \phi_1 e_1 + \phi_2 e_2 + \phi_3 e_3$$

and which has a quaternionic conjugate,

$$\overline{\phi} = \phi_0 - \phi_1 e_1 - \phi_2 e_2 - \phi_3 e_3$$

obtained by simply reversing the signs of all three quaternion units. That's an analog of complex conjugation. (AUDIENCE: Excuse me, of the ϕ_a's, can they be complex, or are they real?) The ϕ_a's are real. And the reason they're real is that complex quaternions don't form a division algebra because

$$(1 + ie_3) \times (1 - ie_3) = 0,$$

so you can have non-zero elements of the algebra multiplying to give zero. However, the Birkhoff-von Neumann construction says that you really want to have a division algebra. A simple heuristic way of seeing this is, if you want to get propagators, you've got to have everything in your algebra of scalars having an inverse, so you're not going to be able to form propagators if you don't have a division algebra. So we must deal with the quaternions over the real numbers. The ψ's are similarly Grassman quaternions. In other words, ψ is a quaternion whose coefficients are Grassman elements of just the form I discussed before,

$$\psi = \psi_0 + \psi_1 e_1 + \psi_2 e_2 + \psi_3 e_3,$$

$$\overline{\psi} = \psi_0 - \psi_1 e_1 - \psi_2 e_2 - \psi_3 e_3.$$

Now if we want to do real-time quantum mechanics, and develop the quaternion analog of the familiar complex case, the analog of the complex e^{iL} will be $e^{\check{L}}$, where $\check{L}$ is a quaternion imaginary Lagrangian. $\check{L}$ will be

$$\check{L} = e_1 L_1 + e_2 L_2 + e_3 L_3,$$

with $L_{1,2,3}$ real. So to do the analog of the complex, real-time path integral, we need to consider oscillatory quaternion Gaussian integrals. To evaluate such integrals we need an integration measure. It turns out that the natural integration measure that replaces $d\overline{z}\, dz$

is $d\phi$, where $d\phi$ is simply the product of the differentials of the real components,

$$\int d\bar{z}\, dz \rightarrow \int d\phi \equiv d\phi_0\, d\phi_1\, d\phi_2\, d\phi_3,$$

and similarly, the analog of $d\bar{\psi}\, d\psi$ is

$$\int d\bar{\psi}\, d\psi \rightarrow \int d\psi \equiv \int d\psi_0\, d\psi_1\, d\psi_2\, d\psi_3.$$

The reason that this is the right measure is that you can show—and this is a little bit of work—that the measure defined this way is invariant under arbitrary quaternion unitary transformations.

From here on things get more complicated, and I'm only going to state a result. Because the e_a's, the quaternion units, don't commute, when you complete the square you find that you cannot then factor the source term out of the integral, because when a and b are not commuting operators,

$$e^{a+b} \neq e^a e^b.$$

So you find that the source dependence is not exponential in general, when you figure out how to do the integrals. It's not just an analog of the $e^{\bar{u}A^{-1}u}$ or $e^{\bar{\xi}A^{-1}\xi}$ of the complex case. But now comes the first surprise. When you do the integrals, what you find is that when the number of bosonic integration variables is equal to the number of fermionic integration variables, the sources do re-exponentiate, and you get the following formula,

$$\lim_{\varepsilon \to 0} \left(\prod_{i=1}^{M} \int d\phi^i\, d\psi^i \right) (4\pi^2)^{-M} \exp(-\bar{\phi}A\phi - \bar{\psi}B\psi + \bar{u}\phi - \bar{\phi}u$$

$$+ \bar{\xi}\psi + \bar{\psi}\xi + \kappa - \varepsilon\bar{\phi}\phi) = \det^2 B \det^{-1}(A^\dagger A) \exp(-\bar{u}A^{-1}u$$

$$+ \bar{\xi}B^{-1}\xi + \kappa).$$

Here κ is a quaternion-imaginary constant, and apart from the infinitesimal convergence term $\varepsilon\bar{\phi}\phi$, everything in the exponents is quaternion imaginary, provided one makes the choice that A is a quaternion anti-Hermitian matrix, and B is a quaternion Hermitian matrix. So you find the surprising feature that the formulas which are valid for bosons and fermions separately in the complex case are valid for quaternions only when the bosons and fermions are combined in equal numbers. I don't know the deep reason for this;

the way I derived this result was by getting a recursion relation which describes what happens as I add one integral at a time. And then, although the computations look very different, it turns out that the recursion relations for bosons and fermions are precisely inverse to each other.

Now let's go back to complex quantum mechanics and sketch how Dirac and Feynman got the path integral. Let me write down a timeline for a system with a single coordinate, x, and divide the segment of the timeline from t to $t + T$ into N little intervals of width Δt:

$$t_N \quad t_{j+1} \quad t_j \quad t_0$$

$$\vdots \quad \vdots \quad \vdots \quad \vdots \qquad \text{Timeline}$$

$$t + T \qquad\qquad\qquad t$$

$$t_{j+1} - t_j = \Delta t = T/N$$

Now, Dirac, in the '30's, made the famous observation that the transformation function from time t_j to time t_{j+1}, in the limit as to Δt goes to 0, is simply related to the classical Lagrangian,*

$$\langle x_{j+1}, t_{j+1} \mid x_j, t_j \rangle = \frac{1}{\zeta} e^{i\Delta t L(x_{j+1/2}, \dot{x}_{j+1/2}, t_{j+1/2})}.$$

Here $L(x, \dot{x}, t)$ is a classical Lagrangian, and $x_{j+1/2}$, $\dot{x}_{j+1/2}$ and $t_{j+1/2}$ are just the position, the velocity and the time evaluated by the trapezoidal rule at the midpoint of the interval,

$$x_{j+1/2} = 1/2(x_{j+1} + x_j)$$

$$\dot{x}_{j+1/2} = (x_{j+1} - x_j)/\Delta t$$

$$t_{j+1/2} = t_j + \Delta t/2.$$

What Feynman did to make the path integral—it's partly implicit in Dirac's paper—what Feynman really did beyond this was to show that you could derive the Schrödinger equation from it, and essentially erect the whole quantum mechanics apparatus from the path integral. Feynman put together many infinitesimal time factors; for each little infinitesimal interval, we know the transformation function: it's just $e^{i\Delta t L}$, with an error that's order $(\Delta t)^2$. That allows us to do a Riemann sum, and so therefore the finite time transformation function, $\langle x', t + T \mid x, t \rangle$, is just the product over all the intermediate

---

* Note: $L(x_{j+1/2}, \dot{x}_{j+1/2}, t_{j+1/2})$ is abbreviated below to $L(j + 1/2)$.

time slices, $j = 1$ to $N - 1$, of integrals of the infinitesimal transformation function,

$$\langle x', t + T \,|\, x, t \rangle$$

$$= \left(\prod_{j=1}^{N-1} \int dx_j \right) \zeta^{-1} e^{i \Delta t L (N - 1/2)} \zeta^{-1} e^{i \Delta t L (N - 3/2)} \ldots \zeta^{-1} e^{i \Delta t L (1/2)}.$$

And in the limit $N \to \infty$, this gives you the Feynman path integral.

This construction of quantum mechanics invites a quaternionic
analog. Let's simply take as our postulate that the transformation
function from a quaternionic state, where the coordinates ϕ and ψ
have values ϕ_j and ψ_j at time t_j, to one where they have values ϕ_{j+1},
ψ_{j+1} at time t_{j+1}, is equal to some constant times the exponential
of Δt, times something which I'll call my quaternion Lagrangian,

$$\langle \{\phi_{j+1}, \psi_{j+1}\}, t_{j+1} \,|\, \{\phi_j, \psi_j\}, t_j \rangle$$

$$= \zeta^{-1} \exp[\Delta t \tilde{L}(\{\phi_{j+1/2}, \dot{\phi}_{j+1/2}, \psi_{j+1/2}, \dot{\psi}_{j+1/2}\}, t_{j+1/2})].$$

But now $\tilde{L}$ has to be quaternion imaginary, so it will be a sum of
three terms,

$$\tilde{L} = L_1 e_1 + L_2 e_2 + L_3 e_3,$$

with $L_{1,2,3}$ real. Thus $\tilde{L}$ will be some function that specifies the dynamics—and again, I'll evaluate it at the midpoint of the interval.
But again, just by combining infinitesimal time steps, I can get a
finite time formula that says that the transformation function from
t_0 and some initial state at time t_0 to t_N and some final state at time
t_N is just the product over all intermediate states, integrating over
all the intermediate state variables, times the infinitesimal transformation functions,

$$\langle \{\phi_N, \psi_N\}, t_N \,|\, \{\phi_0, \psi_0\}, t_0 \rangle$$

$$= \left[\prod_{j=1}^{N-1} \left(\prod_{i=1}^{M} \int d\phi_j^i \, d\psi_j^i \right) \right] \zeta^{-1} e^{\Delta t \tilde{L}(N - 1/2)}$$

$$\times \zeta^{-1} e^{\Delta t \tilde{L}(N - 3/2)} \ldots \zeta^{-1} e^{\Delta t \tilde{L}(1/2)}.$$

In taking the product now the ordering is important, because the
infinitesimal transformation functions don't commute, but we saw
that the Dirac notation gives us a natural factor ordering, and in the
limit $N \to \infty$ we get what physicists call a time-ordered product. This

recipe automatically satisfies the principle of probability amplitude superposition, because if I break the product at some intermediate time, t_I, the product of infinitesimal factors from t_0 to t_I is just a transformation function from t_0 to t_I, there is an integration over the variables at time t_I, and then the remaining string of infinitesimal factors is a transformation function from t_I to t_N,

$$\langle\{\phi_N, \psi_N\}, t_N \mid \{\phi_0, \psi_0\}, t_0\rangle$$

$$= \left(\prod_{i=1}^{M} \int d\phi_I^i \, d\psi_I^i\right) \langle\{\phi_N, \psi_N\}, t_N \mid \{\phi_I, \psi_I\}, t_I\rangle$$

$$\times \langle\{\phi_I, \psi_I\}, t_I \mid \{\phi_0, \psi_0\}, t_0\rangle.$$

So the quaternionic generalization of the path integral satisfies the principle of superposition of probability amplitudes, which I stated at the beginning is, I believe, the fundamental statement of what we mean by quantum mechanics.

We are ready now to take the next step. We've constructed a path integral; the question now is, can we derive a Schrödinger equation from the path integral; and vice versa: from that Schrödinger equation, can we go back and mimic the standard textbook derivation and rederive the path integral? Let us first recall that in the complex case, to derive the Schrödinger equation, one has to specialize the kinetic term by assuming it's a quadratic form in time derivatives. You write

$$L = 1/2\dot{x}^2 - V(x)$$

and then from this Feynman derived the Schrödinger equation. Similarly, in the quaternionic case, if you make an analogous specialization that's suggested by the structure of the Gaussian integral formula, you can derive a Schrödinger equation. The specialization is to write $\tilde{L}$ as

$$\tilde{L} = \tilde{L}_{kin} - \tilde{V}(\{\phi, \psi\}, t),$$

where $\tilde{V}$ is a completely arbitrary quaternion imaginary potential. So I have interactions between my modes. $\tilde{L}_{kin}$ is a quadratic form

$$\tilde{L}_{kin} = \sum_{i=1}^{M} (1/2\bar{\dot{\phi}}^i e_3 \dot{\phi}^i + 1/2\bar{\dot{\psi}}^i \dot{\psi}^i),$$

which is suggested by simply taking my Gaussian integral formula and taking the Hermitian matrix B to be 1, and taking the anti-Hermitian matrix A to be any arbitrary quaternion unit, say e_3. The boson structure is essentially conventional, while the fermion structure is unconventional because it has a second-order wave operator (and this implies an indefinite metric Hilbert space, when you look in detail at the fermion state structure). But it seems to be necessary to use this second-order fermion wave operator in order to get the construction to work. That could be a problem when you try to apply it to physics, because what we see are first-order fermions. If the quaternionic path integral is to describe physics, there'll have to be some mechanism for splitting half the states off, and that's something I don't know yet how to do. The e_3 in the first term is basically arbitrary, since under the quaternion gauge transformation of replacing ϕ by $q\phi$, where q is a unit quaternion with $\bar{q}q = 1$, e_3 gets replaced by $\bar{q}e_3q$, and that allows you to rotate e_3 to any other imaginary quaternion direction. That is, I need the quaternion imaginary unit there to get an imaginary $\bar{L}$, but the choice is completely arbitrary, and can be simply rotated by a change in variable, and this change of variable is an invariance of the integration measure.

Having specialized $\bar{L}$ as described, now we find our second surprise: Using only the Gaussian integral formula, one can derive the Schrödinger equation from the functional integral, and vice versa, in the quaternionic case. To derive the Schrödinger equation, for example, you define a wave function, Ψ, of the coordinates at time t, as

$$\Psi(\{\phi, \psi\}, t) = \langle\langle\{\phi, \psi\}, t \mid \psi\rangle,$$

and then you write

$$\Psi(\{\phi, \psi\}, t + \Delta t) = \left(\prod_i \int d\phi_0^i \, d\psi_0^i\right) \zeta^{-1} e^{\Delta t \bar{L}(1/2)} \Psi(\{\phi_0, \psi_0\}, t),$$

and Taylor-expand in Δt. This is the standard Feynman derivation. And what you find is that the Feynman derivation goes through when you use the Gaussian integral formula. The reason is that the Gaussian formula, when there are equal numbers of fermions and bosons, allows you to integrate out all the modes when you're talking about the potential energy, but the potential still re-exponentiates. Or, you can integrate out all the modes save one, when you're looking at the kinetic energy term, and the kinetic energy of the one mode you're looking at re-exponentiates, and so you get a universal structure independent of how many modes there are. This re-exponentiation feature is essential for the derivation to go through, and what you find is that you get a Schrödinger equation whose form is totally independent of the number of modes. In the process, the normalization constant, ζ, is fixed to be $\zeta = (4\pi^2)^M$. Because of the use of a second-order fermion wave operator, there is no Δt in the path integral normalization; it's purely a phase-space factor. The Δt's from the bosons cancel the Δt's from the fermions, and so you get something that looks mathematically better behaved than you do in the standard path integral.

The Schrödinger equation that you get has the form

$$\frac{d\Psi}{dt} = -\tilde{H}\Psi,$$

where $\tilde{H}$ is the quaternion imaginary Hamiltonian—it's the analog of iH in the complex case. $\tilde{H}$ contains a kinetic term plus a potential term, and the kinetic term has the form

$$\tilde{H}_{kin} = -\frac{1}{6} \sum_{i=1}^{M} (\overline{D}_{\phi^i} e_3 D_{\phi^i} + \overline{D}_{\psi^i} D_{\psi^i}),$$

where D_ϕ and D_ψ are derivative operators that are constructed in a way analogous to the construction of ϕ and ψ. The one peculiar feature is that in the complex analog of this formula, instead of getting 1/6 as the coefficient in the kinetic term, you get 1/2. The extra factor of 3 comes from averaging over quaternionic rotations, and it came as a surprise. That was why I wanted to be very careful to check that I could do the inverse derivation and get back to my original Lagrangian formula, which does have a coefficient of 1/2 in it, to make sure that the 1/6 wasn't a mistake.

Now the converse involves proving that

$$\langle\{\phi, \psi\}, t \mid e^{-\Delta t \tilde{H}} \mid \{\phi_0, \psi_0\}, t\rangle \leftrightarrow (4\pi^2)^{-M} e^{\Delta t(\tilde{L}_{kin} - \tilde{V})},$$

where $\leftrightarrow$ indicates equivalence inside the functional integral. And, in fact, a complicated derivation shows that you can in fact rederive the functional integral from the transformation theory, again using just the Gaussian integral formula as a tool.

Let me now show how in one respect, the standard quantum mechanics familiar from the complex case breaks down in the quaternionic generalization. In the quaternionic case, you don't have a correspondence principle leading to quaternionic classical equations as a correspondence limit. There are many ways of seeing this—

one is to show that the Ehrenfest theorems break down—but let me show you that if you attempt to do a stationary phase approximation, you don't get quaternionic classical equations as a stationary phase approximation. Well, there are two reasons. One is that those infinitesimal factors, $\Delta t\tilde{L}$, are non-commuting. Therefore, their product is not simply the Riemann sum of the $\Delta t\tilde{L}$'s,

$$\prod e^{\Delta t\tilde{L}} \neq e^{\Sigma\Delta t\tilde{L}}.$$

Suppose that we ignore this, and say that in some approximation we can neglect the path ordering, and call the sum of $\Delta t\tilde{L}$'s the classical quaternion action, $\tilde{S}$, and ask: What happens if we require that $\tilde{S}$ be stationary? Well, we then get three variational principles, because the three components of $\tilde{S}$, along the three quaternionic directions, individually have to be stationary,

$$\delta\tilde{S} = \delta\sum \Delta t\tilde{L} = 0 \Rightarrow \delta S_1 = \delta S_2 = \delta S_3 = 0,$$

and three variational principles are not in general stationary on the same orbits. So if I write down a general quaternion imaginary classical action, I can't in general demand that it be stationary, because I get three different sets of Euler equations. I can write down special classical actions, where the stationary orbits of the three components are the same, but in general it's not true. On the other hand, it makes perfect sense to use these $\tilde{L}$'s as infinitesimal phase factors, even if I can't require that there be corresponding classical orbits. Hence the correspondence limit of quaternionic quantum mechanics is not a quaternionic classical mechanics. If there's a correspondence principle, which I believe there is, it's that quaternionic quantum mechanics will map into complex quantum mechanics, and then complex quantum mechanics will have a correspondence principle of the usual sort.

Let me then come to a conclusion. One can construct a quaternionic quantum mechanics with arbitrary numbers of degrees of freedom, and in particular this means one can take the limit of M going to infinity—which is field theory—and so there's no difficulty in this construction in making a quaternionic quantum field theory. The construction depends on having complete boson-fermion symmetry, because the Gaussian integral formula has an especially simple form when the numbers of bosons and fermions are equal. One has a Schrödinger equation, a Dirac transformation theory, and a functional integral. On the other hand, a lot of the apparatus of complex quantum mechanics is absent. You don't have a commuting tensor product, asymptotic states, or an S-matrix, except in a complex specialization. You don't have a canonical formalism, coherent states or a Euclidean continuation. All of these nice things that we associate with complex quantum mechanics are present only if the quaternionic theory is first specialized to a complex one. So the conclusion is that there seems to be a new kind of quantum mechanics. There are many interesting formal questions to study, and hopefully the analysis of these questions will eventually allow us to settle whether it's relevant for particle physics. Thank you.

Light and Life: Niels Bohr's Legacy to Contemporary Biology

Gunther S. Stent

UNIVERSITY OF CALIFORNIA, BERKELEY

On August 15th, 1932 Niels Bohr gave the Opening Address of the International Congress on Light Therapy held in Copenhagen. Bohr's title was "Light and Life," and his purpose was to draw attention to the epistemological implications for the life sciences of the fundamental changes that the quantum theory had brought to the conception of natural law (Bohr, 1933). He thought that these changes, which extend to the very idea of the nature of scientific explanation, were important not only for a full appreciation of the new situation brought to physics by the quantum theory. According to Bohr, these developments had created also an entirely new background for viewing the problems of biology, especially as concerns "our views on the position of living organisms in the realm of natural science." In particular Bohr proposed that the notion of complementarity, which he had first presented five years earlier at the International Congress of Physics in Como, is relevant also for the physiological and psychological aspects of life.

Bohr began his address by stating what is nowadays stigmatized as the "reductionist" credo, namely that the subtle character of the riddle of life notwithstanding, any scientific explanation of biological phenomena "necessarily must consist of reducing the description of more complex phenomena to that of simpler ones." But, Bohr continued, the discovery of an essential limitation of the mechanical description of natural phenomena, as revealed by the recent development of the atomic theory, has lent new interest to the old problem of how far the program of reductive biological explanations can be carried. Bohr then presented a brief historical account of the rise of the quantum theory in connection with the study of light, which he thought "is perhaps the least complex of all physical phenomena."

Yet, it turned out that, despite its lack of complexity, the phenomenon of light cannot be given a coherent mechanical explanation, inasmuch as from the viewpoint of classical physical theories, "the quantum of action . . . must be considered as an irrational notion." The spatial continuity of light propagation, on the one hand, and the atomicity of light effects on the other hand, must, therefore, be considered as *complementary* aspects of one reality, in the sense that each expresses an important feature of the phenomena of light. These aspects are complementary, because, although they are conceptually irreconcilable from a mechanical point of view, they can never be shown to be in direct, or empirical, contradiction. The reason for this is that any closer analysis of either the wave aspect or the particle aspect of light demands mutually exclusive observational arrangements.

In fact, these mutually exclusive observational arrangements needed for the study of complementary aspects turn out to be at the root of a fundamental limitation of our analysis of natural phenomena, since in a physical measurement it is never possible to take directly into account the interaction between object and measuring instruments. Or, as Bohr put it, "the instruments cannot be included in the investigation while they are [also] serving as means of observation." But the admission of a fundamental limitation of empirical knowledge inhering in complementarity puts in doubt our ingrained idea of phenomena existing independently of the means by which they are observed. This claim must have come as a shock to most members of Bohr's audience, which included also the Danish Crown Prince Frederik, since the unlimited validity of that ingrained idea—usually referred to as realism—is the traditional epistemological point of departure of scientific research in the first place. After all, if phenomena did not exist independently of the means by which we observe them, then as far as reality is concerned, there would be no there there, as Gertrude Stein once said of her hometown Oakland, California.

So what novel insights into the problems of biology did Bohr think can be gained by taking into account the concept of complementarity? Bohr, first of all, made clear what it was that he did *not* mean to imply: He did not wish to suggest that at the microscopic quantum level we encounter phenomena which show a closer resemblance to the properties of life than do ordinary, macroscopic physical phenomena. On the contrary, at first sight the essentially indeterministic, statistical character of quantum phenomena is very difficult to reconcile with the highly structured organization of living creatures, which have all the characteristics of their species implanted in tiny genes. It goes without saying that quantum mechanics applies to the chemical behavior of all atoms, be they part of living or of non-living matter. But, so Bohr declared, regarding life as a chemical phenomenon will not explain it any better than had the ancient comparison of life with fire, or the more recent comparison of living organisms with mechanical engines, such as clockworks. Rather, so Bohr said, "an understanding of the essential characteristics of living beings must be sought . . . in their peculiar organization, in which features that may be analyzed by the usual mechanics are interwoven with typically atomistic traits in a manner having no counterpart in inorganic matter." To illustrate this point, Bohr provided a brief and insightful discussion of the human eye and pointed out that the absorption of a single light quantum by the retina suffices to cause a deterministic, macroscopic effect, namely a visual experience by the subject.

So the question at issue is not whether physics can explain some features of the functions of living organisms, which it clearly can, but "whether some fundamental traits are still missing in the analysis of natural phenomena, before we reach an understanding of life on the basis of physical experience." Bohr hastened to point out that among those missing fundamental traits he does *not* include the mysterious "vital force," which vitalist biologists call on for the governance of organic life. He thought that "we all agree with Newton that the real basis of science is that Nature under the same conditions will always exhibit the same regularities. Therefore, if we were able to push the analysis of the mechanism of living organisms as far as that of atomic phenomena, we should scarcely expect to find any features differing from the properties of inorganic matter." What kind of missing fundamental traits did Bohr then have in mind? He was thinking of traits which are still missing because the conditions holding for biological and physical research are not directly comparable, since the necessity of keeping the object of biological investigations alive imposes a restriction on biological research, which finds no counterpart in physical research. According to Bohr, "we should doubtless kill an animal if we tried to carry the investigations of its organs so far that we could describe the role played by single atoms in vital functions. In every experiment on living organisms there must remain an uncertainty as regards the physical conditions to which they are subjected, and the idea suggests itself that the minimal freedom we must allow the organism in this respect is just large enough to permit it, so to say, to hide its ultimate secrets from us.''

Because its ultimate secrets will remain hidden from us, Bohr suggested that "the existence of life must be considered as an elementary fact that cannot be explained. It has to be accepted as a starting point of biology, just as the existence of the elementary particles and of the quantum of action have to be accepted as starting points of atomic physics. The asserted impossibility of a physical or chemical explanation of the function peculiar to life would in this sense be analogous to the insufficiency of the mechanical analysis for the understanding of the stability of atoms."

But if the existence of life itself is to remain an unexplained elementary fact of biology, where are we to look for the missing fundamental traits that will allow us to come to terms with life on the basis of physical experience? Bohr hinted that these traits are likely to reside in "the complexity of the material systems with which we are concerned in biology," in contrast to the properties of matter in its simplest forms, which are the primary focus of interest of atomic physics.

Finally, Bohr extended his considerations from the physiological features of life, which make living organisms appear merely as material objects, albeit highly complex ones, to its psychological features, which make living organisms appear as more than material, namely also as mental objects. He pointed out that the recognition of the epistemological limitation presented by mutually exclusive observational arrangements is suited also to reconcile the apparently contrasting points of view which separate physiology from psychology. "Indeed the necessity of considering the interaction between the measuring instruments and the object under investigation in atomic mechanics corresponds closely to the peculiar difficulties met with in psychological analyses, which arise from the fact that mental content is invariably altered when the attention is concentrated on any single feature of it." However, just as Bohr had warned that the extension of the complementarity argument from atomic phenomena to living organisms should not be regarded as supporting vitalism in the physiological realm so did he warn also that complementarity should not be regarded as supporting spiritualism in the psychological realm. Bohr said that from his point of view, "the feeling of the freedom of the will must be considered as a trait peculiar to conscious life . . . Without entering into metaphysical speculations, I may perhaps add that any analysis of the very concept of an explanation would, naturally, begin and end with a renunciation of an explanation of our own conscious activity.''

Bohr concluded his address by emphasizing that none of his remarks should be taken as intending to express any kind of skepticism as to the future development of physical and biological sciences. For just as the recognition of the limited character of our most fundamental epistemological concepts brought forth by the quantum theory had resulted in far-reaching developments in physics, so was ''the necessary renunciation as regards an explanation of life itself [no] hindrance to the wonderful advances which have been made in recent times in all branches of biology and have, not least, proved so beneficial in the art of medicine.''

Bohr's ''Light and Life'' lecture was to have momentous consequences for the future of 20th century biology, not because anything he said made much of an impact on his contemporary biologists, but because, in addition to Prince Frederik, the audience happened to include also the young physicist Max Delbrück. Delbrück had arrived in Copenhagen that very morning by the nighttrain from Berlin. Leon Rosenfeld had picked up Delbrück at the railway station with a message from Bohr, namely to hurry up and be sure to get to the lecture hall by 10 A.M. (Fischer, 1985). In the preceding year Delbrück had spent six months as a Rockefeller Fellow in Bohr's Institute of Theoretical Physics, and he was now working as an assistant to Lise Meitner at the Kaiser Wilhelm Institute for Chemistry in Berlin-Dahlem. Having entered physics just as the tremendously exciting developments brought by quantum mechanics seemed to be settling down to what Thomas Kuhn has called ''normal science,'' Delbrück had begun to wonder whether he was in the right field. In Berlin he had made some contacts with biologists and he had begun to take an interest in their work. Since he had mentioned that interest to Bohr, Bohr was particularly anxious that Delbrück should listen to the ''Light and Life'' lecture. Delbrück was tremendously excited by the ideas put forward by Bohr, and he decided to try to apply the lesson of complementarity to the deep problem posed by the existence of living matter. One point made by Bohr was to hold a special fascination, not only for Delbrück, but also for other physical scientists who were later to join the movement Delbrück started, namely that we may still be missing some fundamental traits in the analysis of natural phenomena which are needed to fathom living matter. This notion held out the fantastic prospect that by studying living organisms a physicist might make contributions, not just to biology, but to physics itself, since the analysis of complex living ensembles might uncover hitherto unknown properties of matter which are not manifest in its simpler, non-living ensembles. Moreover, another point of Bohr's made a strong impression on Delbrück, namely that at first sight, the seeming lack of logical coherence of attributing to light the properties of both particles and waves might appear very deplorable. But, said Bohr, "as has often happened in the history of science, when new discoveries have revealed an essential limitation of ideas the universal applicability of which had never been disputed, we have been rewarded by getting a wider view and a greater power of correlating phenomena which before might even have appeared as contradictory." The quixotic notion that the search for paradoxical findings is the highroad to deep scientific insights was to remain one of the most characteristic, life-long features of Delbrück's outlook on research strategy (Delbrück 1949, 1970).

Not long after returning to Berlin from this visit to Copenhagen, Delbrück decided that genetics was, in fact, a domain of biological inquiry in which physical and chemical explanations might turn out to be "insufficient" in Bohr's sense (Timofeef-Ressovsky, et al, 1935). Hence there seemed hope that in trying to fathom the then still utterly mysterious nature of the gene, one might encounter one of those paradoxes that could lead to "other laws of physics," as Erwin Schrödinger (1945), called them in his later popular account of Delbrück's project. In 1938 Delbrück received another Rockefeller Fellowship, to enable him to go to Cal Tech to start full-time work in genetics. In Pasadena he chose bacterial viruses as his experimental material, because he realized that these simplest of living structures should make ideal objects for the study of biological self replication, and hence for the physical basis of heredity. Within a few years Delbrück had collected a small group of people (many of them trained in the physical rather than biological sciences) who were interested in the very same problem. Out of Delbrück's group developed what I have previously called the "Informational School" of proto-molecular biologists. It was upon the wedding of Delbrück's Informational School with the Structural School of X-ray crystallographers in 1953, occasioned by the discovery of the DNA double helix by James Watson and Francis Crick, that molecular biology became the recognizable discipline which totally transformed nearly all parts of biology in the following three decades (Stent, 1968).

I give it as my view that Delbrück played a crucial role in the rise of molecular biology, not unlike that which Bohr, whom Delbrück had obviously chosen as a role model, had played in the development of atomic theory. Indeed, I hold that in Delbrück Bohr eventually found his most influential discipline outside the domain of physics. It was through Delbrück that Bohr's epistemology became the intellectual infrastructure of molecular biology. It provided for molecular biologists the conceptual guidance for navigating between the Scylla of crude biochemical reductionism, inspired by 19th century physics, and the Charybdis of obscurantist vitalist holism, inspired by 19th century romanticism. An additional factor, in my opinion also of paramount importance for the development of molecular biology, was Delbrück's successful transplantation of Bohr's "Copenhagen Spirit" to Pasadena. Just as did the international group of atomic theorists who had been associated with Bohr at one time or another of their careers, so did the Informational School constitute a close-knit family whose members were linked by personal ties, and among whom competition never exceeded that of sibling rivalry. For Delbrück was the *pater familias*, whose personal praise was the highest reward anyone could hope for, and he would, of course, see through all dirty tricks (Cairns, et al, 1966).

In 1962, a few months before his death, Bohr accepted Delbrück's invitation to speak at the inauguration of the Institute of Genetics at the University of Cologne, of which Delbrück had been the founding director while on leave from Cal Tech. Bohr entitled this address, which was to be his last public lecture, "Light and Life Revisited" (Bohr, 1963). By that time the initial research goals of molecular biology had been reached, and much more was known about the physical basis of heredity than anyone could have had reason to expect in the 1930's. The gene had been shown to consist of a stretch of the DNA double helix a thousand or so nucleotides in length, encoding the amino acid sequence of a particular protein molecule and self-replicating directly by making and breaking hydrogen bonds between purine and pyrimidine base pairs. No paradoxes had come into focus, no missing fundamental traits of matter and no "other laws of physics" had turned up. Physical and chemical explanations were found to be quite sufficient to account for the phenomena of how, in the living world, like begets like. Thus, Bohr's conjecture that one needs to kill an organism in order to study it at the atomic level and that this is bound to hide the ultimate secrets of life from us, turned out to be wrong.

Throughout these years, Bohr had maintained his interest in biology and was fully aware of the enormous advances which had occurred meanwhile in genetics and which his disciple Delbrück had set in motion. In his "Light and Life Revisited" lecture Bohr first summarized the message he had meant to convey to the light therapists in 1932. He then mentioned a number of important biological discoveries which had been made meanwhile by resort to physical methods and viewpoints, such as the fine structure of muscles and the activity of nerves. Bohr had evidently not given up completely on the idea of unknown fundamental traits of matter, because he said that "these discoveries . . . point to the possibility of physical mechanisms which hitherto have escaped notice." As for genetics, Bohr said that "a turning point in this whole field came . . . about ten years ago with Crick and Watson's ingenious proposal for an interpretation of the structure of DNA molecules. I vividly remember how Delbrück, in telling me about the discovery, said that it might lead to a revolution in microbiology comparable with the development of atomic physics, initiated by Rutherford's nuclear model of the atom." Bohr then recalled that Rutherford's discovery of the atomic nucleus completed the knowledge about the structure of the atom to such an unexpected degree that it challenged the young Bohr and his contemporaries to find out how it could be used for ordering the accumulated information about the physical and chemical properties of matter. "This goal, Bohr said, was largely achieved within a few decades by the cooperation of a whole generation of physicists, which in intensity and scope resembles that taking place these years in genetics and molecular biology."

I think it is fair to say that the ideas set forth by Bohr in his 1932 lecture turned out to have little relevance for the eventual understanding of the gene, except in so far as they had inspired Max Delbrück to work on the problem. But in another, much later lecture entitled "Atoms and Human Knowledge" given in Copenhagen in October, 1955 before the Royal Danish Academy of Sciences, Bohr put forward an idea which provided a deep insight for the currently highly active discipline at the interface between biology and philosophy designated as "evolutionary epistemology" (Bohr, 1958). This discipline is rooted in the demonstration by Immanuel Kant that our knowledge of the world arises by an active, or constructive, process. According to Kant, our sensory impressions become experience, that is gain meaning for us, only after we interpret them in terms of a priori categories—such as time, space and object—

that we bring to rather than derive from experience. Tacit resort to propositions whose validity is similarly accepted a priori, such as "Some A are B; therefore all A are B" (induction) or "The occurrence of a set of conditions A is both necessary and sufficient for the occurrence of B (causation by A of B), then allows our mind to construct an orderly reality from our experience. Kant referred to these a priori categories as "transcendental," because they transcend experience and were thought by him to be beyond the scope of scientific inquiry.

But how can it be that, if we bring these categories to the outer world a priori, they happen to fit that world so well? To this question, which had already troubled Kant, a Darwinist answer was provided only in the 1940's, by Konrad Lorenz (1941): what is a priori for the individual is a posteriori for his species. That is to say, the success of our cerebral circuitry in constructing a version of reality that matches the outer world is merely another product of the process of natural selection which guided our evolutionary history: any early hominids who happen to think that before is after, or that near is far, or who failed to apprehend that A is the cause of B, perished without issue. Characterization of the Kantian categories as a priori for the individual does not mean, however, that they are present already, full-blown at birth. Instead, as the cognitive developmental studies that Jean Piaget initiated in the 1920's have shown, the Kantian categories arise post-natally, as the result of a dialectic interaction between the developing human nervous system and the outer world. Piaget showed that the Kantian categories, though immanent in the mind, are constructed gradually during childhood, via a succession of distinct developmental stages, by a process which Piaget called "genetic epistemology" (Piaget, 1971).

Bohr's contribution to this subject was to draw attention in his "Atoms and Human Knowledge" lecture to the epistemological limitations which the evolutionary origins of our categorical endowment place on our understanding of nature. Bohr did not refer to the contributions of Lorenz and Piaget, nor did he mention Kant, the central figure of modern epistemology, whose name rarely appears in any of Bohr's writings. Rather, in this lecture Bohr seemed to work out from scratch the epistemological roots of the cognitive difficulties presented by the notion of complementarity.

To make his analysis of the source of these limitations, Bohr emphasized the critical relation between language and knowledge. He said that "as the goal of science is to augment and order our experience, every analysis of the conditions of human knowledge must rest on considerations of the character and scope of our means of communication. Our basis [of communication] is, of course, the language developed for orientation in our surroundings and for the organization of human communities." And even though we can vary our experimental conditions in many ways, we must be able to communicate to others what we have done and what we have learned in the terms of the concepts embedded in our language. Foremost among these linguistic concepts are, it goes without saying, the Kantian categories of time, space, object and causality. Accordingly, the models that modern science offers as descriptions of reality are linguistic pictorial representations drawn by use of these a priori categories.

These linguistic procedures were eminently satisfactory as long as knowledge was sought only about phenomena that are commensurate with events which are the subject of our everyday experience (give or take a few orders of magnitude). But this situation began to change when, at the turn of the 20th century, physics had progressed to the stage where problems could be addressed involving either tiny subatomic or immense cosmic events, on scales of time, space and mass billions of times smaller or larger than those of our direct experience. Now, according to Bohr, there arose "difficulties of orienting ourselves in a domain of experience far from that to the description of which our means of expression are adapted." For it turned out that descriptions in ordinary everyday language of phenomena belonging to those transcendent domains lead to contradictions, or mutually incompatible pictures of reality. In order to resolve these contradictions, time, space and mass have to be denatured into generalized concepts, by eliminating from them some hidden presuppositions. It is this procedure which is responsible for the difficulties in orienting ourselves, because the meaning of these denatured concepts no longer matches that inherent in the Kantian categories.

In the 1970's nearing retirement, Delbrück began a project aimed at working out the implications of Bohr's contribution to evolutionary epistemology, not only for physics but also for a wide spectrum of scientific disciplines, ranging from cosmology through organic evolution, sensory perception, cognition and number theory all the way to linguistics. Delbrück intended to show that in whatever direction we push our search for knowledge too far, way beyond the world of the middle dimension of our direct experience to which our conceptual equipment is adapted, we are bound to encounter deep paradoxes. For instance, he considered Gödel's proof of the existence of undecidable propositions in number theory as one of the important examples of our cognitive limitations: the theory of numbers is admirably coherent when applied to counting the apples and pears of our everyday world, but it generates incoherences when pushed too far in the analysis of abstruse sets. Delbrück presented this attempt to expand Bohr's philosophy as a series of twenty lectures at Cal Tech under the title *Mind from Matter?*, but his fatal illness prevented him from preparing them for publication. The text of a transcript of these lectures was edited and expanded by a group of his friends, colleagues and students for posthumous publication (Delbrück, 1986). Quite independently, Gerhard Vollmer (1984) had developed views on evolutionary epistemology which are remarkably similar to Delbrück's and in which Bohr's world of middle dimensions is designated as the "mesocosm."

I conclude my review of Bohr's legacy to contemporary biology with an application of the notion of complementarity to a problem which seems to have been of little interest to Delbrück, namely the relation of biology to the foundations of ethics. In recent years, the ancient quarrel between the adherents of the idealistic ethics advocated by Plato and the naturalistic ethics advocated by Aristotle has flared up again, ignited by the claims of sociobiologists that human moral behavior is to be explained in terms of selective, evolutionary forces (Stent, 1980). According to the sociobiologists' naturalistic argumentation, not only the existence of a moral code, but also its actual content (such as the proscription of incest) is legitimized by the adapative value, or fitness, that the moral code brings to *Homo sapiens* in the evolutionary struggle for survival. To appreciate the philosophical shortcomings of these claims, we may once more turn to Kant, one of whose fundamental insights had been that we live in two metaphysically distinct worlds. One of these worlds is the world of science (including biology), whose natural objects are seen as being governed by laws of causal determination. The other world is the world of morals, whose rational human subjects are seen to be governed by laws of freedom that each person imposes on his or her own actions. Thus from the Kantian perspective the main task for understanding the nature of morality would be to discover and explicate the metaphysical principles presupposed by moral judgments, rather than to discover biological generalizations about the chances for survival or about evolutionary mechanisms. These metaphysical principles can be discovered only by examining the facts of moral experience and by analyzing the logic of moral judgments. This Kantian view is, of course, not very congenial to biologists. How can they accept the Darwinian view of the biological nature of man, while granting Kant's insistence that belief in an autonomous free will, independent of natural law, is a necessary presupposition of morality? In any case, it is obvious that from the Kantian viewpoint the sociobiological project of grounding morals in Darwinism is *radically* ill-conceived. The biology of morals would fail because it brings a categorical confusion to the problem of ethics. Biologists are thus faced with a paradox. On the one hand it must seem self-evident to any biologist, as an heir to Aristotle, the founder of biology as a branch of natural science, that the sociobiological approach to morals has *some* merit. In particular, to any biologist it must seem unreasonable to deny that human morality is not somehow related to human welfare, and hence to the biology of man. Moreover, since man's biological roots reach into the animal kingdom, it would seem equally unreasonable to deny that observations on the social behavior of animals cannot in some way illuminate the nature of human morals. Yet it seems equally apparent that the idealistic, anti-naturalistic notion of an autonomous free will and of a categorical obligation to obey universal moral law cannot be dismissed as so much nonsense in any attempt to fathom the nature of morals.

Kant thus recognized that the relation between biology and ethics is just what Bohr referred to as complementary. For, our intuitive attribution of freedom of the will to the person contradicts our intuitive attribution of causal necessity to nature, just as our attribution of a particulate character to light contradicts our attribution of an undulatory character to the same phenomenon. Yet, as rational beings we cannot abandon the idea of moral freedom of the person any more than we can abandon the idea of scientific causal necessity governing the behavior of human beings. Thus the relation of the person to his or her actions has to be observed from two fundamentally different, mutually exclusive points of view. From one of these points of view the person is seen as a moral agent while from the other the person is seen as a natural thing. As a biologist, I must regard the person as a material object forming part of the causally determined events of the natural world; but as an ethicist, I must regard the person as an intelligent subject forming part of a world of thought independent of the laws of nature. Hence the moral nature of man is formally analogous to the physical nature of light, which is seen either as a particle or as a wave, depending on which one of two mutually exclusive arrangements are used for observing it.

It would appear, therefore, that the need for resorting to mutually exclusive observational setups presents a fundamental limitation in our analysis of ethical phenomena. Just as is the case for our analysis of natural phenomena, it is true also for ethical phenomena that the observational arrangements (i.e., the mutually exclusive viewpoints from which moral behavior is analyzed) cannot be included in the investigation while they are serving as means of observation. Hence, as applied to the problem of the foundational relations between biology and ethics, Bohr's concept of complementarity can help us to understand that the existence of moral behavior must be considered as an elementary fact of life which cannot be explained, and which must be taken as the starting point of ethics. So, in helping us to appreciate the impossibility of accounting for the moral nature of man, the most important lesson for contemporary, meta-molecular biology to be learned from Bohr's "Light and Life" lecture may be that, epistemologically-speaking, the sociobiological approach to the relation of biology to the foundations of ethics is half-baked.

References

Bohr, N.: 1933. Light and Life, *Nature* **131**, 421–423, 457–459.

Bohr, N.: 1958. Atoms and Human Knowledge, in *Atomic Physics and Human Knowledge*, Science Editions, New York.

Bohr, N.: 1963. Light and Life Revisited, in *Essays 1958–1962 in Atomic Physics and Human Knowledge*. Interscience, New York.

Cairns, J., G.S. Stent, J.D. Watson (eds.): 1966. *Phage and the Origins of Molecular Biology*. Cold Spring Harbor Lab. Quant. Biol., Cold Spring Harbor, New York.

Delbrück, M.: 1949. A Physicist Looks at Biology *Trans. Conn. Acad. Arts. Sci.* **38**, 173–190. (Reprinted in Cairns et al, 1966).

Delbrück, M.: 1970. A Physicist's Renewed Look at Biology: Twenty years later. *Science* **168**, 1312–1315.

Delbrück, M.: 1986. *Mind from Matter?* Blackwell Scientific Pubs., Palo Alto, Calif.

Fischer, E.P.: 1985. *Licht und Leben. Ein Bericht über Max Delbrück.* Universitsätsverlag, Konstanz.

Lorenz, K.: 1944. Kant's Lehre vom apriorischen im Lichte gegenwärtiger Biologie. *Blätter für Deutsche Philosophie*, **15**, 94–125.

Piaget, J.: 1971. *Genetic Epistemology* (E. Duckworth, tr.) Norton, New York.

Schrödinger, E.: 1945. *What is Life?* Cambridge University Press, Cambridge.

Stent, G.S.: 1968. That Was the Molecular Biology That Was. *Science* **160**, 390–395.

Stent, G.S. (ed.) 1980. *Morality as a Biological Phenomenon.* University of California Press, Berkeley.

Timofeef-Ressovsky, N.W., K.G. Zimmer, M. Delbrück.: 1935. Über die Natur der Genmutation und Genstruktur. *Nach. Ges. Wiss. Goettingen,* Math-Phys. Kl. Fachgruppe 6, Nr. 13, 190–245.

Vollmer, G.: 1984. Mesocosm and Objective Knowledge, in *Concepts and Approaches in Evolutionary Epistemology.* F.W. Wuketits (ed.) Reidel, Amsterdam.

Bertlmann's Socks and the Nature of Reality

J.S. Bell

CERN

1. Introduction

The philosopher in the street, who has not suffered a course in quantum mechanics, is quite unimpressed by Einstein-Podolsky-Rosen correlations /1/. He can point to many examples of similar correlations in everyday life. The case of Bertlmann's socks is often cited. Dr. Bertlmann likes to wear two socks of different colours. Which colour he will have on a given foot on a given day is quite unpredictable. But when you see (Fig. 1) that the first sock is pink you can be already sure that the second sock will not be pink. Observation of the first, and experience of Bertlmann, gives immediate information about the second. There is no accounting for tastes, but apart from that there is no mystery here. And is not the EPR business just the same?

Consider for example the particular EPR gedanken experiment of Bohm /2/ (Fig. 2). Two suitable particles, suitably prepared (in the 'singlet spin state'), are directed from a common source towards two widely separated magnets followed by detecting screens. Each time the experiment is performed each of the two particles is deflected either up or down at the corresponding magnet. Whether either particle separately goes up or down on a given occasion is quite unpredictable. But when one particle goes up the other always goes down and vice-versa. After a little experience it is enough to look at one side to know also about the other.

So what? Do we not simply infer that the particles have properties of some kind, detected somehow by the magnets, chosen à la Ber-

Reprinted by permission of *Journal de Physique*

Figure 1. Les chaussettes de M. Bertlmann et la nature de la réalité. Fondation Hugot, juin 17, 1980.

tlmann by the source—differently for the two particles? Is it possible to see this simple business as obscure and mysterious? We must try.

To this end it is useful to know how physicists tend to think intuitively of particles with 'spin', for it is with such particles that we are concerned. In a crude classical picture it is envisaged that some internal motion gives the particle an angular momentum about some axis, and at the same time generates a magnetization along that axis. The particle is then like a little spinning magnet with north and south

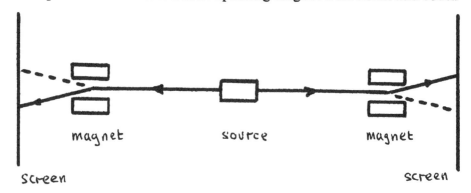

Figure 2. Einstein-Podolsky-Rosen-Bohm gedanken experiment with two spin 1/2 particles and two Stern-Gerlach magnets.

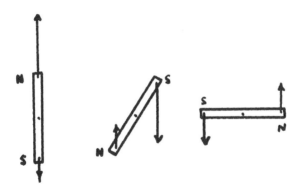

Figure 3. Forces on magnet in non-uniform magnetic field. The field points towards the top of the page and increases in strength in that direction.

poles lying on the axis of rotation. When a magnetic field is applied to a magnet the north pole is pulled one way and the south pole is pulled the other way. If the field is uniform the net force on the magnet is zero. But in a non-uniform field one pole is pulled more than the other and the magnet as a whole is pulled in the corresponding direction. The experiment in question involves such non-uniform fields—set up by so-called 'Stern-Gerlach' magnets. Suppose that the magnetic field points up, and that the strength of the field increases in the upward direction. Then a particle with south-north axis pointing up would be pulled up (Fig. 3). One with axis pointing down would be pulled down. One with axis perpendicular to the field would pass through the magnet without deflection. And one oriented at an intermediate angle would be deflected to an intermediate degree. (All this is for a particle of zero electric charge; when a charged particle moves in a magnetic field there is an additional force which complicates the situation).

A particle of given species is supposed to have a given magnetization. But because of the variable angle between particle axis and field there would still be a range of deflections possible in a given Stern-Gerlach magnet. It could be expected then that a succession of particles would make a pattern something like Figure 4 on a detecting screen. But what is observed in the simplest case is more like Figure 5, with two distinct groups of deflections (i.e., up or down) rather than a more or less continuous band. [This simplest case, with just two groups of deflections, is that of so-called 'spin-1/2' particles; for 'spin-j' particles there are $(2j + 1)$ groups.]

Figure 4. Naive classical expectation for pattern on detecting screen behind Stern-Gerlach magnet.

The pattern of Figure 5 is very hard to understand in naive classical terms. It might be supposed for example that the magnetic field first pulls the little magnets into alignment with itself, like compass needles. But even if this were dynamically sound it would account for only one group of deflections. To account for the second group would require 'compass-needles' pointing in the wrong direction. And anyway it is not dynamically sound. The internal angular momentum, by gyroscopic action, should stabilize the angle between particle axis and magnetic field. Well then, could it not be that the source for some reason delivers particles with axes pointing just one way or the other and not in between? But this is easily tested by turning the Stern-Gerlach magnet. What we get (Fig. 6) is just the same split pattern as before, but turned around with the Stern-Gerlach magnet. To blame the absence of intermediate deflections on the source we would have to imagine that it anticipated somehow the orientation of the Stern-Gerlach magnet.

Phenomena of this kind /3/ made physicists despair of finding any consistent space-time picture of what goes on the atomic and subatomic scale. Making a virtue of necessity, and influenced by positivistic and instrumentalist philosophies /4/, many came to hold not only that it is difficult to find a coherent picture but that it is wrong to look for one—if not actually immoral then certainly unprofessional. Going further still, some asserted that atomic and subatomic particles do not *have* any definite properties in advance of observation. There is nothing, that is to say, in the particles approaching

Figure 5. Quantum mechanical pattern on screen, with vertical Stern-Gerlach magnet.

Figure 6. Quantum mechanical pattern with rotated Stern-Gerlach magnet.

the magnet, to distinguish those subsequently deflected up from those subsequently deflected down. Indeed even the particles are not really there.

For example /5/, 'Bohr once declared when asked whether the quantum mechanical algorithm could be considered as somehow mirroring an underlying quantum reality: "There is no quantum world. There is only an abstract quantum mechanical description. It is wrong to think that the task of physics is to find out how Nature *is*. Physics concerns what we can say about Nature" '.

And for Heisenberg /6/ '. . . in the experiments about atomic events we have to do with things and facts, with phenomena that are just as real as any phenomena in daily life. But the atoms or the elementary particles are not as real; they form a world of potentialities or possibilities rather than one of things or facts'.

And /7/ 'Jordan declared, with emphasis, that observations not only *disturb* what has to be measured, they *produce* it. In a measurement of position, for example, as performed with the gamma ray microscope, "the electron is forced to a decision. We compel it *to assume a definite position*; previously it was, in general, neither here nor there; it had not yet made its decision for a definite position . . . If by another experiment the *velocity* of the electron is being measured, this means: the electron is compelled to decide itself for some exactly defined value of the velocity . . . we ourselves produce the results of measurement" '.

It is in the context of ideas like these that one must envisage the discussion of the Einstein-Podolsky-Rosen correlations. Then it is a little less unintelligible that the EPR paper caused such a fuss, and that the dust has not settled even now. It is as if we had come to deny the reality of Bertlmann's socks, or at least of their colours, when not looked at. And as if a child had asked: How come they always choose different colours when they *are* looked at? How does the second sock know what the first has done?

Paradox indeed! But for the others, not for EPR. EPR did not use the word "paradox". They were with the man in the street in this business. For them these correlations simply showed that the quantum theorists had been hasty in dismissing the reality of the microscopic world. In particular Jordan had been wrong in supposing that nothing was real or fixed in that world before observation. For after observing only one particle the result of subsequently observing the other (possibly at a very remote place) is immediately predictable. Could it be that the first observation somehow fixes what was unfixed, or makes real what was unreal, not only for the near particle but also for the remote one? For EPR that would be an unthinkable 'spooky action at a distance' /8/. To avoid such action at a distance they have to attribute, to the space-time regions in question, *real* properties in advance of observation, correlated properties, which *predetermine* the outcomes of these particular observations. Since these real properties, fixed in advance of observation, are not contained in quantum formalism /9/, that formalism for EPR is *incomplete*. It may be correct, as far as it goes, but the usual quantum formalism cannot be the whole story.

It is important to note that to the limited degree to which *determinism* plays a role in the EPR argument, it is not assumed but *inferred*. What is held sacred is the principle of "local causality"— or "no action at a distance". Of course, mere *correlation* between distant events does not by itself imply action at a distance, but only correlation between the signals reaching the two places. These signals, in the idealized example of Bohm, must be sufficient to *determine* whether the particles go up or down. For any residual undeterminism could only spoil the perfect correlation.

It is remarkably difficult to get this point across, that determinism is not a *presupposition* of the analysis. There is a widespread and erroneous conviction that for Einstein /10/ determinism was always *the* sacred principle. The quotability of his famous "God does not play dice" has not helped in this respect. Among those who had great difficulty in seeing Einstein's position was Born. Pauli tried to help him /11/ in a letter of 1954:

". . . I was unable to recognize Einstein whenever you talked about him in either your letter or your manuscript. It seemed to me as if you had erected some dummy Einstein for yourself, which you then knocked down with great pomp. In particular Einstein does not consider the concept of "determinism" to be as fundamental as it is frequently held to be (as he told me emphatically many times) . . . he *disputes* that he uses as a criterion for the admissibility of a theory the question: "Is it rigorously deterministic?" . . . he was not at all annoyed with you, but only said you were a person who will not listen".

Born had particular difficulty with the Einstein-Podolsky-Rosen argument. Here is his summing up, long afterwards, when he edited the Born-Einstein correspondence /12/:

"The root of the difference between Einstein and me was the axiom that events which happens in different places A and B are independent of one another, in the sense that an observation on the states of affairs at B cannot teach us anything about the state of affairs at A".

Misunderstanding could hardly be more complete. Einstein had no difficulty accepting that affairs in different places could be correlated. What he could not accept was than an intervention at one place could *influence*, immediately, affairs at the other.

These references to Born are not meant to diminish one of the towering figures of modern physics. They are meant to illustrate the difficulty of putting aside preconceptions and listening to what is actually being said. They are meant to encourage *you*, dear listener, to listen a little harder.

Here, finally, is a summing-up by Einstein himself /13/:

'If one asks what, irrespective of quantum mechanics, is characteristic of the world of ideas of physics, one is first of all struck by the following: the concepts of physics relate to a real outside world . . . It is further characteristic of these physical objects that they are thought of as arranged in a space time continuum. An essential aspect of this arrangement of things in physics is that they lay claim, at a certain time, to an existence independent of one another, provided these objects "are situated in different parts of space".

'The following idea characterizes the relative independence of objects far apart in space (A and B): external influence on A has no direct influence on B. . .

'There seems to me no doubt that those physicists who regard the descriptive methods of quantum mechanics as definitive in principle would react to this line of thought in the following way: they would drop the requirement . . . for the independent existence of the physical reality present in different parts of space; they would be justified in pointing out that the quantum theory nowhere makes explicit use of this requirement.

'I admit this, but would point out: when I consider the physical phenomena known to me, and especially those which are being so successfully encompassed by quantum mechanics, I still cannot find any fact anywhere which would make it appear likely that (that) requirement will have to be abandoned.

'I am therefore inclined to believe that the description of quantum mechanics . . . has to be regarded as an incomplete and indirect description of reality, to be replaced at some later date by a more complete and direct one'.

2. Illustration

Let us illustrate the *possibility* of what Einstein had in mind in the context of the particular quantum mechanical predictions already cited for the EPRB gedanken experiment. These predictions make it hard to believe in the completeness of quantum formalism. But of course outside that formalism they make no difficulty whatever for the notion of local causality. To show this explicitly we exhibit a trivial ad hoc space-time picture of what might go on. It is a modification of the naive classical picture already described. Certainly something must be modified in that, to reproduce the quantum phenomena. Previously, we implicitly assumed for the net force in the direction of the field gradient (which we always take to be in the same direction as the field) a form

$$F \cos \Delta \tag{1}$$

where Δ is the angle between magnetic field (and field gradient) and particle axis. We change this to

$$F \cos \Delta / \mid \cos \Delta \mid . \tag{2}$$

Whereas previously the force varied over a continuous range with Δ, it takes now just two values, $\pm F$, the sign being determined by whether the magnetic axis of the particle points more nearly in the direction of the field or in the opposite direction. No attempt is made to explain this change in the force law. It is just an ad hoc attempt to account for the observations. And of course it accounts immediately for the appearance of just two groups of particles, deflected either in the direction of the magnetic field or in the opposite direction. To account then for the Einstein-Podolsky-Rosen-Bohm correlations we have only to assume that the two particles emitted by the source have oppositely directed magnetic axes. Then if the magnetic axis of one particle is more nearly along (than against) one

Stern-Gerlach field) the magnetic axes of the other particle will be more nearly against (than along), a parallel Stern-Gerlach field. So when one particle is deflected up, the other is deflected down, and vice versa. There is nothing whatever problematic or mind-boggling about these correlations, with parallel Stern-Gerlach analyzers, from the Einsteinian point of view.

So far so good. But now go a little further than before, and consider *non*-parallel Stern-Gerlach magnets. Let the first be rotated away from some standard position, about the particle line of flight, by an angle a. Let the second be rotated likewise by an angle b. Then if the magnetic axis of either particle separately is randomly oriented, but if the axes of the particles of a given pair are always oppositely oriented, a short calculation gives for the probabilities of the various possible results, in the ad hoc model,

$$P(\text{up,up}) = P(\text{down,down}) = \frac{|a - b|}{2\pi}$$

$$P(\text{up,down}) = P(\text{down,up}) = \frac{1}{2} - \frac{|a - b|}{2\pi} \qquad (3)$$

where "up" and "down" are defined with respect to the magnetic fields of the two magnets. However, a quantum mechanical calculation gives

$$P(\text{up,up}) = P(\text{down,down}) = \frac{1}{2}\left(\sin \frac{a - b}{2}\right)^2$$

$$P(\text{up,down}) = P(\text{down,up}) = \frac{1}{2} - \frac{1}{2}\left(\sin \frac{a - b}{2}\right)^2. \qquad (4)$$

Thus the ad hoc model does what is required of it (i.e., reproduces quantum mechanical results) only at $(a - b) = 0$, $(a - b) = \pi/2$ and $(a - b) = \pi$, but not at intermediate angles.

Of course this trivial model was just the first one we thought of, and it worked up to a point. Could we not be a little more clever, and devise a model which reproduces the quantum formulae completely? No. It cannot be done, so long as action at a distance is excluded. This point was realized only subsequently. Neither EPR nor their contemporary opponents were aware of it. Indeed the discussion was for long entirely concentrated on the points $|a - b| = 0, \pi/2$, and π.

3. Difficulty with Locality

To explain this denouement without mathematics I cannot do better than follow d'Espagnat /14,15/. Let us return to socks for a moment. One of the most important questions about a sock is "will it wash"? A consumer research organization might make the question more precise: could the sock survive one thousand washing cycles at 45°C? Or at 90°C? Or at 0°C? Then an adaptation of the Wigner-d'Espagnat inequality /16/ applies. For any collection of new socks:

(the number that could pass at 0° and not at 45°)

plus

(the number that could pass at 45° and not at 90°)

is not less than

(the number that could pass at 0° and not at 90°)   (5)

This is trivial, for each member of the third group either could survive at 45°, and so is also in the second group, or could not survive at 45°, and so is also in the first group.

But trivialities like this, you will exclaim, are of no interest in consumer research! You are right; we are straining here a little the analogy between consumer research and quantum philosophy. Moreover, you will insist, the statement has no empirical content. There is no way of deciding that a given sock could survive at one temperature and not at another. If it did not survive the first test it would not be available for the second, and even if it did survive the first test it would no longer be new, and subsequent tests would not have the original significance.

Suppose, however, that the socks come in pairs. And suppose that we know by experience that there is little variation between the members of a pair, in that if one member passes a given test then the other also passes that same test *if it* is performed. Then from d'Espagnat's inequality we can infer the following:

(the number of pairs in which one could pass at 0° and the other not at 45°)

plus

(the number of pairs in which one could pass at 45° and the other not at 90°)

is not less than

(the number of pairs in which one could pass at 0°
and the other not at 90°)   (6)

This is not yet empirically testable, for although the two tests in
each bracket are now on different socks, the different brackets involve different tests on the same sock. But we now add the random
sampling hypothesis: if the sample of pairs is sufficiently large and
if we choose at random a big enough subsample to suffer a given
pair of tests, then the pass/fail fractions of the subsample can be
extended to the whole sample with high probability. Identifying such
fractions with *probabilities* in a thoroughly conventional way, we
now have

(the probability of one sock passing at 0° and the
other not at 45°)

plus

(the probability of one sock passing at 45° and the
other not at 90°)

is not less than

(the probability of one sock passing at 0° and the
other not at 90°)   (7)

Moreover this is empirically meaningful in so far as probabilities can
be determined by random sampling.

   We formulated these considerations first for pairs of socks, moving with considerable confidence in those familiar objects. But why
not reason similarly for the pairs of particles of the EPRB experiment? By blocking off the "down" channels in the Stern-Gerlach
magnets, allowing only particles deflected "up" to pass, we effectively subject the particles to tests which they either pass or do not.
Instead of temperatures we now have angles a and b at which the
Stern-Gerlach magnets are set. The essential difference, a trivial
one, is that the partices are paired à la Bertlmann—if one were to
pass a given test the other would be sure to fail it. To allow for this
we simply take the converse of the second term in each bracket:

(the probability of one particle passing at 0°
and the other at 45°)

plus

(the probability of one particle passing at 45°
and the other at 90°)

is not less than

(the probability of one particle passing at 0°
and the other at 90°)   (8)

In case any one finds the detour by socks a little long, let us look directly at this final result and see how trivial it is. We are assuming that particles have properties which dictate their ability to pass certain tests—whether or not these tests are in fact made. To account for the perfect anticorrelation when identical tests (parallel Stern-Gerlach magnets) are applied to the two members of a pair, we have to admit that the pairing is a generalized à la Bertlmann—when one has the ability to pass a certain test, the other has not. Then the above assertion about pairs is equivalent to the following assertion about either member:

(the probability of being able to pass at 0° and not able
at 45°)

plus

(the probability of being able to pass at 45° and not able
at 90°)

is not less than

(the probability of being able to pass at 0° and not able
at 90°)   (9)

And this is indeed trivial. For a particle able to pass at 0° and not at 90° [and so contributing to the third probability in (9)] is either able to pass at 45° (and so contributes to the second probability) or not able to pass at 45° (and so contributes to the first probability).

However, trivial as it is, the inequality is not respected by quantum mechanical probabilities. From (4) the quantum mechanical probability for one particle to pass a magnet with orientation a and the other to pass a magnet with orientation b (called $P(\text{up,up})$) in (4) is

$$\frac{1}{2}\left(\sin\frac{a - b}{2}\right)^2$$

Inequality (9) would then require

$$\tfrac{1}{2}(\sin 22.5°)^2 + \tfrac{1}{2}(\sin 22.5°)^2 \geqslant \tfrac{1}{2}(\sin 45°)^2$$

or

$$0.1464 \geqslant 0.2500,$$

which is not true.

Let us summarize once again the logic that leads to the impasse. The EPRB correlations are such that the result of the experiment on one side immediately foretells that on the other, whenever the analyzers happen to be parallel. If we do not accept the intervention on one side as a causal influence on the other, we seem obliged to admit that the results on both sides are determined in advance anyway, independently of the intervention on the other side, by signals from the source and by the local magnet setting. But this has implications for non-parallel settings which conflict with those of quantum mechanics. So we *cannot* dismiss intervention on one side as a causal influence on the other.

It would be wrong to say 'Bohr wins again' (Appendix 1); the argument was not known to the opponents of Einstein, Podolsky and Rosen. But certainly Einstein could no longer write so easily, speaking of local causality' . . . I still cannot find any fact anywhere which would make it appear likely that requirement will have to be abandoned'.

4. General Argument

So far the presentation aimed at simplicity. Now the aim will be generality /17/. Let us first list some aspects of the simple presentation which are not essential and will be avoided.

The above argument relies very much on the perfection of the correlation (or rather anticorrelation) when the two magnets are aligned ($a = b$) and other conditions also are ideal. Although one could hope to approach this situation closely in practice, one could not hope to realize it completely. Some residual imperfection of the set-up would spoil the perfect anticorrelation, so that occasionally

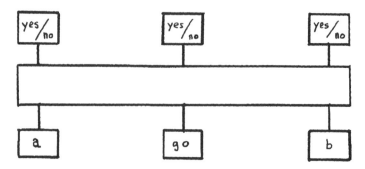

Figure 7. General EPR set-up, with three inputs below and three outputs above.

both particles would be deflected down, or both up. So in the more sophisticated argument we will avoid any hypothesis of perfection.

It was only in the context of perfect correlation (or anticorrelation) that *determinism* could be inferred for the relation of observation results to preexisting particle properties (for any indeterminism would have spoiled the correlation). Despite my insistence that the determinism was inferred rather than assumed, you might still suspect somehow that it is a preoccupation with determinism that creates the problem. Note well then that the following argument makes no mention whatever of determinism.

You might suspect that there is something specially peculiar about spin-1/2 particles. In fact there are many other ways of creating the troublesome correlations. So the following argument makes no reference to spin-1/2 particles, or any other particular particles.

Finally you might suspect that the very notion of particle, and particle orbit, freely used above in introducing the problem, has somehow led us astray. Indeed did not Einstein think that fields rather than particles are at the bottom of everything? So the following argument will not mention particles, nor indeed fields, nor any other particular picture of what goes on at the microscopic level. Nor will it involve any use of the words "quantum mechanical system", which can have an unfortunate effect on the discussion. The difficulty is not created by any such picture or any such terminology. It is created by the predictions about the correlations in the visible outputs of certain conceivable experimental set-ups.

Consider the general experimental set-up of Figure 7. To avoid inessential details it is represented just as a long box of unspecified equipment, with three inputs and three outputs. The outputs, above in the figure, can be three pieces of paper, each with either "yes" or "no" printed on it. The central input is just a "go" signal which sets the experiment off at time t_1. Shortly after that the central output says "yes" or "no". We are only interested in the "yes" 's, which confirm that everything has got off to a good start (e.g., there are no "particles" going in the wrong directions, and so on). At time $t_1 + T$ the other outputs appear, each with "yes" or "no" (depending for example on whether or not a signal has appeared on the "up" side of a detecting screen behind a local Stern-Gerlach magnet). The apparatus then rests and recovers internally in preparation for a subsequent repetition of the experiment. But just before time $t_1 + T$, say at time $t_1 + T - \delta$, signals a and b are injected at the two ends. (They might for example dictate that Stern-Gerlach magnets be rotated by angles a and b away from some standard position). We can arrange that $c\delta \ll L$, where c is the velocity of light and L the length of the box; we would not then expect the signal at one end to have any influence on the output at the other, for lack of time, whatever hidden connections there might be between the two ends.

Sufficiently many repetitions of the experiment will allow tests of hypotheses about the joint conditional probability distribution

$$P(A,B \mid a,b)$$

for results A and B at the two ends for given signals a and b.

Now of course it would be no surprise to find that the two results A and B are correlated, i.e., that P does not split into a product of independent factors:

$$P(A,B \mid a,b) \neq P_1(A \mid a) \, P_2(B \mid b)$$

But we will argue that certain particular correlations, realizable according to quantum mechanics, are *locally inexplicable*. They cannot be explained, that is to say, without action at a distance.

To explain the "inexplicable" we explain "explicable". For example the statistics of heart attacks in Lille and Lyons show strong correlations. The probability of M cases in Lyons and N in Lille, on a randomly chosen day, does not separate:

$$P(M,N) \neq P_1(M) \, P_2(N).$$

In fact when M is above average N also tends to be above average. You might shrug your shoulders and say "coincidences happen all the time", or "that's life". Such an attitude is indeed sometimes advocated by otherwise serious people in the context of quantum philosophy. But outside that particular context, such an attitude would be dismissed as unscientific. The scientific attitude is that correlations cry out for explanation. And of course in the given example explanations are soon found. The weather is much the same in the two towns, and hot days are bad for heart attacks. The day of the week is exactly the same in the two towns, and Sundays are especially bad because of family quarrels and too much to eat. And so on. It seems reasonable to expect that if sufficiently many such causal factors can be identified and held fixed, the *residual* fluctuations will be independent, i.e.,

$$P(M,N \mid a,b,\lambda) = P_1(M \mid a,\lambda) \, P_2(N \mid b,\lambda), \tag{10}$$

where a and b are temperatures in Lyons and Lille respectively, λ denotes any number of other variables that might be relevant, and $P(M,N \mid a,b,\lambda)$ is the conditional probability of M cases in Lyons and N in Lille for *given* (a,b,λ). Note well that we already incorporate in (10) a hypothesis of "local causality" or "no action at a distance". For we do not allow the first factor to depend on a, nor the second on b. That is, we do not admit the temperature in Lyons as a causal influence in Lille, and vice versa.

Let us suppose then that the correlations between A and B in the EPR experiment are likewise "locally explicable". That is to say we suppose that there are variables λ, which, if only we knew them, would allow decoupling of the fluctuations:

$$P(A,B \mid a,b,\lambda) = P_1(A \mid a,\lambda) \, P_2(B \mid b,\lambda) \tag{11}$$

We have to consider then some probability distribution $f(\lambda)$ over these complementary variables, and it is for the averaged probability

$$P(A,B \mid a,b) = \int d\lambda \, f(\lambda) \, P(A,B \mid a,b,\lambda) \tag{12}$$

that we have quantum mechanical predictions.

But not just any function $p(A,B \mid a,b)$ can be represented in the form (12).

To see this it is useful to introduce the combination

$$E(a,b) = \left(\begin{array}{c} P(\text{yes,yes}) \mid a,b) + P(\text{no,no} \mid a,b) \\ -P(\text{yes,no} \mid a,b) - P(\text{no,yes} \mid a,b) \end{array} \right) \tag{13}$$

Then it is easy to show (Appendix 1) that if (12) holds, with however many variables λ and whatever distribution $\rho(\lambda)$, then follows the Clauser-Holt-Horne-Shimony /18/ inequality

$$| E(a,b) + E(a,b') + E(a',b) - E(a',b') | \leqslant 2 \qquad (14)$$

According to quantum mechanics, however, for example with some practical approximation to the EPRB gedanken set-up, we can have approximately [from (4)]

$$E(a,b) = \left(\sin \frac{a - b}{2} \right)^2 - \left(\cos \frac{a - b}{2} \right)^2$$
$$= - \cos (a - b) \qquad (15)$$

Taking for example

$$a = 0°, \, a' = 90°, \, b = 45°, \, b' = -45° \qquad (16)$$

We have from (15)

$$E(a,b) + E(a,b') + E(a',b) - E(a',b')$$
$$= - 3 \cos 45° + \cos 135° = -2 \sqrt{2} \qquad (17)$$

This is in contradiction to (14). Note that for such a contradiction it is not necessary to realize (15) accurately. A sufficiently close approximation is enough, for between (14) and (17) there is a factor of $\sqrt{2}$.

So the quantum correlations are locally inexplicable. To avoid the inequality we could allow P_1 in (11) to depend on b or P_2 to depend on a. That is to say we could admit the input at one end as a causal influence at the other end. For the set-up described this would not be only a mysterious long range influence—a non-locality or action at a distance in the loose sense—but one propagating faster than light (because $c\delta \ll L$)—a non-locality in the stricter and more indigestible sense.

It is notable that in this argument nothing is said about the locality, or even localizability, of the variable λ. These variables could well include, for example, quantum mechanical state vectors, which have no particular localization in ordinary space time. It is assumed only that the outputs A and B, and the particular inputs a and b, are well localized.

5. Envoi

By way of conclusion I will comment on four possible positions that might be taken on this business—without pretending that they are the only possibilities.

First, and those of us who are inspired by Einstein would like this best, quantum mechanics may be *wrong* in sufficiently critical situations. Perhaps Nature is not so queer as quantum mechanics. But the experimental situation is not very encouraging from this point of view /19/. It is true that practical experiments fall far short of the ideal, because of counter inefficiencies, or analyzer inefficiencies, or geometrical imperfections, and so on. It is only with added assumptions, or conventional allowance for inefficiencies and extrapolation from the real to the ideal, that one can say the inequality is violated. Although there is an escape route there, it is hard for me to believe that quantum mechanics works so nicely for inefficient practical set-ups and is yet going to fail badly when sufficient refinements are made. Of more importance, in my opinion, is the complete absence of the vital *time* factor in existing experiments. The analyzers are not rotated during the flight of the particles. Even if one is obliged to admit some long range influence, it need not travel faster than light—and so would be much less indigestible. For me, then, it is of capital importance that Aspect /19,20/ is engaged in an experiment in which the time factor is introduced.

Secondly, it may be that it is not permissible to regard the experimental settings a and b in the analyzers as independent variables, as we did /21/. We supposed them in particular to be independent of the supplementary variables λ, in that a and b could be changed without changing the probability distribution $\rho(\lambda)$. Now even if we have arranged that a and b are generated by apparently random radioactive devices, housed in separate boxes and thickly shielded, or by Swiss national lottery machines, or by elaborate computer programmes, or by apparently free willed experimental physicists, or by some combination of all of these, we cannot be *sure* that a and b are not significantly influenced by the same factors λ that influence A and B /21/. But this way of arranging quantum mechanical correlations would be even more mind boggling that one in which causal chains go faster than light. Apparently separate parts of the world would be deeply and conspiratorially entangled, and our apparent free will would be entangled with them.

Thirdly, it may be that we have to admit that causal influences *do* go faster than light. The role of Lorentz invariance in the completed theory would then be very problematic. An "ether" would be the cheapest solution /22/. But the unobservability of this ether would be disturbing. So would the impossibility of "messages" faster than light, which follows from ordinary relativistic quantum mechanics in so far as it is unambiguous and adequate for procedures we can actually perform. The exact elucidation of concepts like 'message' and 'we', would be a formidable challenge.

Fourthly and finally, it may be that Bohr's intuition was right—in that there is no reality below some 'classical' 'macroscopic' level. Then fundamental physical theory would remain fundamentally vague, until concepts like 'macroscopic' could be made sharper than they are today.

Appendix 1—The Position of Bohr

While imagining that I understand the position of Einstein /23,24/, as regards the EPR correlations, I have very little understanding of the position of his principal opponent, Bohr. Yet most contemporary theorists have the impression that Bohr got the better of Einstein in the argument and are under the impression that they themselves share Bohr's views. As an indication of those views I quote a passage /25/ from his reply to Einstein, Podolsky and Rosen. It is a passage which Bohr himself seems to have regarded as definitive, quoting it himself when summing up much later /26/. Einstein, Podolsky and Rosen had assumed that '. . . if, without in any way disturbing a system, we can predict with certainty the value of a physical quantity, then there exists an element of physical reality corresponding to this physical quantity'. Bohr replied: '. . . the wording of the above mentioned criterion . . . contains an ambiguity as regards the meaning of the expression "without in any way disturbing a system". Of course there is a case like that just considered no question of a mechanical disturbance of the system under investigation during the last critical stage of the measuring procedure. But even at this stage there is essentially the question of *an influence on the very conditions which define the possible types of predictions regarding the future behaviour of the system* . . . their argumentation does not justify their conclusion that quantum mechanical description is essentially incomplete . . . This description may be characterized as a rational utilization of all possibilities of unambiguous interpretation of measurements, compatible with the finite and uncontrollable interaction between the objects and the measuring instruments in the field of quantum theory'.

Indeed I have very litte idea what this means. I do not understand in what sense the word 'mechanical' is used, in characterizing the disturbances which Bohr does not contemplate, as distinct from those which he does. I do not know what the italicized passage means—'an influence on the very conditions. . . .' Could it mean just that different experiments on the first system give different kinds of information about the second? But this was just one of the main points of EPR, who observed that one could learn *either* the position *or* the momentum of the second system. And then I do not understand the final reference to 'uncontrollable interactions between measuring instruments and objects', it seems just to ignore the essential point of EPR that in the absence of action at a distance, only the first system could be supposed disturbed by the first measurement and yet indefinite predictions become possible for the second system. Is Bohr just rejecting the premise—'no action at a distance'—rather than refuting the argument?

Appendix 2—The Clauser-Holt-Horne-Shimony Inequality

From (13) and (11),

$$
\begin{aligned}
E(a,b) &= \int d\lambda \, f(\lambda) \, \{P_1(\text{yes}|a,\lambda) \\
&\quad - P_1(\text{no}|a,\lambda), \{P_2(\text{yes}|b,\lambda) - P_2(\text{no}|b,\lambda)\} \\
&= \int d\lambda \, f(\lambda) \, \bar{A}(a,\lambda) \, \bar{B}(b,\lambda)
\end{aligned}
\tag{18}
$$

where $\bar{A}$ and $\bar{B}$ stand for the first and second curly brackets. Note that since the P's are probabilities,

$$
0 \leqslant P_1 \leqslant 1, \qquad 0 \leqslant P_2 \leqslant 1
$$

and it follows that

$$
|\bar{A}(a,\lambda)| \leqslant 1, \, |\bar{B}(b,\lambda)| \leqslant 1
\tag{19}
$$

From (18),

$$
E(a,b) \pm E(a,b') \leqslant \int d\lambda \, f(\lambda) \, \bar{A}(a,\lambda)[\bar{B}(b,\lambda) \pm \bar{B}(b',\lambda)]
$$

so from (19),

$$
|E(a,b) \pm E(a,b')| \leqslant \int d\lambda \, f(\lambda) \, |\bar{B}(b,\lambda) \pm \bar{B}(b',\lambda)|;
$$

likewise

$$
|E(a',b) \mp E(a',b')| \leqslant \int d\lambda \, f(\lambda) \, |\bar{B}(b,\lambda) \mp \bar{B}(b',\lambda)|
$$

Using (19) again,

$$
|\bar{B}(b,\lambda) \pm \bar{B}(b,\lambda)| + |\bar{B}(b,\lambda) \mp \bar{B}(b',\lambda)| \leqslant 2
$$

and from

$$\int d\lambda\ f(\lambda) = 1$$

it follows that

$$| E(a,b) \pm E(a,b')| + | E(a',b) \mp E(a',b')| \leq 2, \qquad (20)$$

which includes (14).

References

/1/ A. Einstein, B. Podolsky and N. Rosen, Phys. Rev. 46 (1935) 777. For an introduction see the accompanying paper of F. Laloë.

/2/ D. Bohm, quantum Theory, (Englewood Cliffe, New Jersey, 1951).

/3/ Note, however, that these *particular* phenomena were actually inferred from other quantum phenomena in advance of observation.

/4/ And perhaps romanticism. See P. Forman, "Weimar culture, causality and quantum theory, 1918–1927", in Historical Studies in the Physical Sciences, R. Mc Cormach, ed. (University of Pennsylvania Press, Philadelphia, 1971) vol. 3, p. 1–115.

/5/ M. Jammer, The Philosophy of Quantum Mechanics, John Wiley (1974), p. 204, quoting A. Petersen, Bulletin of the Atomic Scientist 19 (1963) 12.

/6/ M. Jammer, ibid, p. 205, quoting W. Heisenberg, Physics and Philosophy (Allen and Unwin, London 1958) p. 160.

/7/ M. Jammer, ibid, p. 161, quoting E. Zilsel, "P. Jordans Versuch, den Vitalismus quanten mechanisch zu retten", Erkenntnis 5 (1935) 56–64.

/8/ The phrase is from a 1947 letter of Einstein to Born (Ref. 11), p. 158.

/9/ The accompanying paper of F. Laloë gives an introduction to quantum formalism.

/10/ And his followers. My own first paper on this subject [Physics 1 (1965) 195] starts with a summary of the EPR argument *from locality to* deterministic hidden variables. But the commentators have almost universally reported that it begins with deterministic hidden variables.

/11/ M. Born (editor), The Born-Einstein-Letters, (Macmillan, London, 1971), p. 221.

/12/ M. Born, ibid, p. 176.

/13/ A. Einstein, Dialectia (1948) 320, included in a letter to Born (Ref. 11). p. 168.

/14/ B. d'Espagnat, Scientific American, November 1979, p. 158.

/15/ B. d'Espagnat, A la Recherche du Réel (Gautheir-Villars, Paris, 1979).

/16/ "The number of young women is less than or equal to the number of woman smokers plus the number of young non-smokers." (Ref. 15, p. 27). See also E.P. Wigner, Am. J. of Physics 38 (1970) 1005.

/17/ Other discussions with some pretension to generality are: J.F. Clauser and M.A. Horne, Phys. Rev. 10D (1974) 526; J.S. Bell, CERN preprint TH-2053 (1975), reproduced in Epistemological Letters (Association Ferd. Gonseth, CP 1081, CH-2051, Bienne) 9 (1976) 11; H.P. Stapp, Foundations of Physics 9 (1979) 1, and to appear; J.S. Bell, Comments on Atomic and Molecular Physics, to appear. Many other references are given in the reviews of Clauser and Shimony /19/ and Pipkin /19/.

/18/ J.F. Clauser, R.A. Holt, M.A. Horne and A. Shimony, Phys. Rev. Letters 23 (1969) 880.

/19/ The experimental situation is surveyed in the accompanying paper of A. Aspect. See also: J.F. Clauser and A. Shimony, Reports on Progress in Physics 41 (1978) 1881; F.M. Pipkin, Annual Reviews of Nuclear Science, (1978).

/20/ A. Aspect, Phys. Rev. 14D (1976) 1944.

/21/ For some explicit discussion of this, see contributions of Shimony, Horne, Clauser and Bell in Epistemological Letters (Association Ferdinand Gonseth, CP 1081, CH-2051, Biennne) 13 (1976) p. 1; 15 (1977) p. 79, and 18 (1978) p. 1. See also Clauser and Shimony /19/.

/22/ P.H. Eberhard, Nuovo Cimento 46B (1978) 392.

/23/ But Max Jammer thinks that I misrepresent Einstein (Ref. 5, p. 254). I have defended my views in Ref. 24.

/24/ J.S. Bell, in Frontier Problems in High Energy Physics, in honour of Gilberto Bernardini, Scuola Normale, Pisa, 1976.

/25/ N. Bohr, Phys. Rev. 48 (1935) 696.

/26/ N. Bohr, In Albert Einstein, Philosopher-Scientist (P.S. Schlipp, Ed., Tudor, N.Y., 1949).

Fifty Years Later: When Gedanken Experiments Become Real Experiments

Alain Aspect* and Philippe Grangier

INSTITUT D'OPTIQUE THÉORIQUE ET APPLIQUÉE

1. Introduction

I would like to report here on sort of a laboratory-class experiment, on the question of wave-particle duality. As Professor Weisskopf recounts in these interesting movies, Niels Bohr enjoyed reasoning on thought experiments. And he was very careful to draw all the details of the apparatus, specially the screws. As you have already seen with the Einstein-Podolsky-Rosen Gendanken—experiment we can, fifty years later, realize these thought experiments. But our screws are lasers, and our field is not Quantum Mechanics but rather Quantum Optics. So, wave-particle duality will be illustrated with photons.

Wave particle duality for the photon is considered a typical example of quantum behaviour. As so, it is often introduced at the very beginning of elementary courses in quantum mechanics, where it is presented as a well established fact. More generally, it is a widely spread belief that all the famous historical "single photon interference experiments" are a definite experimental evidence for the wave-particle duality of light.

We will first show that the situation is certainly not so clear. There is no doubt that these experiments have shown interferences, even with very weak light, evidencing the wave-like behaviour. But we will argue that there was no unquestionable evidence of a particle— like behaviour. More, in the light of modern discussions about non-classical effects in the statistical properties of light [1,2], one can claim that it would have been impossible to observe any particle—

---

* Now with Collège de France, Paris.

like behaviour. This claim is related to the type of source used in all these experiments, producing chaotic light, for which it has now been recognized that there is no non-classical effect (a non-classical effect is an effect that cannot be interpreted by a model in which the light is treated by a classical field, represented by c. numbers).

We will then develop the idea that light produced in "single photon states" would exhibit a clear particle—like behaviour, anticorrelation on a beam-splitter. A quantitative criterium will be produced, allowing to show that such a behaviour is incompatible with any classical model of the light.

We have been able to build a source of single photon pulses of light. This source has allowed to observe the particle-like behaviour, characteristic of single photon states. We have then used the same source for an interference experiment, that we thus think correct to call a single photon interference experiment [3].

2. Feeble Light Interference Experiments

Very early in the development of quantum mechanics, the question was raised of knowing whether "a photon interferes with itself". Many experiments have been carried out in this context (see Table 1), after the first experiment by Taylor (1909), where he made a photograph of a diffraction pattern with six months of exposition time. Although some conflicting results exist, almost all these experiments have shown that the visibility of the light is strongly attenuated. The various authors have thus concluded that "a photon interferes with itself" [4].

Their argument generally amounts to evaluating the energy flux ϕ in the interferometer, from which a photon flux $\phi/h\nu$ is calculated. They then show that, in the average, there is less than one photon at a time in the interferometer. Assuming the statistical independence between the emissions, the probability that two photons are simultaneously present in the system is thus weaker than the probability of presence of one photon. The conclusion is that the observed effect is essentially due to single photons.

This argument is correct in its great lines, as soon as the concept of photons is accepted, i.e. if one admits that a light beam is constituted of quanta of energy $h\nu$. But the question is still open to know whether it is necessary to introduce this concept of photon. It is generally admitted that the discrete nature of the detection pro-

Table I. Feeble light interference experiments. All these experiments have been realized with attenuated light from a usual source (atomic discharge).

Author	Date	Experiment	Detector	Photon flux (s^{-1})	Interferences
Taylor (a)	1909	Diffraction	Photography	10^6	Yes
Dempster et al. (b)	1927	(i) Grating	Photography	10^2	Yes
		(ii) Fabry Pérot	Photography	10^5	Yes
Janossy et al. (c)	1957	Michelson Interferometer	Photomultiplier	10^5	Yes
Griffiths (d)	1963	Young slits	Image Interferometer	2×10^3	Yes
Scarl et al. (e)	1968	Young slits	Photomultiplier	2×10^4	Yes
Donstov et al. (f)	1967	Fabry Pérot	Image Intensifier	10^3	No
Reynolds et al. (g)	1969	Fabry Pérot	Image Intensifier	10^2	Yes
Bozec et al. (h)	1969	Fabry Pérot	Photography	10^2	Yes
Grishaev et al. (i)	1969	Jamin Interferometer	Image Intensifier	10^3	Yes

(a) G.I. Taylor, Proc. Cambridge Philos. Soc. *15*, 114 (1909).
(b) A.J. Dempster and H.F. Batho, Phys. Rev. *30*, 644 (1927).
(c) L. Janossy and Z. Naray, Acta Phys. Hungaria 7, 403 (1967).
(d) H.M. Griffiths, Princeton University Senior Thesis (1963).
(e) G.T. Reynolds et al., Advances in electronics and electron physics *28 B*, Academic Press, London (1969).
(f) Y.P. Dontsov and A.I. Baz, Sov. Phys. JETP *25*, 1 (1967).
(g) G.T. Reynolds, K. Spartalian and D.B. Scarl. Nuovo Cim. *B 61*, 355 (1969).
(h) P. Bozec, M. Cagnet and G. Roger, C.R. Acad. Sci. *269*, 883 (1969).
(i) A. Grishaev et al., Sov. Phys. JETP *32*, 16 (1969).

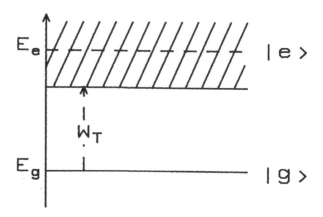

Figure 1. Model of detector for the photoelectric effect. The atom has a ground state $| g \rangle$, and a continuum of ionized states. In state $| e \rangle$, the free electron has a kinetic energy $E_e - E_g - W_T$.

cess is an evidence of the concept of photon. More specifically, Einstein's interpretation of the photoelectric effect is often considered a proof of the necessity of describing the light as composed of photons. In fact, it is an elementary exercise in quantum mechanics to show that all the characteristics of the photoelectric effect can be obtained with a model where a quantized detector interacts with a classical electromagnetic field [1,5].

Let us for instance consider a simple model where the detector is an atom with a ground state $| g \rangle$ and a continuum of ionized excited states $| e \rangle$, with a gap W_T (Fig. 1). This atom interacts with a classical electromagnetic field via an electric dipole interaction, represented by the hamiltonian $\epsilon \cdot \hat{D}$, where the electric field ϵ is a number, and $\hat{D}$ is the atomic electric dipole operator. If ϵ is a field with an amplitude ϵ_0 oscillating at the angular frequency ω, we can easily find the transition rate from state $| g \rangle$ to state $| e \rangle$, by use of Fermi's Golden Rule:

$$\frac{d}{dt} \mathcal{P}_{g \to e} = \frac{\Pi}{2\hbar} |\langle e|\hat{D}|g \rangle|^2 \cdot \epsilon_0^2 \cdot \rho(e) \, \delta(E_e - E_g - \hbar\omega). \quad (1)$$

Integration over the final states would yield the total ionization rate, but formula (1) embodies all the features of the photoelectric effect. The existence of a threshold is related to the fact that the density of excited states, $\rho(e)$, is zero if $E_e \langle E_g + W_T$. The probability of detection is proportional to the intensity of ϵ_0^2. The final energy of the system is $E_e = E_g + \hbar\omega$, i.e. the kinetic energy of the electron is $\hbar\omega - W_T$. Finally, the fact that photodetection is constituted by discrete events can be attributed to the quantization of the detector.

To conclude this discussion, we can say that there is no logical necessity to introduce the concept of photon to describe these experiments of interference with weak light. It is enough to use a model where the light is treated as a classical electromagnetic field (for which we are not surprised to have interference) and where the detector is quantized.

One may argue that we know now that there exist some effects—called non-classical properties of light [1], [6,9]—that cannot be interpreted by such semi-classical models. But these effects, observable on higher order intensity correlation functions, can only be observed for light emitted in certain particular states. And it is well known that the light emitted by usual sources—such as used in the interference experiments under discussion—has absolutely no non-classical feature, even in weak intensity beams. In other words, it is important to realize that a classical light (emitted by a usual source, or a laser) has no non-classical features, even if strongly attenuated. In particular, it is definitely different from light emitted in single-photon states, as we discuss it now.

3. Single Photon Behaviour: Anticorrelation on a Beam-Splitter

All the observed non-classical properties of light are related to second order (in intensity) coherence properties of light. But there had still been no test of the conceptually very simple situation dealing with single-photon states of the light impinging on a beam splitter. In this case, the quantum theory of light predicts a perfect anticorrelation for detections on both sides of the beam splitter (a single photon can only be detected once). On the other hand, any description involving classical waves would predict some amount of coincidences, since a wave is divided on a beam splitter. It is thus possible to characterize experimentally single photon behaviour, by the observation of an anticorrelation between the detections on both sides of the beam splitter.

In order to find a more quantitative criterium, we consider a source emitting light pulses, which impinge on a beam splitter (Fig. 2). Synchronized with each pulse, we have a gate which enables the

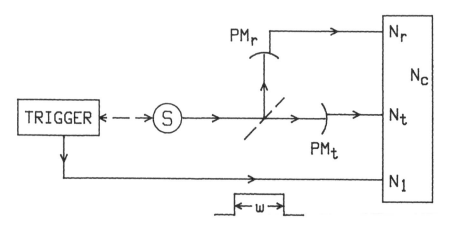

Figure 2. Study of detection correlations after a beam-splitter. The source s emits light pulses, which impinge on a beam-splitter. A triggering system produces a gate of duration W, synchronized with the light pulse. The detections are authorized only during the gates. A coincidence is counted when both PM_r and PM_t register a detection during the same gate.

photodetectors for a detection time w overlapping the arrival of the corresponding pulse. During that gate, one monitors singles detections in the transmitted or reflected channels, and a coincidence if both channels register a count during the same gate. Denoting N_1 the rate of gates, N_r and N_t the rates of singles counts, and N_c the rate of coincidences, we obtain the probabilities for a single count during a gate

$$P_t = \frac{N_t}{N_1} \qquad P_r = \frac{N_r}{N_1} \tag{2.a}$$

and the probability for a coincidence

$$P_c = \frac{N_c}{N_1}. \tag{2.b}$$

For a classical-wave description of the light, the intensity $I(t)$ is divided on the beam splitter into a reflected and a transmitted part. The probabilities of photodetection during the n-th gate are proportional to the average intensity during this gate

$$i_n = \frac{1}{w} \int_{t_n}^{t_n+w} I(t)dt. \tag{3}$$

Denoting by brackets an average over the ensemble of gates, we have

$$P_t = \alpha_t w \langle i_n \rangle \qquad P_r = \alpha_r w \langle i_n \rangle, \qquad (4.\text{a})$$

where α_r and α_t are overall detection efficiencies, including the splitting coefficient on the beam splitter.

Restraining this study to the case when the singles probabilities are small compared to one, the coincidence probability then writes

$$P_c = \alpha_t \alpha_r w^2 \langle i_n^2 \rangle. \qquad (4.\text{b})$$

The standard Cauchy-Schwartz inequality

$$\langle i_n^2 \rangle \geqslant \langle i_n \rangle^2 \qquad (5)$$

holds for the bracket average. Therefore, for any classical-wave description of the experiment of Figure 2, we expect

$$P_c \geqslant P_r P_t \qquad (6.\text{a})$$

or equivalently

$$\alpha \geqslant 1 \qquad \text{with } \alpha = \frac{N_c N_1}{N_r N_t}. \qquad (6.\text{b})$$

The intuitive meaning of this inequality is clear. For a classical wave divided on the beam splitter, there is a minimum rate of coincidences, corresponding to the "accidental coincidences".

We have thus obtained a criterium for empirically characterizing a single particle behaviour of light pulses. The violation of inequality (6) will indicate that the light pulses should not be described as wave packets divided on a beam splitter but rather as single photons that cannot be detected simultaneously on both sides of the beam splitter.

4. Anticorrelation on a Beam-Splitter: Experiments

We have built an experimental setup corresponding to the scheme of Fig. 2, i.e. allowing to measure singles and coincidences on the two sides of a beam-splitter, during gates triggered by events synchronous with the light pulses. This system has been used to study light pulses emitted by a classical source (4.a). Then, we have studied pulses from a source designed to emit single-photons wave packets, as explained in 4b.

Table II. Anticorrelation experiment for light-pulses from an attenuated photodiode (0.01 photon/pulse). The last column corresponds to the expected number of coincidences for $\alpha = 1$. All the measured coincidences are compatible with $\alpha = 1$; there is no evidence of anticorrelation.

Trigger rates $N_1 (s^{-1})$	Single rates		Duration $T(s)$	Measured coincidences $N_c T$	Expected coincidences for $\alpha = 1$ $\frac{N_{2r} \cdot N_{2t}}{N_1} T$
	$N_{2r} (s^{-1})$	$N_{2t} (s^{-1})$			
4 760	3.02	3.76	31 200	82	74.5
8 880	5.58	7.28	31 200	153	143
12 130	7.90	10.2	25 200	157	167
20 400	14.1	20.0	25 200	341	349
35 750	26.4	33.1	12 800	329	313
50 800	44.3	48.6	18 800	840	798
67 600	69.6	72.5	12 800	925	955

Note that the singles rates are similar to the ones of table III.

4a. Attenuated Classical Source

In order to confirm experimentally our arguments, and also to test the photon counting system, we have first studied light from a pulsed photodiode. It produced light pulses with a rise time of 1.5 ns and a fall time about 6 ns. The gates, triggered by the electric pulses driving the photodiode, were 9 ns wide (as in 4b) and overlapped almost completely the light pulses.

The source was attenuated to a level corresponding to one detection per 1,000 emitted pulses. With a detector quantum efficiency about 10%, the average energy per pulse can be estimated to about 0.01 photon. In the context of Table I, this source would certainly have been considered a source of single photons.

Table II shows the results of the anticorrelation measurements. The quantity α (of inequality (6)) is consistently found equal to 1, i.e. no anticorrelation is observed.

This experiment thus support the claim that light emitted by an attenuated classical source does not exhibit a single photon behaviour on a beam-splitter. This has been found true, even with well separated light pulses, with an average energy by pulse much less than one photon.

4b. Single Photon Pulses

An excited atom emits a single photon, because of energy conservation. In classical sources, many atoms are simultaneously in view of the detectors, and the number of excited atoms fluctuates. As a

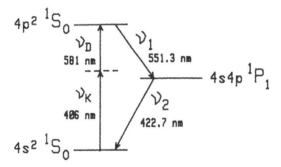

Figure 3. Radiative cascade of Calcium used to produce the single photons wave packets. The detection of photon v_1 is used as trigger (cf. Fig. 2). The photon v_2 is sent onto the beam-splitter. The calcium cascade used in this experiment was excited by two photon absorption from stabilized c.w. lasers, and the cascade rate was held constant to a few percent.

consequence, the emitted light is described by a density matrix reflecting these fluctuations, including the possibility that several photons are emitted simultaneously. For a Poisson fluctuation of the number of emitting atoms, one can show that the statistical properties of the light cannot be distinguished from the one of classical light.

In order to observe non-classical properties in fluorescence light, it is thus necessary to isolate single atom emission. This was realized by Kimble et al. [8] who had only one atom in their observation region when they demonstrated antibunching. In our experiment, we have been able to isolate single atom emission not in space but in time. Our source is composed of atoms that we excite to the upper level of a two-photon radiative cascade (Fig. 3) [10], emitting two photons at different frequencies v_1 and v_2. The time intervals between the detections of v_1 and v_2 are distributed according to an exponential law, corresponding to the decay of the intermediate state with a life-time $\tau_s = 4.7$ ns. By choosing the rate of excitation of the cascades much smaller than $(\tau_s)^{-1}$, we have cascades well separated in time. We can use the detection of v_1 as a trigger for a gate of duration $\omega \simeq 2\,\tau_s$, corresponding to the scheme of Fig. 2. During a gate, the probability for the detection of a photon v_2 coming from the same atom that emitted v_1 is much bigger than the probability of detecting a photon v_2 coming from any other atom in the source. We are then in a situation close to an ideal single-photon pulse [11], and we expect the corresponding anticorrelation behaviour on the beam splitter.

The expected values of the counting rates can be obtained by a straight-forward quantum mechanical calculation. Denoting N the rate of excitation of the cascades, and ϵ_1, ϵ_t and ϵ_r the detection efficiencies of photon v_1 and v_2 (including the collection solid angles, optics transmissions, and detector efficiencies) we obtain

$$N_1 = \epsilon_1 N \tag{7.a}$$

$$N_t = N_1 \epsilon_t [f(w) + Nw] \tag{7.b}$$

$$N_r = N_1 \epsilon_r [f(w) + Nw] \tag{7.b'}$$

$$N_c = N_1 \epsilon_t \epsilon_r [2f(w) Nw + (Nw)^2]. \tag{7.c}$$

The quantity $f(w)$, very close to 1 in this experiment, is the product of the factor $[1 - \exp(-w/\tau_s)]$ (overlap between the gate and the exponential decay) by a factor somewhat greater than 1 related to the angular correlation between v_1 and v_2 [12].

The quantum mechanical prediction for α (eq. (6)) is thus

$$\alpha_{QM} = \frac{2f(w) Nw + (Nw)^2}{[f(w) + Nw]^2}, \tag{8}$$

which is smaller than one, as expected. The anticorrelation effect will be stronger (α small compared to 1) if Nw can be chosen much smaller than $f(w)$. This condition corresponds actually to the intuitive requirement that N is smaller than $(\tau_s)^{-1}$ (w is of the order of τ_s).

The counting electronics, including the gating system, was a critical part in this experiment. The gate w was actually realized by time-to-amplitude converters followed by threshold circuits. These single-channel analyzers are fed by shaped pulses from PM1 (detecting v_1) on the START input, and from PM$_r$ or PM$_t$ on the STOP input. This allows to adjust the gates with an accuracy of 1 ns. A third time-to-amplitude converter measures the delays between the various detections, and allows to build the various time delay spectra, useful for the control of the system.

Table III shows the measured counting rates, for different values of the excitation rate of the cascade. The corresponding values of α have been plotted on Fig. 4, as a function of Nw. As expected, the violation of inequality (6) increases when Nw decreases, but the

Table III. Anticorrelation experiment with single-photon pulses from the radiative cascade. The last column corresponds to the expected number of coincidences for $\alpha = 1$. The measured coincidences show a clear anticorrelation effect.

Trigger rates N_1 (s^{-1})	Single rates		Duration $T(s)$	Measured coincidences $N_c T$	Expected coincidences for $\alpha = 1$ $\dfrac{N_{2r} \cdot N_{2t}}{N_1} T$
	N_{2r} (s^{-1})	N_{2t} (s^{-1})			
4 720	2.45	3.23	1 200	6	25.5
8 870	4.55	5.75	17 200	9	50.8
12 100	6.21	8.44	14 800	23	64.1
20 400	12.6	17.0	19 200	86	204
36 500	31.0	40.6	13 200	273	456
50 300	47.6	61.9	8 400	314	492
67 100	71.5	95.8	3 600	291	367

These data can be compared to table II

signal decreases simultaneously, and it becomes necessary to accumulate the data for long periods of time to achieve a reasonable statistical accuracy. A maximum violation of more than 13 standard deviations has been obtained for a counting time of 5 hours. The value of α is 0.18 ± 0.06, corresponding to a total number of coincidences of 9, instead of a minimum value of 50 expected for a classical model of the light.

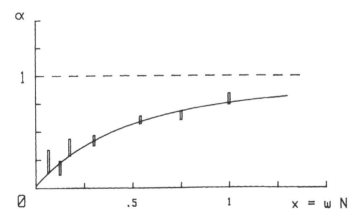

Figure 4. Anticorrelation parameter α as a function of wN (number of cascades excited during a gate). The indicated error is ± 1 standard deviation. The full line is the theoretical prediction of Quantum Mechanics. The inequality $\alpha < 1$ characterizes the quantum domain.

4c. Conclusion

In these experiments, we have thus confirmed that light pulses emitted by a classical source do not exhibit a single-photon behaviour on a beam splitter, even when the average energy by pulse is much less than one photon.

We have also demonstrated a source that produces single photon wave-packets, with a synchronized triggering signal. On a beam splitter, these single photon pulses exhibit a very clear anticorrelation, that is to say a behaviour characteristic of single photons, as predicted by quantum mechanics.

With such a source, it is thus possible to revisit the question of single photon interferences.

5. Single-Photon Interference

Starting with the same source and beam-splitter we have built a Mach-Zehnder interferometer (Fig. 5). The detectors in the two outputs Z1 and Z2 are gated as in the previous experiments, so that we measure the probabilities that a single photon is transmitted to output Z1 or Z2. According to quantum mechanics, these probabilities P_{Z1} and P_{Z2} are modulated with a visibility unity when the path difference δ is varied.

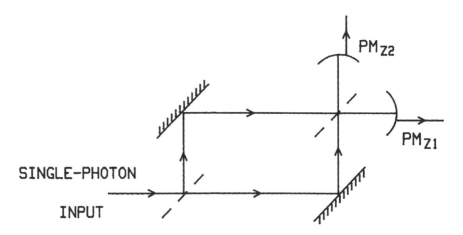

Figure 5. Mach-Zehnder interferometer. The photomultipliers PM_{Z1} and PM_{Z2} are gated as in the previous experiments, so that the interferences are due to single photon wave packets. The path difference is controlled by moving the mirrors.

The interferometer has been designed to accept the large optical spread of the beam from the source (diameter 40 mm, divergence 25 mrad), without altering the visibility of the fringes. The interference is thus observed in the focal planes of lenses in the outputs Z1 and Z2, and δ is varied around zero. The planeities of the beam splitters and of the mirrors are close to $\lambda/50$ on the whole beam diameter. Their orientations are controlled by mechanical stages at the same precision. A translation system, piezzo driven, allows to maintain the orientation of the mirrors while changing the path difference.

The interferometer has first been checked using light from the cascade source, but without the gating system (in order to have a greater signal). We define the fringe visibility as

$$V = \frac{N_{Z1}^{\text{Max}} - N_{Z1}^{\text{Min}}}{N_{Z1}^{\text{Max}} + N_{Z1}^{\text{Min}}} \qquad (9)$$

where N_{Z1}^{Max} and N_{Z1}^{Min} are the maximum and minimum counting rates on output Z1 when δ is varied (dark counts of the photomultipliers are subtracted for this calculation). This fringe visibility (without gating) was found equal to 98.7% $\pm$ 0.5%, the uncertainty corresponding to variations in the alignment from day to day.

For a single-photon interference experiment, the gating system is used, and δ is varied around $\delta = 0$ in 256 steps of $\lambda/50$ each, with a counting time of 1 s at each step. All the data are stored in a computer, that also controls the path difference. Various sweeps can be compiled to improve the signal to noise ratio. A single sweep and the compiled results are shown on Fig. 6. This experiment was performed with the source running in the regime corresponding to an anticorrelation parameter $\alpha = 0.18$, measured in the previous experiment (4b). The visibility of the fringes is clearly very close to 1.

The exact measurement of the visibility of the fringes is a delicate question. One can use the definition (9), but this procedure uses only a small fraction of the information stored, since only the maximum and minimum counting rates are used. Another method, using all the data, consists in searching the linear regression between the gated rates and the non-gated rates (that were also monitored and stored). Standard statistical methods yield the coefficients of this regression, which allows to express the visibility of the gated fringes as a function of the visibility of the non-gated fringes for which the uncertainty is negligible.

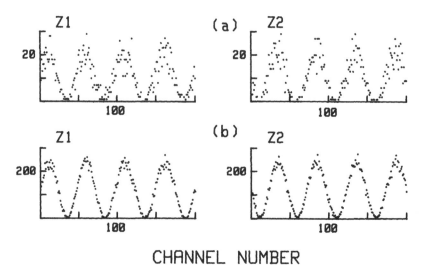

Figure 6. Number of counts in outputs Z1 and Z2 as a function of the path difference δ (one channel corresponds to $\lambda/50$). (a) 1 s counting time per channel; (b) 15 s counting time per channel (compilation of 15 elementary sweeps). This experiment corresponds to an anticorrelation parameter $\alpha = 0.18$.

Both procedures have given consistent results, but the second one yields results with a better accuracy. These results are presented on Fig. 7, where it is clear that the visibility appears constant when the anticorrelation increases (the error bars increase because of the diminution of the signal). We see for instance that for $WN = 0.06$, the visibility is found equal to 0.98 ± 0.01. This regime corresponds to an anticorrelation parameter $\alpha = 0.11$, i.e. a clear single-photon situation.

6. Conclusion—Wave Particle Duality for a Single Photon

We have presented two complementary experiments, bearing on the light pulses emitted by our source. The same gating procedure, synchronized in the same way, has been used in both experiments. In the first experiment, we have observed a clear anticorrelation on both sides of the beam splitter. We have concluded that the source emits single-photon wave packets, as predicted by our analysis. In

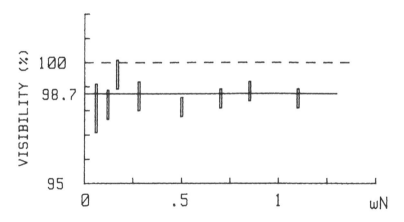

Figure 7. Visibility of the fringes in the single-photon regime, as a function of wN. A correction (less than 0.3%) has been made for the dark counts of the photomultipliers.

the second experiment, we have observed interference with a visibility close to 1, in the single photon regime. We think that this last experiment is a genuine single photon interference experiment.

These two experiments illustrate in a direct way the notion of wave particle duality for a single photon. If we want to use classical pictures (or concepts) to interpret these experiments, we must use a particle picture for the first one ("a photon is not split on a beam-splitter"). On the contrary, the second experiment can only be understood in the framework of a wave theory ("the electromagnetic field is coherently split on the first beam splitter, and recombined on the second, and this recombination depends on the path difference"). The logical conflict between these two classical pictures is at the very basis of famous difficulties of interpretation of quantum mechanics.

Not pretending to solve these difficulties, we can anyway make some remarks which are the application to this particular situation of general statements appearing in the old discussions about the interpretation of wave-particle duality. It is true that the two classical pictures (wave or particle) corresponding to the two experiments are mutually incompatible, but the two corresponding experiments are also mutually exclusive, in the sense that they could not be performed simultaneously. One has to choose between the measurements described on Fig. 2 or on Fig. 5. The impossibility of performing simultaneously experiments corresponding to com-

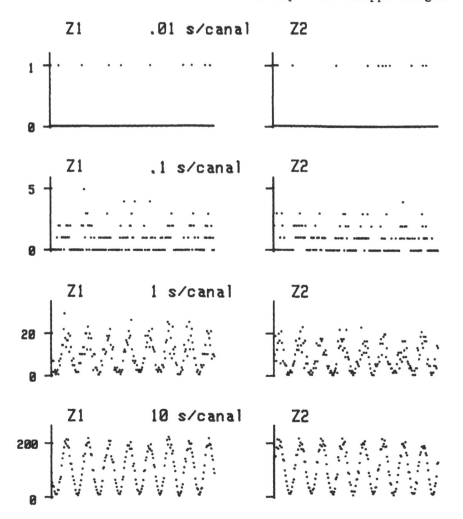

Figure 8. Apparition of the fringes "photon by photon". The figure shows the number of counts in the outputs Z1 and Z2 as a function of the path difference δ (one channel corresponds to $\lambda/25$). The counting time per channel is varied from 0.01 s to 10 s. The regime of the source corresponds to an anticorrelation parameter $\alpha = 0.2$.

plementary descriptions was an often repeated argument of Bohr, who concluded that Quantum Mechanics is a coherent and complete theory. It may also be worth noting that the problem of incompatible descriptions arises only if we insist on using classical concepts such as waves or particles. But if we stand to the Quantum Mechanical description, there is a unique description of the light, by the same state vector (or density-matrix) for experiments of Fig. 2 or Fig. 5.

It is the observed quantity which changes, according to the measurement performed, but not the description of the light.

To conclude, I cannot resist presenting the apparition of the interferences, "photon by photon", since this is a classic of this topics. Figure 8 shows how the interference pattern is built when the data are accumulated. In spite of the excitation that we felt when observing this apparition of the fringes, we hope to have convinced you that it is not the most significant result. Wave-particle duality is really illustrated by the combination of experiments of Figs. 2 and 5.

References

[1] R. Loudon, Rep. Progr. Phys. 43,913 (1980).

[2] R. Loudon, The Quantum Theory of Light, 2nd edition (Clarendon, Oxford, 1983).

[3] P. Grangier, G. Roger and A. Aspect, Europhysics Lett. 1, 173 (1986). P. Grangier, thèse de doctorat d'état, Université Paris XI, Orsay (1986).

[4] P.A.M. Dirac, "The principles of Quantum Mechanics", Oxford University Press (1958).

[5] W.E. Lamb and M.O. Scully, in "Polarisation, Matière et Rayonnement", ed. Société Française de Physique, Presses Universitaires de France, Paris (1969).

[6] D.C. Burnham and D.L. Weinberg, Phys. Rev. Lett. 25, 84 (1970). S. Friberg, C.K. Hong and L. Mandel, Phys. Rev. Lett. 54, 2011 (1985).

[7] J.F. Clauser, Phys. Rev. D 9, 853 (1974).

[8] H.J. Kimble, M. Dagenais and L. Mandel, Phys. Rev. Lett. 39, 691 (1977). J.D. Cresser, J. Hager, G. Leuchs, M. Rateike and H. Walther, in "Topic in current Physics", Vol 27, Springer Verlag, Berlin (1982).

[9] R. Short and L. Mandel, Phys. Rev. Lett. 51, 384 (1983). M.C. Teich and B.E.A. Saleh, J.O.S.A. B 2, 275 (1985). All the recent observations of squeezing also constitute an evidence for "non-classical states of light".

[10] A. Aspect, P. Grangier and G. Roger, Phys. Rev. Lett. 47, 460 (1981).

[11] Since our experiment, single photon pulses have been produced by a different technique, which consists in producing pairs of photons emitted in parametric splitting, and using one of these photons for the gating system. A theoretical advantage of these techniques is a better time and angular correlation between the two photons of a pair. See C.K. Hong and L. Mandel, Phys. Rev. Lett. 56, 58 (1986). J.G. Walker and E. Jakeman, Optica Acta 32, 1303 (1985).

[12] E.S. Fry, Phys. Rev. A 8, 1219 (1973).

Physical and Philosophical Issues in the Bohr-Einstein Debate

Abner Shimony

BOSTON UNIVERSITY

1. Aspects of the Debate

The debate between Niels Bohr and Albert Einstein concerning the interpretation of quantum mechanics extended from the fifth Solvay Conference in 1927 until the end of Einstein's life. The most dramatic exchange occurred in 1935, when Einstein, in collaboration with B. Podolsky and N. Rosen, published a paper in *Physical Review* entitled "Can quantum-mechanical description of physical reality be considered complete?"[1], concluding with a negative answer, and Bohr replied in the same journal, with a paper of the same title but giving a positive answer.[2] Their arguments were restated, in some respects with greater clarity, in the Library of Living Philosophers volume on Einstein in 1949.[3] The disagreements between Bohr and Einstein concerned not only the physical question expressed in the common title of their papers, but also philosophical questions about physical reality and human knowledge.

Much light was thrown upon the Bohr-Einstein debate by the theorem of J.S. Bell (1964)[4] and the experiments which it inspired.[5] Bell showed that any physical theory which applies to a pair of spatially separated systems (as considered in the thought experiment of Einstein, Podolsky, and Rosen) and satisfies a certain locality condition will, in certain circumstances, disagree statistically with quantum mechanics. Consequently, the supplementation of the quantum mechanical description envisaged by Einstein and his collaborators (usually referred to as a "hidden variables theory") must either violate the locality condition or clash with quantum mechanical predictions. Almost certainly this theorem would have surprised Einstein, who believed in locality in the relativistic sense but also treasured the predictive power of quantum mechanics as a statistical theory of the atomic domain. A series of experiments, culminating with that of A. Aspect, J. Dalibard, and G. Roger (1982), strongly supported quantum mechanics against the whole family of local hidden variables theories. In the experimental arrangement of Aspect et al., locality in the sense of relativity theory implies that Bell's locality condition is satisfied, and hence the results refuted the kind of hidden variables theory which Einstein seems to have had in mind. The physical question in the debate between Bohr and Einstein was thus answered in favor of Bohr. It should be said that there are some loopholes in the experiments performed so far, which have kept alive the hopes of dedicated advocates of local hidden variables theories. Most students of the subject, however, do not regard the exploitation of these loopholes to be promising.[6]

The lectures of Bell and Aspect in the present volume provide quite detailed accounts of the arguments of Einstein, Podolsky and Rosen, of Bell's theorem, and of the consequent experiments. Hence I shall not expand the condensed account given in the preceding paragraph. Instead, I shall be concerned primarily with the philosophical differences between Bohr and Einstein.

I shall maintain that the correctness of Bohr's answer to the question, "Can quantum-mechanical description of physical reality be considered complete?", does not by itself constitute a victory over Einstein in their philosophical disagreements. It will be argued in Section 2 that Einstein's physical realism can be suitably generalized to accommodate the results of the experiments inspired by Bell's theorem. The generalization consists in recognizing that a new modality of reality is implicit in quantum mechanical description, which may appropriately be called "potentiality," and in acknowledging a peculiar kind of quantum mechanical nonlocality. Einstein's fundamental philosophical thesis that the physical world has an existence independent of human knowledge is preserved by this generalization, even though there is obviously a retrenchment from some of his characterizations of physical reality. Furthermore, it will be argued in Section 3 that the experimental results do not support the most radical of Bohr's philosophical innovations, and therefore other considerations are needed in order to assess his philosophy. I shall propose that Bohr's thought can be related to a certain philosophical tradition (including Hume and Kant) in which a theory of knowledge is worked out without commitment to a theory of existence. The placement of Bohr's thought in a well explored philosophical tradition helps to exhibit some serious lacunae in the realization of his philosophical program, and questions can be raised about the prospects of filling these lacunae.

2. Nonlocality, Potentiality, and Einstein's Realism

As preparation for some critical remarks about Einstein's philosophical views, it will be very useful to analyze the concept of locality in some detail.

Figure 1 is an abstract representation of the kind of physical system studied by Bell, and earlier by Einstein, Podolsky, and Rosen. Two particles, labelled 1 and 2, propagate to the right and left respectively from a common source. At the moment of their departure from the source the *complete state* of the composite system $1 + 2$ is denoted by λ. For the present, no assumption is made that λ is the quantum mechanical state of $1 + 2$ or a state described by hidden variables. It is only assumed that λ provides a complete specification of the properties of $1 + 2$ when they leave the source. Particle 1 impinges upon an analyzer which has an adjustable parameter a, and the particle emerges from the analyzer in one of two channels, which are labelled $+$ and $-$ respectively. Likewise, particle 2 impinges on an analyzer with adjustable parameter b, and it too emerges in one of two channels labelled $+$ and $-$. Let x_a be the outcome of analysis of particle 1, hence either $+$ or $-$, and let x_b be the outcome of analysis of particle 2, again either $+$ or $-$. The following probability notation will be used:

Figure 1. Particles 1 and 2 propagate from a common source, the complete state of the particles at the moment of departure being λ. The analyzer of particle 1 has an adjustable parameter a, and the analyzer of particle 2 has an adjustable parameter b. Each particle can emerge from its respective analyzer in one of two channels, labeled $+$ and $-$.

$p_\lambda(x_a,x_b \mid a,b)$ = the probability of joint outcomes x_a,x_b, provided that λ is the complete state of $1 + 2$ when they leave the source and a and b are the settings of the respective adjustable parameters.

In terms of $p_\lambda(x_a,x_b \mid a,b)$ we can define probability of single outcomes, and also conditional probability in a standard manner:

$$p_\lambda^1(x_a \mid a,b) = p_\lambda(x_a, + \mid a,b) + p_\lambda(x_a, - \mid a,b),$$

$$p_\lambda^2(x_b \mid a,b) = p_\lambda(+,x_b \mid a,b) + p_\lambda(-,x_b \mid a,b),$$

$$p_\lambda^1(x_a \mid a,b,x_b) = p_\lambda(x_a,x_b \mid a,b)/p_\lambda^2(x_b \mid a,b),$$

$$p_\lambda^2(x_b \mid a,b,x_a) = p_\lambda(x_a x_b \mid a,b)/p_\lambda^1(x_a \mid a,b).$$

In terms of these probability expressions two distinct independence conditions can be defined, as proposed by J. Jarrett.[7]
1. Parameter Independence: $p_\lambda^1(x_a \mid a,b)$ is independent of b,
   $p_\lambda^2(x_b \mid a,b)$ is independent of a.
2. Outcome Independence: $p_\lambda^1(x_a \mid a,b,x_b) = p_\lambda^1(x_a \mid a,b),$
   $p_\lambda^2(x_b \mid a,b,x_a) = p_\lambda^2(x_b \mid a,b).$
The names of these independence conditions are self-explanatory. Furthermore, if the compound event consisting of the choice of parameter a and the occurrence of outcome x_a has space-like separation from the compound event consisting of the choice of parameter b and the occurrence of outcome x_b (as is the case in the experiment of Aspect et al.) then both independence conditions are *prima facie* required by the limitations upon direct causal connectedness imposed by relativity theory. It should be noted that the two independence conditions permit correlations between x_a and x_b, for the complete state λ may imply a deterministic or a probabilistic correlation between the outcomes. But precisely because λ is the complete state of 1 and 2 at the moment of their separation, the specification of the parameter b and the outcome x_b should not, from the standpoint of relativistic locality, have any effect upon the outcome x_a that is not already implicit in λ (and likewise, if the letters a and b are interchanged).

Jarrett showed that the conjunction of Parameter Independence and Outcome Independence is equivalent to Bell's locality condition, which says essentially that

$$p_\lambda(x_a,x_b \mid a,b) = p_\lambda^1(x_a \mid a)p_\lambda^2(x_b \mid b).$$

Since the experimental result of Aspect et al. contradicts Bell's locality condition (provided the loopholes mentioned above are set aside), it follows that at least one of the two independence conditions must be false. It turns out that the consequences of violation of Parameter Independence are very different from the consequences of violation of Outcome Independence, as the following analysis shows.

1. Suppose that Parameter Independence fails, e.g., because $p_\lambda^2(x_b \mid a,b) \neq p_\lambda^2(x_b \mid a',b)$. Then at a moment when particles 1 and 2 are well separated and are about to impinge upon their respective analyzers, an experimenter can make a choice between the parameter values a and a', thereby affecting the probability of the outcome $x_b = +$ for the analysis of particle 2; and if an ensemble of pairs of replicas of the pairs $1 + 2$ is prepared in a sufficiently short interval of time, then the frequency of $+$ outcomes will be affected with virtual certainty by the choice between a and a'. In this way a bit of information is conveyed from the experimenter to an observer of the output of the second analyzer. Most important, *this bit of information can be transmitted faster than light.* Consequently, the failure of Parameter Independence implies the possibility of an unequivocal violation of the special theory of relativity.

2. Suppose that Outcome Independence fails, e.g., because $p_\lambda^2(x_b \mid a,b,x_a) \neq p_\lambda^2(x_b \mid a,b)$ for some choice of a,b, from which it follows that $p_\lambda^2(x_b \mid a,b,+) \neq p_\lambda^2(x_b \mid a,b,-)$. Again a message can be sent from an experimenter associated with the analyzer of particle 1 to an observer of the analysis of particle 2, but the procedure must be different from the foregoing. The experimenter must monitor each of the particles 1 of an ensemble of replicas of $1 + 2$, and decide whether to block the propagation of the particle 2 if $x_a = +$, or else to block its propagation if $x_a = -$. The particle 2 must be placed "on hold" (for instance, a photon may be trapped in a light guide) until the monitoring of its partner has been completed. An observer of the output of the second analyzer can infer from the statistics of x_b which decision the experimenter made. In this way a bit of information is conveyed from the experimenter to the observer. Clearly, however, *this information cannot be transmitted faster than light.* Time is required to monitor the outcome x_a, to perform the operation of taking particle 2 "off hold," and to permit the propagation of particle 2 to its analyzer; and the total time required is greater than the time required for a direct transmission of a light signal between the two analyzers.

Since, as noted previously, the experimental results of Aspect et al. agree with quantum mechanics, the Bell locality condition must be violated by the quantum mechanical predictions concerning the test situation. Which, then, of the two conditions, Parameter Independence and Outcome Independence, is violated by quantum mechanics? If the quantum mechanical polarization state of the two photons in their experiment is

$$\Psi = \frac{1}{\sqrt{2}} \left[u_x(1) \otimes u_x(2) + u_y(1) \otimes u_y(2) \right],$$

where $u_x(1)$ represents polarization of photon 1 along the x-axis, $u_x(2)$ represents polarization of photon 2 along the x-axis, and $u_y(1)$ and $u_y(2)$ have analogous meanings. Then it can easily be seen as follows that the quantum mechanical predictions based upon Ψ violate Outcome Independence. Let the analyzers of photons 1 and 2 be idealized sheets of polaroid placed perpendicularly to the paths of the photons, with the transmission axes of each sheet oriented along the x-axis. In the notation introduced above, the parameters a and b of the two analyzers are each chosen to be the angle 0. The outcome $x_a = +$ will be taken to mean that photon 1 passes through the polaroid sheet oriented at an angle a to the x-axis, and $x_a = -$ means that it fails to pass; and the values $+$ and $-$ of x_b have analogous meanings. The quantum mechanical prediction based upon Ψ yields the probability $\frac{1}{2}$ that photon 1 will pass through its polaroid sheet, because the absolute square of the coefficient of the term $u_n(1) \otimes u_x(2)$ in Ψ is $\frac{1}{2}$. If, however, in addition to knowing that the initial quantum state of the two photons is Ψ, one also knows that photon 2 has failed to pass through its sheet of polaroid, prior to a polarization analysis of photon 1, then the term $u_y(1) \otimes u_y(2)$ is picked out from Ψ, thereby ensuring that photon 1 will not pass through its polaroid sheet. If we express these quantum mechanical probabilities in the foregoing notation, taking the complete state λ to be Ψ, we obtain

$$p_\Psi^1(+ \mid 0,0) = \tfrac{1}{2},$$

$$p_\Psi^1(+ \mid 0,0,-) = 0,$$

in contradiction to Outcome Independence.

It can be shown, though a somewhat elaborate argument is needed,[8] that quantum mechanics does not violate Parameter Independence. As pointed out above, a violation of Bell's locality condition (and hence the possibility of agreement with the results of Aspect et al.) is ensured by the failure of either Parameter Independence or Outcome Independence. But a greater strain in current physical theory would result from failure of the former than of the latter, because only failure of Parameter Independence permits a signal to be transmitted faster than light. We conclude that quantum mechanics is a non-local theory, in the sense of violating Bell's locality condition, but its non-locality "peacefully coexists" with the relativistic prohibition of superluminal signals. Indeed, it may be appropriate to introduce a notation which is familiar in a political context in order to summarize the relation between quantum mechanics (QM) and special relativity theory (SR):

$$QM \ O \ SR.$$

An examination of the structure of the quantum mechanical description Ψ of the pair of photons $1 + 2$ will throw much light upon this peaceful coexistence. First of all, according to Ψ the polarization of photon 1 with respect to the $x-y$ axes is not definite, and the same is true of photon 2. If the state were represented only by $u_x(1) \otimes u_x(2)$, then both photons would have definite polarization along the x-axis; and if the state were represented only by $u_y(1) \otimes u_y(2)$, then both would have definite polarization along the y-axis. But since both terms are present in Ψ, there is a probability $\frac{1}{2}$ that photon 1 will exhibit a polarization along x and a probability $\frac{1}{2}$ that it will exhibit a polarization along y, and similarly for photon 2. It is tempting to regard Ψ as merely a description of the state of the scientist's *knowledge* of the two photons, or alternatively, as a description of an inhomogeneous ensemble of photon pairs, the individual members of which have definite properties that are not described by Ψ. In fact, the temptation is to supplement the quantum mechanical description of the pair of photons by a hidden variables description. But that is precisely what Bell's theorem and the associated experimental results preclude, unless one is willing to accept a hidden variables theory which violates Bell's locality condition. If, however, we concede that Ψ is a complete description of the polarization state of the pair of photons, then we must accept the *indefiniteness* of the polarization of each with respect to the $x-y$ axes as an objective fact, not as a feature of the knowledge of one scientist or of all human beings collectively. We must also acknowledge *objective chance* and *objective probability*, since the outcome of the polarization analysis of each photon is a matter of probability. It is convenient to use a term of Heisenberg to epitomize objective indefiniteness together with the objective determination of probabilities of the various possible outcomes; the polarizations of the photons are *potentialities*.[9] The work initiated by Bell has the consequence of making virtually inescapable a philosophically radical interpretation of quantum mechanics: that there is a modality of existence of physical systems which is somehow intermediate between bare logical possibility and full actuality, namely, the modality of potentiality.

A further peculiarity of the quantum mechanical view of the physical world is exhibited by Ψ, the existence of an n-particle state (with $n \geqslant 2$) in which the individual particles are not in definite states. It can be shown that there is no pair of one-photon polarization states w and z such that Ψ is equivalent to the conjoined attribution of w to photon 1 and z to photon 2. Schrödinger called an n-particle state with this peculiarity an "entangled" state. When the photon pair is in the state Ψ, the polarizations of both 1 and 2 with respect to the x–y axes (and actually with respect to all other pairs of orthogonal axes) are potentialities, but the actualization of either one of them automatically ensures the actualization of the other. If, as a result of a polarization analysis (or other appropriate process), photon 1 exhibits a definite polarization along x or along y, then photon 2 would certainly exhibit the same definite polarization if it is subjected to a subsequent polarization analysis; and likewise if 1 and 2 are interchanged. The entanglement of states of an n-particle system thus rests upon the fact that a quantum mechanical state of a system is a network of potentialities, the content of which is not exhausted by a catalogue of the actual properties which are assigned to the system.

The conclusions just reached have a direct bearing upon the correctness of Einstein's physical world view. In a late and careful paper of 1948 Einstein says the following:

"the concepts of physics refer to a real external world, i.e., ideas are posited of things that claim a 'real existence' independent of the perceiving subject (bodies, fields, etc.), and these ideas are, on the other hand, brought into as secure a relationship as possible with sense impressions. Moreover, it is characteristic of these physical things that they are conceived as being arranged in a space-time continuum. Further, it appears to be essential for this arrangement of the things introduced in physics that, as a specific time, these things claim an existence independent of one another, insofar as these things 'lie in different parts of space'."[10]

The entanglement of the states of the spatially separated photons 1 and 2 conflicts with Einstein's thesis that "these things claim an existence independent of one another." Of course, Einstein was well aware of the entanglement in the quantum mechanical description of composite physical systems, and in fact his paper with Podolsky and Rosen made essential use of entanglement. But Einstein wished to interpret quantum mechanical entanglement epistemically, as expressing the scientist's knowledge of correlations among the properties of spatially separated system, and entanglement would not characterize the complete states which he envisaged. As we have seen, Bell's theorem and subsequent experimental work seem to preclude this epistemic interpretation, thereby confirming that entanglement is an objective fact about the physical world.

The first of Einstein's theses in the passage quoted, and the one which appears to stand highest in his philosophical hierarchy, is that physical things "claim a 'real existence' independent of the perceiving subject." This thesis *is* consistent with all the conclusions which we drew from an analysis of Bell's theorem, the relevant experiments, and the formalism of quantum mechanics. However, this thesis of physical realism, when separated from the rest of Einstein's theses, leaves open the character of the real existence of physical things. The foregoing analysis led to radical conclusions regarding the character of physical existence: i.e., that there are objective indefiniteness, objective chance, and objective probability, in short, that there is a modality of existence which has been designated as potentiality. It is hard to resist speculating about how Einstein would have reacted to Bell's theorem and the experiments of Aspect et al., but of course we must acknowledge candidly that any guesses about this hypothetical reaction are sheer speculation. A good starting point for speculation is the remarkable passage at the conclusion of the posthumously published Fifth Edition of *The Meaning of Relativity*:

From the quantum phenomena it appears to follow with certainty that a finite system of finite energy can be completely described by a finite set of numbers (quantum numbers). This does not seem to be in accordance with a continuum theory, and may lead to an attempt to find a purely algebraic theory for the description of reality. But nobody knows how to obtain the basis of such a theory.[11]

This passage acknowledges the empirical success of quantum theory and indicates a willingness to incorporate some of the implications of quantum mechanics into his world view. In the spirit of this passage might Einstein have been receptive of the ideas of potentiality and entanglement and might he have been reconciled to the abandonment of the mutually independent existence of spatially separated things? We cannot answer this question, but we can say that one coherent option which remains open to a sympathetic but independent-minded follower of Einstein is to accept his highest thesis, that of physical realism, but to give it a sense which is derived from an analysis of quantum mechanics.

One qualification to the last sentence is needed: that the "coherence" of the option is not established beyond all doubt. The problem of the actualization of potentialities—known also as the problem of the reduction of the wave packet and the problem of measurement—is a dark cloud on the horizon. A consequence of the linear dynamical law of quantum mechanics (the Schrödinger equation) is that in the final physical stage of a measurement process, the indexical property of the apparatus (e.g., the direction of a pointer on a dial) may be objectively indefinite—apparently in gross disagreement with laboratory experience. The most dramatic exposition of this difficulty is the famous "cat paradox" of Schrödinger,[12] but the difficulty was well known to other pioneers of quantum mechanics, including von Neumann and Einstein himself. Until this problem is solved one cannot claim that an extended version of physical realism, taking the complete state of a physical system to be a network of potentialities, is a coherent view of the physical world.

3. On the Philosophy of Bohr

The version of physical realism arrived at in Section 2 recognizes Bohr as the victor in the debate with Einstein, Podolsky, and Rosen over the *physical* question, "Can quantum-mechanical description of physical reality be considered complete?". It does not follow that Bohr would be satisfied with this kind of physical realism, despite its incorporation of ideas inspired by quantum mechanics. Furthermore, Bohr's own reasons for defending the completeness of quantum mechanical descriptions are entirely different from the considerations of Bell's theorem and the experiments which it inspired.

The heart of Bohr's answer to Einstein and his collaborators is the following passage:

Of course there is in a case like that just considered no question of a mechanical disturbance of the system under investigation during the last critical stage of the measuring procedure. But even at this stage there is essentially the question of *an influence on the very conditions which define the possible types of predictions regarding the future behavior of the system.* Since these conditions constitute an inherent element of the description of any phenomenon to which the term 'physical reality' can be properly attached, we see that the argumentation of the mentioned authors does not justify their conclusion that quantum-mechanical description is essentially incomplete.[13]

There is nothing in this passage that suggests a commitment to objective entanglement or to potentiality as a new modality of physical existence. In fact, the thrust of the passage is away from ontological questions and towards questions about knowledge and language—e.g., questions of "possible types of predictions regarding the future behavior of the system" and of "the description of any phenomenon to which the term 'physical reality' can be properly attached." In order to see how Bohr's answer applies to the pair of photons which we have been considering, it is useful to note that the quantum mechanical polarization state Ψ studied in Section 2 can be expressed in different ways, which are mathematically equivalent:

$$\Psi = \frac{1}{\sqrt{2}}\left[u_x(1) \otimes u_x(2) + u_y(1) \otimes u_y(2) \right]$$
$$= \frac{1}{\sqrt{2}}\left[u_{x'}(1) \otimes u_{x'}(2) + u_{y'}(1) \otimes u_{y'}(2) \right],$$

where the $x'-y'$ axes are obtained by rotating the $x-y$ axes by an arbitrary angle about the line of propagation of the two photons. The first of the expressions for Ψ shows that a polarization analysis of photon 1 with respect to the $x-y$ axes would yield polarization information concerning photon 2 with respect to the same axes, without disturbing photon 2; while the second expression for Ψ shows the same with respect to the $x'-y'$ axes. Since polarization analysis of photon 1 with respect to the $x-y$ axes and with respect to $x'-y'$ axes are mutually exclusive, and since the choice determines what kind of prediction can be made concerning photon 2, Bohr maintains that it is not proper to attach the term "physical reality" to polarization of photon 2 with respect to both $x-y$ and $x'-y'$. The limitation upon the attribution of two distinct properties to photon 2 in no way stems from action at a distance, because an analysis performed upon photon 1 does not physically disturb photon 2.

There is only a limitation upon possible predictions which an experimenter can make.

In the answer to Einstein et al. just quoted, and repeatedly elsewhere in his essays, Bohr insists that the unequivocal application of terms of theoretical physics to physical situations depends upon the careful specification of experimental arrangements, and he illustrates his dicta with penetrating analyses of interactions between objects and apparatus. Bohr's discussion constitutes an important contribution to one of the central problems of modern philosophy of science, that of the relation between theoretical terms and experience. It also constitutes a contribution to the part of the theory of language called "pragmatics," which—in contrast to syntactics and semantics—specifically studies the circumstances of the language user. Bohr is one of the most remarkable writers on the pragmatics of the languages of the exact sciences, even though this aspect of his thought has not been extensively studied.[14]

Although Bohr's writing is obviously philosophical in the sense of exploring fundamental questions concerning nature and human knowledge, it is far from clear how close he came to formulating a systematic philosophy. Some crucial passages will be cited, however, which at least will indicate the direction in which he was attempting to achieve systematization.

In the treatment of atomic problems, actual calculations are most conveniently carried out with the help of a Schrödinger state function, from which the statistical laws governing observations obtainable under specified conditions can be deduced by definite mathematical operations. It must be recognized, however, that we are here dealing with a purely symbolic procedure, the unambiguous physical interpretation of which in last resort requires a reference to a complete experimental arrangement. Disregard of this point has sometimes led to confusion, and in particular the use of phrases like 'disturbance of the phenomena by observation' and 'creation of physical attributes of objects by measurements' is hardly compatible with common language and practical definition.[15]

It is not relevant that experiments involving an accurate control of the momentum or energy transfer from atomic particles to heavy bodies like diaphragms and shutters would be very difficult to perform, if practical at all. It is only decisive that, in contrast to the proper measuring instruments, these bodies together with the particles would constitute the system to which the quantum-mechanical formalism has to be applied.[16]

While in the mechanical conception of nature, the subject-object distinction was fixed, room is provided for a wider description through the recognition that the consequent use of our concepts requires different placing of such a separation.[17]

Such considerations point to the epistemological implications of the lesson regarding our observational position, which the development of physical science has impressed on us. In return for the renunciation of accustomed demands on explanation, it offers a logical means of comprehending wider fields of experience, necessitating proper attention to the placing of the object-subject separation. Since, in philosophical literature, reference is sometimes made to different levels of objectivity or subjectivity or even of reality, it may be stressed that the notion of an ultimate subject as well as conceptions like realism and idealism find no place in objective description as we have defined it.[18]

Without entering into metaphysical speculations, I may perhaps add that an analysis of the very concept of explanation would, naturally, begin and end with a renunciation as to explaining our own conscious activity."[19]

There have been many attempts to fit these passages and others like them—which are characteristically penetrating, suggestive, and elliptical—into a coherent philosophy.[20] It is beyond the scope of the present paper to assess these efforts of exposition and systematization of Bohr's thought. Some judgments can be made with confidence, however. The second passage is clearly discordant with the interpretation of Bohr as a macro-realist, who attributes objective existence to macroscopic bodies but treats microphysics only as an instrument for predicting observable behavior of macroscopic bodies. This passage shows that Bohr does not reserve a quantum mechanical description for microscopic entities and a classical description for macroscopic ones. Rather, he treats quantum mechanically any physical system which is the object of investigation and uses a classical description for parts of the apparatus of measurement, which are situated on the subject's side of the subject-object separation. The fourth passage is discordant with the interpretation of Bohr as an idealist who regards the contents of consciousness as the fundamental reality, and all physical discourse as merely an instrument or short-hand for summarizing, systematizing, and anticipating these contents.[21]

A. Petersen suggests a very interesting generalization of the two foregoing negative judgments: not only is it incorrect to attribute to Bohr either a macrophysical or an idealistic ontology, but any ontology whatever is alien to his thought. Here are some of Petersen's statements, which are either explicitly or implicitly presented as expositions of Bohr's philosophical ideas:

Bohr's remarks on quantal and thermodynamical irreversibility illustrate his approach to the description problem in physics. Especially, they indicate that he thought this problem to be a purely conceptual one. The question is not what *is* in an ontological sense, but what can be stated unambiguously in physical terms.[22]

In the course of the interpretation discussion [of quantum mechanics], the irrelevance of ontological ideas became increasingly conspicuous, but their elimination has been a slow and difficult process which may still be far from complete.[23]

When it was objected that reality is more fundamental than language and lies beneath language, Bohr answered, 'We are suspended in language in such a way that we cannot say what is up and what is down.'[24]

Petersen's proposals command attention for the intrinsic reason that they provide a lucid, unforced, and illuminating interpretations of passages like the first, third, fourth, and fifth above, and for the extrinsic reason that he was Bohr's assistant for seven years. What is not convincing, however, is Petersen's claim that Bohr's replacement of ontology by other concerns is a radically new departure in western philosophy. Bohr's ideas belong quite clearly to an important philosophical tradition: that in which epistemology is studied in deliberate abstention from considerations of ontology, and particularly from suppositions about the ontological status of the knowing subject. Classical philosophers in this tradition are Hume and Kant, but some of the analytic philosophers associated with the Vienna Circle and some of the followers of Wittgenstein can reasonably be said to belong to it. An opposing tradition envisages the meshing of epistemology and ontology in two ways: (1) epistemology aims at (and perhaps partially succeeds in) showing how human beings can obtain knowledge of the world as it is, at least to a good approximation and with a reasonably high degree of reliability; and (2) it aims at exhibiting an ontological niche for the knowing subject. In this tradition are found Aristotle, Leibniz, Newton, Locke, and Whitehead, and perhaps Einstein (the last in a fragmentary way, since he did not attempt to formulate a systematic philosophy that is more comprehensive than he needed for understanding physics). This contrast of two philosophical traditions is crude, as grand intellectual classifications usually are, since they group together diverse philosophies, thereby playing down their nuances and individualities. In spite of the crudity of the classification, however, it throws some light upon Bohr's thought and helps to pose some important criticisms.

Particularly instructive is a comparison of Bohr with Kant, since both in their respective ways were critical of claims to knowledge of things in themselves. It should be emphasized that their critiques are directed not only against a "naive realism," according to which perception yields direct knowledge of the existence of things in themselves and their properties, but also of "critical realism," according to which such knowledge can be obtained indirectly and inferentially. Kant's doctrine that space and time are forms of intuition imposed upon appearances by the faculty of sensibility, and that causality, substance, and other fundamental concepts are categories imposed upon experience by the faculty of understanding, undermines the legitimacy of all the inferences upon which critical realism relies in order to achieve indirect knowledge of the existence and properties of the things in themselves. The Kantian doctrine excludes a spatio-temporal theater in which both things in themselves and phenomena are located, and it is incompatible with a causal treatment of the relation between these two types of entities— e.g., it precludes regarding phenomena as mental events that are aroused in the knowing subject (itself a thing in itself) by its interaction with other things in themselves. No claim is being made here that the Kantian doctrine of space, time, and the categories holds up under critical scrutiny; indeed, when the synthetic *a priori* judgments which Kant adduces in geometry and in pure natural science are undermined, his argument for the ideality of space and time and for the imposition of the categories by the understanding loses its force. Nevertheless, the power and coherence of Kant's epistemology may be acknowledged by the following conditional statement: if his doctrine of the origin of space, time, and the categories were correct, then the impossibility in principle of human theoretical knowledge of the properties of the things in themselves would follow. The solidity of this conditional statement makes Kant the exemplary exponent of the tradition in which epistemology is developed in deliberate abstention from ontology.

Since Bohr rejects Kant's affirmations of synthetic *a priori* knowledge, one may properly ask how Bohr can prove in principle the illegitimacy of the essential reasoning of critical realism. Bohr nowhere explicitly confronts this question as it has just been posed. There are, however, many statements, particularly about complementarity, which can be construed as answering his own reformulation of the question. For example, in a late and carefully written article Bohr says,

In quantum physics, however, evidence about atomic objects obtained by different experimental arrangements exhibits a novel kind of complementary relationship. Indeed, it must be recognized that such evidence which appears contradictory when combination into a single picture is attempted, exhausts all conceivable knowledge about the object. Far from restricting our efforts to put questions to nature in the form of experiments, the notion of *complementarity* simply characterizes the answers we can receive by such inquiry, whenever the interaction between the measuring instruments and the objects forms an integral part of the phenomena.[25]

This passage contains no philosophical locutions like "properties of things in themselves," but there is a surrogate for them in Bohr's phrase "combination into a single picture," and the argument which he presents for the impossibility of such a picture can be construed as a reason for rejecting in principle the feasibility of the critical realism.

It must be emphasized, however, that Bohr's argument is limited in scope. It is directed against a "single picture" of the kind that is drawn in classical mechanics, in which all the properties of a physical object are simultaneously assigned definite values. Nothing in his argument precludes a "single picture" of the kind envisaged in the realistic interpretation of quantum mechanics presented at the end of Section 2, as a reasonable extension of Einstein's physical realism. In this picture some properties of a physical object are actual, but others (and indeed most) have the status of potentialities. Bohr never seems to consider this possibility explicitly, and his arguments do not suffice explicitly or implicitly to rule it out. Remarkably, however, Heisenberg—who at one time was close to Bohr—recognizes the possibility of this kind of realistic interpretation of quantum mechanics and somewhat tentatively endorses it:

Kant had pointed out that we cannot conclude anything from the perception about the 'thing-in-itself.' This statement has . . its formal analogy in the fact that in spite of the use of the classical concepts in all the experiments a nonclassical behavior of the atomic objects is possible. The 'thing-in-itself' is for the atomic physicist, if he uses this concept at all, finally a mathematical structure; but this structure is—contrary to Kant—indirectly deduced from experience.[26]

At the conclusion of Section 2 a fundamental difficulty of the realistic interpretation of quantum mechanics was mentioned: that is, the difficulty of explaining the actualization of potentialities within the framework of quantum mechanics. Much effort is currently being devoted to this problem. Among the proposals under examination are nonlinear variants of the dynamical law of quantum mechanics, stochastic variants of quantum mechanical dynamics, non-quantum mechanical behavior of macroscopic bodies, and non-quantum mechanical behavior of the space-time field itself.[27] No decisive progress has been achieved along any of these lines, but it is premature to conclude that these efforts are doomed to fail. One can say conditionally, however, that if sustained efforts to provide a realistic account of the actualization of quantum mechanical potentialities should prove to be unsuccessful, then there would be *a posteriori* reasons to conclude that the program itself is misconceived and to return to Bohr. He would presumably object to the locution "actualization of potentialities," just as he objected, in the first passage above, to locutions like "creation of physical attributes of objects by measurement." It is conceivable that Bohr's ideas will be vindicated in this roundabout way. But a precondition to a full vindication of Bohr's philosophy is to fill the lacunae of his exposition and to transform his suggestions into a systematic and coherent world view.

References

[1] A. Einstein, B. Podolsky, and N. Rosen, "Can quantum-mechanical description of physical reality be considered complete?", Physical Review **47**, 777–780 (1935).

[2] N. Bohr, "Can quantum-mechanical description of physical reality be considered complete?", Physical Review **48**, 696–702 (1935).

[3] P.A. Schilpp, ed., *Albert Einstein: Philosopher-Scientist* (Evanston, IL: The Library of Living Philosophers, Inc., 1949).

[4] J.S. Bell, "On the Einstein Podolsky Rosen paradox," Physics **1**, 195–200 (1964).

[5] S.J. Freedman and J.F. Clauser, "Experimental test of local hidden variable theories," Physical Review Letters **28**, 938–941 (1972); E.S. Fry and R.C. Thompson, "Experimental test of local hidden-variable theories," Physical Review Letters **37**, 465–468 (1976); A. Aspect, J. Dalibard, and G. Roger, "Experimental test of inequalities using variable analyzers," Physical Review Letters **49**, 1804–1807 (1982). Other experiments are reviewed in J. Clauser and A. Shimony, "Bell's Theorem: experimental tests and implications," Reports on Progress in Physics **41**, 1881–1927 (1978).

[6] See, for example, the article of Clauser and Shimony in Ref. 5, and also B. d'Espagnat, "Nonseparability and the tentative description of reality," Physics Reports **110**, 202–264 (1984). For a contrary opinion see T.

Marshall and E. Santos, letter to Physics Today **38**, no. 11, 9–11 (1985) and references given there.

[7] J. Jarrett, "On the physical significance of the locality conditions in the Bell arguments," Nous **18**, 569–589 (1984). The terminology and notation used in the text differ from Jarrett's.

[8] P.H. Eberhard, "Bell's Theorem and the different concepts of locality," Il Nuovo Cimento **46B**, 392–419 (1978); G.C. Ghirardi, A. Rimini, and T. Weber, "A general argument against superluminal transmission through the quantum mechanical measurement process," Lettere al Nuovo Cimento **27**, 293–298 (1980).

[9] W. Heisenberg, *Physics and Philosophy* (New York: Harper and Bros., 1958), 185.

[10] A. Einstein, "Quantenmechanik und Wirklichkeit," Dialectica **2**, 320–324 (1948), translated by Don Howard.

[11] A. Einstein, *The Meaning of Relativity*, fifth edition (Princeton, N.J.: Princeton University Press, 1955), 165–6.

[12] E. Schrödinger, "The present situation in quantum mechanics," in *Quantum Theory and Measurement*, ed. J.A. Wheeler and W. Zurek (Princeton, N.J.: Princeton University Press, 1983), translated from a German article of 1935.

[13] N. Bohr, *Atomic Physics and Human Knowledge* (New York: Science Editions, 1961), 60–61.

[14] This aspect of Bohr's thought is emphasized by P. Zinkernagel, *Conditions for Description* (London: Routledge & Kegan Paul, 1961), and by A. Petersen, *Quantum Physics and the Philosophical Tradition* (Cambridge MA: MIT Press, 1968).

[15] N. Bohr, *Essays 1958–1962 on Atomic Physics and Human Knowledge* (New York: Vintage Books, 1966), 5.

[16] N. Bohr, Ref. 13, 50.

[17] Ibid., 91–92.

[18] Ibid., 78–79.

[19] N. Ibid., 11.

[20] Among these are the books mentioned in Ref. 14; and also P. Feyerabend, "Problems of Microphysics," in *Frontiers of Science and Philosophy*, ed. R. Colodny (Pittsburgh PA: University of Pittsburgh Press, 1962); K.M. Meyer-Abich, *Korrespondenz, Individualität und Komplementarität* (Wiesbaden: Steiner, 1965); L. Rosenfeld, "Strife about complementarity." Science Progress **41**, 393–410 (1953); C.A. Hooker, "The nature of quantum mechanical reality: Einstein versus Bohr," in *Paradigms and Paradoxes*, ed. R. Colodny (Pittsburgh PA: University of Pittsburgh Press, 1972); M. Jammer, *The Philosophy of Quantum Mechanics* (New York: Wiley, 1974); J. Honner, "The transcendental philosophy of Niels Bohr," Studies in History and Philosophy of Science **13**, 1–29 (1982); and H. Folse, *The Philosophy of Niels Bohr* (Amsterdam: North-

Holland, 1985), reviewed by A. Shimony, Physics Today **38**, no. 10, 108–109 (1985).

[21] In Ch. 8, sect. 2 of Folse's book, Ref. 20, there is a good compilation of quotations from Bohr confirming this interpretation.

[22] A. Petersen, Ref. 14, 157.

[23] Ibid., 63.

[24] Ibid., 188.

[25] N. Bohr, Ref. 15, 4.

[26] W. Heisenberg, Ref. 9, 91.

[27] A survey of these proposals is given in A. Shimony, "Events and Processes in the Quantum World," In *Quantum Concepts of Space and Time*, ed. C. Isham and R. Penrose (Oxford: Oxford University Press, 1986).

The Soviet Reaction to Bohr's Quantum Mechanics

Loren R. Graham

MASSACHUSETTS INSTITUTE OF TECHNOLOGY

Soviet attitudes toward Niels Bohr and the Copenhagen School of physics can be divided into two phases which I will call "The Early Dogmatic Phase" and "The Continuing Philosophical Critique." By categorizing Soviet reactions to Bohr in this rather schematic way I am, of course, somewhat simplifying the historical record. Many Soviet physicists would say, for example, that the earlier Stalinist criticism of quantum mechanics has disappeared entirely, and that I am distorting the record by speaking of a "continuing philosophical critique." From the standpoint of the working physicist, this remark is quite understandable; quantum mechanics in the Soviet Union is the same as quantum mechanics everywhere. In fact, Soviet physicists used the mathematical apparatus of quantum mechanics throughout the entire Stalinist period, despite the ideological quarrels that sometimes surrounded the topic. I wish to show in this paper, however, that even today Soviet literature in philosophy of science continues to contain criticisms of some of the leading members of the Copenhagen School of physics, including Bohr, despite the fact that these same Soviet authors accept entirely both the mathematical core and the physical theory at the base of quantum mechanics. Several of the points being made by Soviet authors in this more recent period deserve more serious consideration than they have so far been given.

The Early Dogmatic Phase

Under Stalin, ideological interference with science was a very common phenomenon in the Soviet Union. Starting in the early nineteen thirties Soviet ideologists frequently attacked scientists they described as being under the influence of "bourgeois" interpretations of scientific theories.[1] As materialists, Soviet philosophers sharply opposed any interpretation of nature that could be described as idealistic or subjective; as non-reductionists, they castigated as "vulgar materialists" those scientists who attempted to explain complex phenomena like living organisms in terms of elementary physical laws; as determinists, they were ardent defenders of the principle of causality in nature; and as believers in the existence of unitary objective truth about the world, they criticized as "agnostics" or "relativists" those people who either doubted that scientists could attain truth or those who believed that there exist multiple forms of truth.

As Stalinism deepened before and immediately after World War II, the intellectual content of these discussions became thinner and thinner, and the political element ever more obvious and dominant. Soviet dialectical materialists in this period frequently used ideological arguments for personal advantage rather than for intellectual clarification. The most graphic example of this upsurge of political elements and demotion of intellectual ones was not in physics but in biology, where Trofim Lysenko rose to power in the late nineteen forties. As almost everyone knows, this ideological and political campaign culminated in 1948 with the suppression of genetics in the Soviet Union. Not until 1965 was Lysenko's hold on Soviet biology broken.[2]

There were moments in Soviet history between 1930 and 1953 when it appeared that something similar might happen to quantum mechanics.[3] The fact that it did not is a result of several factors: the greater distance of quantum mechanics than biology from the concerns of political leaders and ideologists; the extremely intelligent and vigorous resistance of a few Soviet physicists to ideological interference; and the dramatic elevation in political influence of physicists in Soviet society after it became apparent that they could build nuclear weapons.

In a book entitled *Science and Philosophy in the Soviet Union* I have described these events in detail, and I do not wish to repeat them here.[4] Instead, I would like to concentrate on topics that I did not cover in my book, especially the critique of some of Niels Bohr's philosophical positions that has continued in the Soviet Union to the present day. Before doing so, however, I would like to summarize the criticisms advanced against quantum mechanics in the heyday of Stalinism, and indicate briefly the successful responses to these criticisms that Soviet physicists and the more enlightened Soviet philosophers developed.

The first Stalinist criticism of quantum mechanics was that it denied causality and therefore deprived any world view based on science of a secure intellectual framework of predictable cause-and-effect relationships.[5] As Stalin's chief ideologist A.A. Zhdanov declared in a now-infamous speech in 1948, "The Kantian vagaries of modern bourgeois atomic physicists led them to inferences about the electron's possessing 'free will' . . . and to other devilish tricks."[6] The response which the Soviet physicists developed to this critique was that, contrary to certain exaggerated statements on causality made by Western physicists like Bohr and Heisenberg, quantum mechanics did not deny causality at all; it merely redefined causality. A true denial of causality, continued the Soviet physicists, would mean that when one tried to predict the future position of an electron in space, all possible future positions would have equal probability. This is far from the case, they noted, as anyone who uses a television set should realize very well, since the cathode ray tube at the heart of a television receiver controls beams of electrons with amazing accuracy. Only at the microlevel, they continued, does quantum mechanics introduce an element of indeterminacy, and this element is so small that it does not interfere with man's control over nature or his ability to construct a world view based on causal principles. The problem, the Soviet physicists continued, is not that quantum mechanics denies causality, but that Western physicists in their enthusiasm for quantum mechanics have not been very interested in explaining to lay people how causality and determinism can be redefined in terms of probability, and how potent causality in nature still is.[7]

Another favorite target of Stalinist critics of quantum mechanics was the concept of complementarity.[8] To the dogmatic philosophers, complementarity was a denial of the principle of unitary truth; they insisted that a microbody could not be both a particle and a wave, but one or the other. What is the truth here? The reply of the Soviet physicists was that truth surely exists in quantum mechanics as it does everywhere else, and that microbodies in quantum mechanics are in truth neither particles nor waves; they are instead material forms of existence for which the terms "particle" and "wave" are inappropriate, but which we have to use nonetheless because we have no better terms available. Complementarity, they continued, is not the denial of the existence of unitary truth, but the assertion that when one uses inappropriate terms to try to describe micro-bodies, understandably one cannot adequately describe them. Microbodies are just as objectively real as all other material bodies in the world, continued the Soviet physicists, and our acquisition of ever-greater unitary truth about them should not be obscured by the fact that our everyday language does not fit them well.

The defense of complementarity in the Soviet Union turned out to be more difficult than the defense of causality; from 1948 to 1960 the term was actually banished from Soviet publications. The one exception to this rule was the eminent Soviet physicist Vladimir Fock, who was a dedicated Marxist who refused to allow professional philosophers to define his Marxism for him; he continued to use the term "complementarity" with reference to physics for a few years after 1948, but even he eventually abandoned the term under pressure. The reason that he gave for his retreat was intellectually interesting. He wrote in 1951:[9]

At first the term complementarity signified the situation that arose directly from the uncertainty relation: Complementarity concerned the uncertainty in coordinate measurement . . . . and the term 'principle of complementarity' was understood as a synonym for the Heisenberg relation. Very soon, however, Bohr began to see in his principle of complementarity a certain universal principle . . . applicable not only in physics but even in biology, psychology, sociology, and all the sciences. . . . To the extent that the term 'principle of complementarity' has lost its original meaning . . . it would be better to abandon it.

We see from the above passage that Fock was willing to defend complementarity if the term were restricted to physics, where it could be explained as the result of trying to describe phenomena in the microworld with terms taken from the macroworld, but he would not defend complementarity if it were extended to social phenomena. On occasion, Bohr did just this, applying complementarity to such subjects as anthropology and religion. He said in 1938, for example, that complementarity provided anthropologists with a way to appreciate cultures with contradictory social practices.[10] At another time he spoke of the complementary one-sided views of spiritual truth and scientific truth.[11] Fock refused to follow Bohr down these paths away from physics because he knew that if Bohr's statements were taken literally, they led away from the concept of unitary social truth that was a part of Soviet Marxism. No Soviet Marxist could agree that religion and science each provide a piece of truth, nor could such a Marxist take the relativist view in anthropology that Bohr's comments seemed to favor.

After 1960, the term complementarity returned to Soviet physics and philosophy of science, but, in line with Fock's objections, it was always applied strictly to quantum mechanics, not to society. Soviet commentators on complementarity often even today refer approvingly to Einstein's jibe that when Bohr begins to think about complementarity "he thinks of himself as a prophet." [12]

The Continuing Philosophical Critique

During the last twenty years Soviet analysts of quantum mechanics have abandoned their criticism of the physical theory itself and they have even accepted most of Bohr's philosophical interpretation. After redefining and defending causality and determinism in terms that take quantum mechanics into account, they have even performed the rather breathtaking maneuver of maintaining that quantum mechanics is a brilliant illustration of dialectical materialism rather than a challenge to it. M.E. Omel'ianovskii maintained in the late seventies that "quantum mechanics is once more confirming all the statements of dialectical materialism." [13] At the same time, however, most of the philosophers have continued to maintain that Bohr and his colleagues made mistakes in their early interpretations of the significance of quantum mechanics. The most important continuing criticism centers on the concept of "uncontrollable influence" during measurement of microbodies.

The idea that the effect of measurement of a microbody cannot be reduced to zero is characteristic of most discussions of quantum mechanics. Often this effect is stated in the following way: The greater the accuracy in determining a particle's position, the less the accuracy in determining its momentum, and conversely. Sometimes the way in which this phenomenon manifests itself is described by saying that measurement exerts an influence on the microbody which is in principle indeterminable and that this influence puts the microbody into a new state.

It is exactly this concept of the "uncontrollable influence of the measuring instrument on the microbody" that has been the target of recent Soviet criticism of the Copenhagen Interpretation. A very common Soviet view is that this approach attributes less reality to the microbody than it does to the measuring instrument and therefore undermines a philosophical principle to which they are committed, namely, the objective reality of all aspects of nature which physicists study; the Soviet critics insist that objective reality is shared equally by both the measuring instrument and the microbody.

The two most prominent Soviet interpreters of quantum mechanics during the last twenty years, the physicist V.A. Fock and the philosopher M.E. Omel'ianovskii, maintained that the first step toward a more philosophically rigorous interpretation of quantum mechanics was the recognition that the term "uncontrollable influence of the measuring instrument" is inaccurate because it often leads one to believe that the measuring instrument actually exerts a physical force on the microbody.[14] They asserted that in the classical instances cited in quantum mechanics, no such force is exerted. They asked us to consider the following situation: If we think of a crystalline lattice as the measuring instrument for an electron, before passing through the lattice the electron is located in a state with a definite momentum and an indefinite position; after passing through the lattice, the electron is in a state with a definite position and an indefinite momentum. Measurement therefore changes the state of the micro-object, but this change is not a result of a force acting on the micro-object. Indeed, no physicist in describing the passage of an electron through a crystalline lattice speaks of the action of a normal force known to physicists, such as gravitational or electromagnetic force, on the electron. The influence of the crystalline lattice is not that it exerts an uncontrollable physical force on the electron, but that it changes the frame of reference within which we view the electron. In several articles and in a 1977 book Omel'iianovskii picked up this concept of "frame of reference" by making an analogy between measurement in quantum mechanics and change of reference frames in classical mechanics. As he wrote, "The change of quantum state under the influence of measurement is similar to the change of mechanical state of a body in classical theory when one makes the transition from one system of reference to another moving relative to the first."[15] Thus, a falling body in a moving train is described as descending in a straight line by a passenger in the train; it is described as descending along a parabolic curve to a person standing outside the train. Analogically, said Omel'ianovskii, an electron that is approaching a crystalline lattice is described by the physicist with a indefinite position and a definite momentum before it reaches the lattice; after passing through the lattice, the electron is described with a definite position and an indefinite momentum. In neither the case of the falling body in the train nor the electron passing through the lattice, observed Omel'ianovskii, were physical forces responsible for the different descriptions; instead a change in reference frames resulted in contrasting descriptions. Furthermore, Omel'ianovskii insisted that there was equal justification for describing both the falling body in the train and the electron as objectively-real entities despite the different physical descriptions that can be assigned to them in different situations.

The Soviet physicist V.A. Fock advanced a similar criticism of the concept of "uncontrollable action of the measuring instrument" on the microbody. In pointing out what he considered to be the unsatisfactory character of Bohr's term "uncontrollable interaction," Fock remarked:[16]

As a matter of fact, it is a matter here not of an interaction in the proper sense of the term but of the logical connection between the quantum and classical methods of description at the junction between that part of the system which is described in quantum-mechanical terms (the object) and that part which is described in classical terms (the measuring instrument).

In February and March 1957 Fock visited Bohr in Copenhagen where they held a series of conversations on the philosophic significance of quantum mechanics. The discussions took place both in Bohr's home and at his Institute of Theoretical Physics. Fock later reported on the conversations in the following way:

From the very beginning Bohr said that he was not a positivist and that he attempted simply to consider nature exactly as it is. I pointed out that several of his expressions have ground for an interpretation of his views in a positivistic sense that he, apparently, did not wish to support. . . . Our views constantly came closer together; in particular it became clear that Bohr completely recognized the objectivity of atoms and their properties . . . ; he further said that the term 'uncontrollable mutual influence' was unsuccessful and that actually all physical processes are controllable.[17]

Soviet writers have declared that from this time onward, Bohr's views became acceptable to them, and they have claimed at least partial credit for the alleged change. Fock, for example, said that "After Bohr's correction of his formulations, I believe that I am in agreement with him on all basic items."[18] And he further claimed that these "corrections" that Fock led Bohr to make "found their reflection in his article 'Quantum Physics and Philosophy' . . ."[19] Omel'ianovskii similarly maintained that Fock's criticisms caused

Bohr to make "a definite advance toward a materialist approach to quantum mechanics."[20]

This version of events is misleading. It portrays Soviet dialectical materialists as having been correct all along about quantum mechanics, and it describes Bohr as having "come over to their side." This interpretation fails to take account of the fact that by far the most significant changes in position on quantum mechanics during the last fifty years have been among the dialectical materialists rather than among the supporters of the Copenhagen School. Not only did the dialectical materialists repudiate the concept of complementarity for many years, they even on occasion denied the legitimacy of quantum mechanics as a science. There were moments in the nineteen thirties and forties when it appeared that the dialectical materialist criticism of quantum mechanics might actually result in the suppression of the theory, just as actually happened in the case of genetics. Fortunately, the intellectual tragedy of genetics was not repeated in physics.

On the specific point of Bohr's attitude toward the concept of "uncontrollable influence," however, the Soviet commentators have made claims that deserve to be examined more carefully than they so far have been. It *does* appear that Bohr changed his mind on this subject. Writing in 1935, Bohr spoke of "the impossibility, in the field of quantum theory, of accurately controlling the reaction of the object on the measuring instruments, i.e., the transfer of momentum in case of position measurements, and the displacement in case of momentum measurements."[21] Elsewhere in the same essay he wrote that "the impossibility of controlling the reaction of the object on the measuring instruments" entails "a final renunciation of the classical ideal of causality and a radical revision of our attitude towards the problem of physical reality."[22]

Bohr's emphasis on the interaction of the measuring instrument with the microbody and his criticism of the concept of causality continued for many years. In his 1949 article "Discussion with Einstein on Epistemological Problems in Atomic Physics," he spoke many times of the "unavoidable" or "uncontrollable" interaction between the instrument and micro-objects, as well as the "momentum exchange" between the electron and the measuring agency.[23] He also wrote that this interaction forces a "renunciation" of causal description.[24]

If we now jump ahead in time from Bohr's 1935 and 1949 essays to his 1958 one entitled "Quantum Physics and Philosophy" (the one Fock claimed to have influenced), we find a different tone. In the later essay, Bohr did not use the concept "uncontrollable interaction." In addition, he added that the description of atomic phenomena "has in these respects a perfectly objective character, in the sense that no explicit reference is made to any individual observer."[25] And, on the subject of causality, he concluded "far from involving any arbitrary renunciation of the ideal of causality, the wider frame of complementarity directly expresses our position as regards the account of fundamental properties of matter presupposed in classical physical description, but outside its scope."[26]

Soviet philosophers have described this essay as a fundamental change of position by Bohr, but Bohr's son Aage has explained it in somewhat different terms.[27] Writing shortly after his father's death, Aage Bohr said that "My father felt that, in this little article (the 1958 one—LRG), he had succeeded in formulating some of the essential points more clearly and concisely than on earlier occasions."[28] The question before us is, therefore, the following: Does the 1958 article represent a change in philosophical position when compared to the 1935 and 1949 ones, as the leading Soviet interpreters of quantum mechanics have maintained, or is it merely a clarification, as indicated by Aage Bohr? Historians and philosophers of science need to investigate this question much more deeply than they have so far done.

More research on this issue is in process, but it already seems rather clear to me that some transition did occur in Bohr's thought, a transition away from emphasis on the interaction of the measuring instrument and the micro-object as the key to quantum mechanics, away from a renunciation of causality, and toward a greater recognition of the physical reality of the microbodies of quantum mechanics.

What is not clear, however, is whether Bohr's discussions with Fock or his knowledge of Marxist critiques of the Copenhagen School had anything to do with his changes of emphases. I am doubtful that any such influence can ever be proved, although it is possible that examination of Bohr's papers at the time of Fock's visit might be helpful in clarifying what happened. In the meantime, we are justified in taking a somewhat skeptical view of the Soviet interpretation that pictures Bohr as correcting his positions under the influence of Soviet Marxist criticism. As early as 1938, long before Bohr's crucial conversations with Fock, Bohr warned against using phrases in discussions of quantum mechanics such as "disturbing of phenomena by observation" and "creating physical attributes to atomic objects by measurement."[29] Throughout this period, however, and on up until at least 1949, Bohr continued to speak of the "uncontrollable interaction between the objects and the measuring instruments," as the Soviet critics have been at pains to point out.[30]

It seems likely that the transitions in Bohr's thought in the last years of his life were the result of a multitude of influences, including the general decline in the persuasiveness of positivism, rather than being the consequence of a particular event or intellectual influence. Furthermore, it was only natural that after the great achievements of quantum mechanics were firmly established that its pioneers would find some of their earlier philosophical statements somewhat overblown. During the heroic days of the establishment of quantum mechanics it was helpful to be willing to question the basic assumptions of most physicists, including causality and objective reality. At one point, in 1924, Bohr even entertained the abandonment of the conservation of energy and momentum in radiation processes, a hypothesis that he later readily admitted was mistaken.[31] In the last decades of his life, after the structure of quantum mechanics was almost universally regarded as complete, Bohr may have seen a useful role in the reverse function of showing how much of the earlier world could still be retained, both in terms of physics and in terms of philosophy.

When Bohr debated Einstein about quantum mechanics at the Solvay conferences in the late twenties and thirties and again in print in the thirties and forties the question at issue was whether quantum mechanics was complete, or whether it would be eventually replaced by another theory permitting more exact description in classical terms.[32] In the context of this debate with Einstein, Bohr emphasized the radical nature of quantum mechanics and the impossibility of a return to classical description.

The criticism from Fock in the late nineteen fifties came from a different direction, since Fock agreed entirely with Bohr that quantum mechanics was complete, physically speaking. What Fock sought was not a revision in quantum mechanics, but a revision of the philosophical terms in which Bohr described quantum mechanics. As Fock observed in 1963, "I never doubted that Bohr was essentially correct" so far as the physics went; Fock merely wanted to separate Bohr's physics from "the positivistic coating that at first glance seemed to be intrinsic to it."[33] Fock wanted Bohr to reformulate his views in a way that would drop emphasis on "uncontrollable interaction'' and would grant the continuing importance of terms like "causality" and "objective reality." Bohr moved in this direction sometime between 1949 and 1958, but for reasons that are still not clear.

The value of the retention of as much of classical physics as possible was at the base of Bohr's "correspondence principle" developed in the early twenties, and therefore obviously not a new thought to him in his later life. To extend the correspondence principle from physics to philosophy was not a major step. It is quite reasonable to think that Bohr's greater attention in the last part of his life to traditional philosophical concepts like objective reality and causality could have occurred without the particular stimulus of Fock's criticism of his views. But the question of just how much Bohr changed and whether Fock or other Marxist critics had anything to do with the changes remains open.

References

1. See David Joravsky, *Soviet Marxism and Natural Science, 1917–1932*, New York, 1961; Gustav Wetter, *Dialectical Materialism: A Historical and Systematic Survey of Philosophy in the Soviet Union*, New York 1958; Loren Graham, *The Soviet Academy of Sciences and the Communist Party, 1927–1932*, Princeton, 1967.

2. See Joravsky, *The Lysenko Affair*, Cambridge, 1970; Zhores Medvedev, *The Rise and Fall of T.D. Lysenko*, New York and London, 1969; and Graham, *Science and Philosophy in the Soviet Union*, New York, 1972, pp. 195–256.

3. One period of particular difficulty was the Cultural Revolution in the years 1929 and 1930, when both quantum mechanics and relativity physics came under heavy attack. A Marxist physicist who resisted these incursions particularly effectively was Boris Hessen. See Loren Graham, "The Socio-Political Roots of Boris Hessen," *Social Studies of Science*, forthcoming. In the years 1947–53 ideological currents were also very strong, but by this time atomic physicists had much greater influence because of their role in the development of atomic weapons.

4. Loren Graham, *Science and Philosophy in the Soviet Union*, New York, 1972. A revised and updated edition of this book is *Science, Philosophy and Human Behavior in the Soviet Union*, New York, 1987. The footnotes in these works give a much fuller representation of Soviet reactions to quantum mechanics than I am able to present in this short essay.

5. See D.I. Blokhintsev, *Printsipal'nye voprosy kvantovoi mekhaniki*, Moscow, 1966; V.A. Fock, "Ob interpretatsii kvantovoi mekhaniki," in P.N. Fedoseev et al., eds, *Filosofskie problemy sovremennogo es-*

testvoznaniia, Moscow, 1959, pp. 212–236; and "Lenin o prichinnosti i kvantovaia mekhanika," *Vestnik akademii nauk SSSR*, No. 4 (1958), pp. 3–12.

6. A.A. Zhdanov, *Vystuplenie na diskussii po knige G.F. Aleksandrova 'Istoriia zapadnoevropeiskoi filosofii' 24 iiunia 1947 g.*, Moscow, 1951, p. 43.

7. See the discussion in Graham, *Science and Philosophy in the Soviet Union*, pp. 69–110.

8. M.E. Omel'ianovskii, "Dialekticheskii materializm i tak nazyvaemyi printsip dopolnitel'nosti Bora," in A.A. Maksimov et al., eds., *Filosofskie voprosy sovremennoi fiziki*, Moscow, 1952; for a later view, see L.B. Bazhenov et al., eds., *Printsip dopolnitel'nosti i materialisticheskaia dialektika*, Moscow, 1976.

9. V.A. Fok, "Kritika vzgliadov Bora na kvantovuiu mekhaniku," *Uspekhi fizicheskikh nauk*, XLV No. 1 (September, 1951), p. 13.

10. N. Bohr, "Natural Philosophy and Human Cultures," *Nature* (February 18, 1939), vol. 143, p. 271. For a discussion of how this essay needs to be seen as a criticism of racist views in anthropology, especially in Nazi Germany, see Loren Graham, *Between Science and Values*, New York, 1981, p. 57.

11. N. Bohr, "Science and the Unity of Knowledge," in Lewis Leary, ed., *The Unity of Knowledge*, Garden City, N.Y., 1955, pp. 58–61; also, see N. Bohr, "Physical Science and the Study of Religions," in *Studia Orientalia Ioanni Pedersen septuagenario A.D. VII ID. NOV. ANNO MCMLIII*, Einar Munksgaard, 1953, p. 389.

12. Quoted in R.S. Shankland, "Conversations with Albert Einstein," *American Journal of Physics* (1963), vol. 31, p. 50.

13. M.E. Omelyanovsky (sic), *Dialectics in Modern Physics*, Moscow, 1979, p. 144.

14. M.E. Omel'ianovskii, "Filosofskie aspekty teorii izmereniia," in *Materialisticheskaia dialektika i metody estestvennykh nauk*, Moscow, 1968, pp. 207–255; V.A. Fok, "Ob interpretatsii kvantovoi mekhaniki," pp. 219–222.

15. M.E. Omel'ianovskii, "Filosofskie aspekty teorii izmereniia," p. 248; also, see his *Dialektika v sovremennoi fizike*, Moscow, 1977.

16. Quoted in Omelyanovsky, *Dialectics in Modern Physics*, p. 312.

17. V.A. Fok, "Nil's Bor v moei zhizni," *Nauka i chelovechestvo 1963*, Vol. II, Moscow, 1963, pp. 518–519.

18. V.A. Fok, "Ob interpretatsii kvantovoi mekhaniki," p. 235. In 1965 Fock wrote in the following way of his approving but nonetheless critical approach to Bohr's interpretation: *Le mérite d'une nouvelle position du problème de la description des phénomènes à l'échelle atomique appartient à Niels Bohr; le point de vue adopté dans le présent article est le résultat de nos recherches et méditations ayant pour but d'appro-*

fondir, de préciser—et si nécessaire de critiquer et de corriger—les idées de Bohr. V. Fock, "La physique quantique et les idéalisations classiques," *Dialectica*, No. 3-4 (1965), p. 223.

19. V.A. Fock, "Zamechaniia k stat'e Bora o ego diskussiakh s Einshteinom," *Uspekhi fizicheskikh nauk*, LXVI, No. 4 (December, 1958), p. 602.

20. Omelyanovsky, *Dialectics in Modern Physics*, p. 50.

21. N. Bohr, "Can Quantum-Mechanical Description of Physical Reality be Considered Complete?", *Physical Review*, Vol. 48 (October 15, 1935), p. 699.

22. *Ibid.*, p. 697.

23. N. Bohr, "Discussion with Einstein on Epistemological Problems in Atomic Physics," in Paul Arthur Schilpp (ed.), *Albert Einstein: Philosopher-Scientist*, Evanston, 1949, pp. 209, 210, 211, *passim*.

24. *Ibid.*, p. 202, 203, 204, 211.

25. Niels Bohr, "Quantum Physics and Philosophy: Causality and Complementarity," in his *Essays 1958-1962 on Atomic Physics and Human Knowledge*, New York, 1963, p. 3.

26. *Ibid.*, p. 6.

27. See, especially, Omelyanovsky, *Dialectics in Modern Physics*, pp. 47, 50, 54, 57, 58, and 311-313.

28. Aage Bohr, "Preface," in N. Bohr, *Essays 1958-1962 on Atomic Physics and Human Knowledge*, p. vi.

29. Quoted in Bohr, "Discussion with Einstein . . . ," in Schilpp, *op. cit.*, p. 237, from *New Theories in Physics*, Paris, 1938, p. 11.

30. Bohr, "Discussion with Einstein . . . ," p. 211.

31. The original article was N. Bohr, H.A. Kramers, and J.C. Slater, "The Quantum Theory of Radiation," *Philosophical Magazine* (May, 1924), 74, pp. 785-802. Bohr acknowledged that this approach was "unpromising" (*aussichtlos*) in "Über der Wirkung von Atomen bei Stössen," *Zeitschrift für Physik*, No. 34, pp. 156-157.

32. See Bohr, *Discussion with Einstein . . .*

33. Fock, "Nil's Bor v moei zhizni," pp. 518-519.

Niels Bohr and the First Principles of Arms Control

Martin J. Sherwin

TUFTS UNIVERSITY

Four months after the end of World War II, in response to a massive lobbying effort by the atomic scientists, President Truman ordered the Department of State to develop a proposal for the international control of atomic energy. That plan, generally known as the Acheson-Lilienthal Report, was formulated, designed, and probably even largely written by J. Robert Oppenheimer, a consultant to the Lilienthal committee. But the basic principles, and even some of the central ideas in that report, were conceived by Niels Bohr.[1]

In 1964 Oppenheimer returned to Los Alamos to give a public address on "Niels Bohr and the Atomic Bomb"—an explicit acknowledgment of the great physicist's magisterial place among the scientific authorities of this century.[2] But Oppenheimer's decision to discuss Bohr's unsuccessful attempts to influence wartime atomic energy policy carried other messages as well: it symbolically connected Oppenheimer's own political activities and defeats during the Cold War to Bohr's failed arms control efforts during World War II, and it acknowledged the intellectual debt that atomic scientists, and Oppenheimer in particular, owed to Bohr in the area of nuclear arms control.

Niels Bohr was the first person not only to advocate nuclear arms control, but also to formulate its fundamental principles and work to have them adopted by Western governments. In ways not generally recognized, he set the intellectual agenda, both theoretically and politically, for the postwar nuclear arms control movement.

Above all else Bohr recognized two characteristics of the nuclear problem that are still imperfectly understood:

The timing of an initiative can be more critical than its details.

The greatest spur to the nuclear arms race—and the deadliest enemy of the arms negotiation process—is not the fear of nuclear attack but the anticipation of nuclear leverage.

After escaping late in 1943 from Nazi-occupied Denmark, Bohr tried to convince President Roosevelt and Prime Minister Churchill to open negotiations with Stalin on the international control of atomic energy. Bohr's proposal was not accepted, and for over 20 years scientists who regretted that decision believed that Roosevelt and Churchill had misunderstood Bohr. As Oppenheimer told his audience at Los Alamos, it was "easy, as history has shown, for even wise men not to know what Bohr was talking about." Oppenheimer explained Bohr's unsuccessful efforts to avert a postwar atomic armaments race between the great powers as a tragedy; his obscure prose and barely audible speech were the tragic flaws that prevented sympathetic statesmen from heeding his advice.

But tragedy is history's uncertainty principle and should never be accepted as its final explanation. Since Oppenheimer prepared his talk in the early 1960s, a great deal has been learned about the wartime political and diplomatic decisions surrounding the atomic bomb. It is now clear that Bohr's proposals were rejected not because they were misunderstood, but rather because Roosevelt and Churchill understood them perfectly—and opposed them absolutely. The temptation to use their emerging nuclear advantage after the war left them cold to Bohr's plan for the international control of atomic energy—a proposal that, in effect, called for the voluntary neutralization of a new weapon of heretofore unimagined potential.

As Oppenheimer said: "Officially and secretly [Bohr] came to [Los Alamos] to help the technical enterprise, [but] most secretly of all . . . he came to advance his case and his cause." In the broadest sense, Bohr's cause was to insure that atomic energy "is used to the benefit of all humanity and does not become a menace to civilization."[3] Specifically, he warned that, "quite apart from the question of how soon the weapon will be ready for use and what role it may play in the present war," some agreement had to be reached with the Soviet Union about the future control of atomic energy.[4]

Bohr's ideas on the international control of atomic energy are significant beyond their immediate historical context. He was universally admired by his colleagues, and his influence among them, even on political and social issues, stemmed from his qualities as a human being as much as from his professional achievements. A leader in nuclear physics, Nobel laureate in 1922, founder and director of an internationally famed research institute in his native Denmark, and social conscience of the international community of physicists during the interwar years, Bohr's judgments on political matters always received a respectful hearing and a respectable following. Whether or not other scientists ultimately agreed with his political proposals, they shared the basic view of science from which his political ideas grew. His proposals therefore reveal more than the insights of an individual scientist; they represent an effort to transfer the scientific ideal into the realm of international politics.

In politics as in science Bohr was an activist who believed that every serious person was obliged to confront the historical process, not merely as an observer but as a participant. As he himself put it: "Every valuable human being must be a radical and a rebel for what he must aim at is to make things better than they are."[5]

Bohr's concerns and his understanding of the discoveries that he had done so much to introduce into the world led him, in 1944, to reject traditional approaches to international relations for the postwar years. Simply put, he believed that the development of the atomic bomb necessitated a new international order. At the heart of his proposal were a scientist's distrust of secrecy and the recognition that the existence of nuclear weapons demanded an environment in which each nation could be confident that no potential enemy was stockpiling these weapons. He therefore urged Roosevelt to consider "any arrangement which can offer safety against secret preparations."[6]

There were no historical precedents to encourage Bohr, but the threat of atomic warfare was also unprecedented. What was necessary, he once remarked to Oppenheimer in jest, referring to the quantum theory, was "another experimental arrangement."[7] But in a deeper sense he was not jesting, for his proposal called for a political quantum leap into the nuclear age. In 1944, such a leap seemed possible to Bohr; indeed, if the world were to survive, he considered it essential.

A close examination of Bohr's plans for international control of atomic energy is necessary to gain a clear understanding of what he proposed to Churchill and Roosevelt during the spring and summer of 1944.

The problem that Bohr foresaw was an atomic arms race after World War II. He never opposed the wartime use of the bomb. "What role it may play in the present war," he wrote to Roosevelt, was a question "quite apart" from the postwar issue.[8] Looking beyond the war, however, Bohr argued that an agreement for international control could be accomplished only by promptly inviting Soviet participation in atomic energy planning, *before the bomb was a certainty* and before the war was over.

What ought to be done during the war, Bohr asked, to make possible the postwar international control of atomic energy? He began his answer with two assumptions: that the bomb was a creation out of proportion to anything else in human experience, and that it could not be monopolized. He therefore concluded that its development would endanger rather than enhance the future security of the United States and Great Britain if it were not effectively neutralized. A world in which rival nations could employ atomic bombs would be in constant danger of total destruction. In such a world traditional concepts of security through military protection no longer applied.

Bohr believed that the bomb limited international alternatives. A world in which each great power could feel confident that no rival nation was producing nuclear weapons was one choice; a world dominated by the constant specter of total destruction was the other. The new weapon was too effective, too destructive to afford a middle ground. Without the guarantee of some form of international control of atomic energy, he concluded, the great powers would inevitably choose to produce nuclear weapons in the expectation of gaining diplomatic advantages. Such a short-sighted policy would plant the seeds of world destruction.

Timing was at the heart of Bohr's proposal. Since the atomic bomb would be the critical factor in the postwar international climate, it was necessary that Stalin be informed about the Manhattan Project before the war ended. The Soviet leader had to be assured that an Anglo-American alliance, based on an atomic monopoly, was not being formed against his country. Discussions had to be initiated before matters proceeded to a point where an approach to the Soviets would appear more coercive than friendly.

While Bohr understood that the initiative he urged did not guarantee the postwar cooperation of the Soviet Union, he also believed that cooperation was impossible unless his proposal, or a similar one, was adopted. He did not ignore the uncertainties, but relative to the stakes at issue, the risks appeared slight. His suggestion was that the Soviets be informed of the existence of the Manhattan Project, but not of the details of the bomb's construction. Should their response to this limited disclosure be positive, then the way was open for further cooperation. "In preliminary consultation [with the Soviets]," he wrote to Roosevelt, "no information as regards important technical developments should, of course, be exchanged: on the contrary, the occasion should be used frankly to explain that all such information must be withheld until common safety against the unprecedented dangers has been guaranteed."[9]

Bohr wanted Roosevelt to offer Stalin a comprehensive arms control agreement: international control of atomic energy in exchange for surrendering traditional secrecy that inhibited verification. A modified version of this concept characterized Secretary of War Henry Stimson's views toward the end of the war. And in June 1946 it emerged again, although in a distorted form, in the U.S. government's plan for international control of atomic energy—the Baruch Plan. Two years earlier, Bohr was emphasizing that such an offer had to be made before the weapon was a certainty, and before the end of the war.

A final point revealed Bohr's faith in the political influence of the international community of scientists, a faith that the Pugwash conferences later justified. He suggested to Roosevelt that "helpful support may perhaps be afforded by the worldwide scientific collaboration which for years had embodied such bright promises for common human striving. On this background personal connections between scientists of different nations might even offer means of establishing preliminary and noncommittal contact." He was certain that among eminent Soviet scientists "one can reckon to find ardent supporters of universal co-operation."[10]

That their influence with the Soviet government on political matters might be nil was a possibility he did not raise. The scientific perspective which led him to predict the course of international affairs also led him to underestimate the extremely difficult political obstacles. These obstacles were not limited to Stalin's suspicions of his Allies; they were first raised by Churchill and Roosevelt, who intended to use the atomic bomb as a diplomatic counter for postwar bargaining with the Soviet Union.

By mid-May 1944 Bohr succeeded in obtaining a meeting with Churchill—but it went very badly. The prime minister had little patience with Bohr's intrusion into his jealously guarded arena of international affairs. The scheduled 30-minute interview was not long underway before Churchill became embroiled in an argument with his science adviser, Lord Cherwell, the only other person present. Bohr was left out of the discussion; frustrated and depressed, he was unable to bring the conversation back to what he considered the most important diplomatic problem of the war. In a last attempt to communicate his anxieties and ideas to the prime minister, Bohr asked if he might forward a memorandum to him. A letter from Bohr, Churchill bitingly replied, was always welcome, but he hoped it would deal with a subject other than politics. Summing up the meeting, Bohr remarked: "We did not even speak the same language."[11]

Churchill was unmoved by Bohr's argument because he rejected its basic assumption: that the bomb could change the very nature of international relations. To accept it the prime minister would have had to substitute Bohr's analysis for his own. Bohr wanted Churchill to reject, in effect, the very axioms that had guided the prime minister to the wartime leadership of his country. The argument that a new weapon invalidated traditional power considerations ignored every lesson Churchill read into history.

Like Bohr, Churchill believed that the atomic bomb would be a major factor in postwar diplomacy, but he did not believe that it would alter the basic nature of international relations. Nor, apparently, did he believe that international control was practicable regardless of how straightforward and cooperative the United States and Britain might be toward the Soviet government. "You can be quite sure," he wrote less than a year later, "that any power that gets hold of the secret will try to make the article [atomic bomb] and this touches the existence of human society. This matter is out of all relation to anything else that exists in the world, and I could not think of participating in any disclosure to third [France] or fourth [Soviet Union] parties at the present time."[12]

Nothing could persuade Churchill that his nation would be better served by exchanging traditional political and military behavior for a scientist's prediction about the effects on international relations of an extraordinary new weapon. Churchill's perceptions of the future were strongly influenced by his knowledge of the past. The monopoly of the atomic bomb that England and the United States would enjoy after the war would be a significant diplomatic advantage in settling postwar geopolitical rivalries with the Soviet Union. Churchill did not believe that anything could be gained by surrendering that advantage.

By mid-June 1944 Niels Bohr had returned to the United States— less optimistic, but still resolute. An interview with Roosevelt was arranged for August 26, 1944, and Bohr spent many long, hot evenings in his Washington apartment summarizing his ideas for presentation to the president.

To Bohr's initial delight, the meeting was in marked contrast to his confrontation with Churchill. The two men talked for over an hour about the atomic bomb, Denmark, and world politics in general. With respect to the postwar importance of atomic energy, Bohr told his son, Roosevelt agreed that contact with the Soviet Union had to be tried along the lines that Bohr suggested. The president said he was optimistic that such an initiative would have a "good result." Stalin was, he thought, enough of a realist to understand the revolutionary importance of this development and its consequences. The president was also confident that Churchill would come to share these views. They had disagreed before, he said, but in the end they had always succeeded in resolving their differences. Another meeting, Roosevelt suggested, might be useful after he had spoken with Churchill about the matter at the Second Quebec Conference scheduled for early September. In the meantime, if Bohr had any further suggestions, the president would welcome a letter.[13]

Roosevelt's enthusiasm for Bohr's proposals turned out to be insubstantial. The president did not mention them to Churchill until the two leaders discussed atomic energy issues on September 18 at Hyde Park, after the Quebec conference. He then endorsed Churchill's point of view.

The decisions made at Hyde Park on atomic energy policy were summarized and documented in an *aide-mémoire* signed by Roosevelt and Churchill on September 19, 1944. It bears all the markings of Churchill's attitude toward the atomic bomb and Niels Bohr. Among other points, it stated: "The suggestion that the world should be informed regarding tube alloys [the atomic bomb], with a view to an international agreement regarding its control and use, is not acceptable. The matter should continue to be regarded as of the utmost secrecy."

But Bohr had never suggested that "the world" be informed—only that the Soviet government should be officially notified about the Manhattan Project before the inevitable moment when any discussion of the matter would appear more coercive than friendly.

The *aide-mémoire* covered three additional subjects. One dealt with maintaining the Anglo-American atomic monopoly into the postwar period. Another offered some insight into how Roosevelt intended to use the weapon in the war: "When a bomb is finally available it might perhaps, after mature consideration, be used against the Japanese, who should be warned that this bombardment will be repeated until they surrender." But it was the last paragraph which revealed Churchill's success in bringing his point of view to dominate: "Enquiries should be made," it noted, "regarding the activities of Professor Bohr and steps taken to ensure that he is responsible for no leakage of information particularly to the Russians."[14]

Whatever caused Roosevelt to suspect Bohr—the physicist's conversations with Supreme Court Justice Felix Frankfurter, who had arranged the meeting between Bohr and Roosevelt; Bohr's correspondence with the Soviet physicist Piotr Kapitza; or Bohr's interest in informing the Soviets about the Manhattan Project—Churchill was undoubtedly the primary force behind the president's suspicions. Clearly the prime minister considered the Danish scientist a security risk. After the Hyde Park meeting he asked Lord Cherwell how Bohr had become involved with the Manhattan Project, stating that both he and Roosevelt were very concerned about Bohr's communications with Frankfurter and Kapitza. "It seems to me," he concluded, that "Bohr ought to be confined or at any rate made to see that he is very near the edge of mortal crimes."[15]

Churchill's opinions about Bohr's proposals were as definite and unwavering as his views of the scientist himself. In March 1944 he curtly rejected a memorandum from Cherwell and Sir John Anderson urging him to inform his own war cabinet about the atomic developments, and to take up the problem in postwar international relations. A month later he refused even to listen to Bohr's opinions. He also dismissed a number of other suggestions, urged upon him by members of his inner circle, to face up to the fact that "plans for world security which do not take account of [the atomic bomb] must be quite unreal."[16]

Churchill's hostility to Bohr's proposals, and to Bohr himself, appear to be related to the prime minister's determination to secure for Britain a position of equality with the United States in a postwar atomic energy partnership. Britain's future position as a world economic and military power, he believed, depended upon that relationship. No one understood better than Churchill how severely the war had weakened Britain. As early as 1943 he had stated that atomic energy was "necessary for Britain's independence in the future," a view confirmed by Cherwell, who told Harry Hopkins, a special assistant to Roosevelt, that Britain was considering "the whole [atomic energy] affair *on an after-the-war military basis*" (emphasis added).[17]

Bohr's proposals were, in fact, vulnerable to the internal logic of Churchill's position. After all, Bohr had not dealt directly with the possibility that *any* discussion of atomic energy matters with the Soviet government was bound to have a coercive element, given the Anglo-American lead. From the standpoint of seeking cooperation, the earlier the Soviets were told, the better. But Roosevelt and Churchill had to choose between the short-range advantages of having the weapon first and alone against the long-range advantages of international control. As has happened so often since, the immediate advantage appeared to be the more attractive choice.

But was international control achievable? Bohr thought it was possible if Roosevelt and Churchill promptly initiated the necessary first step: to inform Stalin of the Manhattan Project's existence and make assurances of an Anglo-American commitment to a British-U.S.-Soviet postwar arms control regime. Without this first step he knew that such an arrangement was impossible. Thus he believed that it was necessary to try. Such an effort, he insisted, was advantageous regardless of the outcome. If Stalin proved cooperative, international control might be achieved; if not, then Roosevelt and Churchill would be better able to assess accurately the nature of international relations in the postwar world. Continued secrecy, he believed, offered the worst of both worlds.

Although suspicions about Bohr were dissipated within the next six months, his chance to influence the policies of the Allied governments was lost. Bohr's colleagues carried his ideas and his formulations forward after the war, but the opportunity Bohr's ideas offered to gauge and perhaps influence the Soviet response to the international control of atomic energy could not be regained. Thus, instead of fostering their most laudable postwar aims, the atomic energy policy pursued by Roosevelt and Churchill made those aims more difficult to achieve.

For more than 40 years their successors have struggled to control the arms race that Bohr predicted. Occasionally, as in 1963 during the negotiations over the nuclear test ban treaty, Bohr's principles of arms control have been rediscovered. But those rediscoveries have been sporadic, and their influence ephemeral. Is it too much to suggest that in 1944 an opportunity was lost?

Indeed, it was more than *an* opportunity; it was the only opportunity to begin the nuclear age with a set of initiatives and accompanying assumptions that would have made possible the postwar control of nuclear weapons. How to recreate that opportunity is one of the great questions upon which survival hangs. But if a new opportunity does present itself, it will be well to remember Bohr's first principles of arms control:

With respect to weapons of mass destruction, short-range advantage is the deadliest enemy of long-term security.

When the opportunity for an initiative presents itself, timing is more critical than detail.

References

1. For Bohr's wartime nuclear arms control proposals and related memoranda see the "Frankfurter-Bohr" file, box 34, J. Robert Oppenheimer papers (hereafter JROP), Library of Congress, Washington, D.C.
2. An abbreviated version of Oppenheimer's talk, "Niels Bohr," was published in the *New York Review of Books* (Dec. 17, 1966).
3. Bohr, memorandum (May 8, 1945), JROP.
4. Bohr to Roosevelt (July 3, 1944), JROP.
5. J. Rud Nielson,"Memories of Niels Bohr," *Physics Today*, vol. 16 no. 10 (Oct. 1963), p. 29.
6. Bohr to Roosevelt (March 24, 1945), JROP.
7. Oppenheimer, "Niels Bohr," p. 8.
8. Bohr to Roosevelt (July 3, 1944), JROP.
9. Bohr (May 8, 1945), JROP.
10. Bohr, "Notes Concerning Scientific Co-operation with the USSR. Written in Connection with Memorandum of July 3, 1944–September 30, 1944," JROP.
11. Margaret Gowing, *Britain and Atomic Energy, 1939–1945* (London: MacMillan, 1964), p. 355; see also Bohr to Churchill (May 22, 1944), JROP.
12. Ibid., p. 360.
13. Aage Bohr, "The War Years and the Prospects Raised by Atomic Weapons," in Stefan Rozental, ed., *Niels Bohr* (New York: Wiley, 1967), p. 197; Franklin D. Roosevelt Papers, Official File 2240, Franklin D. Roosevelt Library, Hyde Park, N.Y. (herafter FDRL). Justice Felix Frankfurter first brought Bohr's concerns to Roosevelt's attention in February 1944 (see Martin J. Sherwin, *A World Destroyed: The Atomic Bomb and the Grand Alliance* (New York: Knopf, 1975), pp. 99–101.
14. For complete text of the *aide-mémoire* see Sherwin, *A World Destroyed*, appendix C.
15. Quoted in Gowing, op. cit., p. 358.
16. Ibid., p. 352.

17. Churchill quotation in Harvey Bundy, "Memorandum of Meeting at 10 Downing Street on July 23, 1943," AEC document no. 312; Cherwell is quoted in Vannevar Bush, "Memorandum of Conference with Mr. Harry Hopkins and Lord Cherwell at the White House," May 25, 1943, in the Harry Hopkins papers, "A-bomb" file, FDRL. See also Bush, "Memorandum of Conference with the President," June 24, 1943, AEC document no. 133.

American Attitudes on Security: Comments

John Steinbruner

THE BROOKINGS INSTITUTION

It is a great honor for a man when friends and colleagues gather to celebrate his memory. It is an additional honor when they are joined by those who did not know him, who are drawn by recognition of the importance of his accomplishments. It is an honor for me to be able to pay this latter tribute on this occasion to a man whose accomplishments are likely to be remembered throughout the course of human history.

Though I did not know him personally, I can readily perceive that Niels Bohr was a very great visionary—an enormously successful one as a physicist and a sadly neglected one as a statesman. I presume that his contributions to quantum theory will forever endure in the history of science even if greater revelations should someday come to supersede them. His appeal for openness among nations, however—an attempt to apply the values and principles of science to the business of diplomacy—did not succeed in his lifetime and at the moment its prospects appear very bleak indeed. That is nonetheless a vision we must not forget and the underlying purpose—a transformation of the relations among nations—is one that we dare not abandon.

In a spirit that I am sure Bohr would endorse, I would like to review why his appeal for a transformation of international relations did not succeed and to note some important consequences of its rejection. In doing so I will focus on the actions of my own country, not because it deserves exclusive blame but rather because I believe that it must carry the primary hope for a better future. If relations among nations are eventually to be developed along the lines of Bohr's vision, then the process must begin in the United States. A necessary step in that process is to recognize the magnitude and implications of what we have done in the thirty-five years that have passed since Bohr advanced his ideas in his open letter to the United Nations.

As all of you are aware, the United States maintains a great diversity of opinion on the issues of security as on most matters of public policy. This diversity is lavishly displayed in our public literature and in the many forms of political action that constitute the workings of a democracy. The kaleidoscopic variations and internal inconsistencies that arise are quite confusing—to participating Americans as well as international observers. They often obscure the fact that over time there has been an inner simplicity and relentless focus to American security policy. Fear, distrust, and political antagonism toward the Soviet Union has prevailed over other international political sentiments; security has been based on immediate preparations for war; nuclear weapons have been established as the central element in our theory of defense.

When Bohr addressed his appeal to the U.N. this pattern had not yet crystallized and he could still have some plausible hope of preventing it. The letter was dated June 9, 1950, however, 16 days before the outbreak of the Korean War. That war became the triggering event for a process that transformed American security in a manner quite inconsistent with the image that he advanced.

Prior to Korea, United States military forces were not prepared for immediate operations on any major scale. Security was based on isolation from any immediate threat, on the inherent capacity of the American economy, and on relative dominance of pertinent military technology. Nuclear weapons were maintained in small numbers under civilian authority. They were not dispersed to military commanders and preparations for their actual use were rudimentary. We believed we could mobilize after a direct military threat had appeared and that in due course we could prevail.

Beginning with the Korean War and proceeding through the decades that followed, we adopted a much more urgent sense of threat and fundamentally altered the terms of our security. Driven by Soviet acquisition of nuclear technology and of ballistic missiles, we invested for the first time in our history in a large peacetime military establishment capable of conducting extensive military operations on short notice anywhere in the world. We dispersed our own nuclear weapons under the immediate control of many thousands of military personnel and now continuously exercised them in preparation for attack on the Soviet Union. At any given moment, United

States nuclear forces are literally capable of initiating an attack within an hour and completing it within a day on a scale that under any reasonable prior assessment must be considered fatal to the military, economic, and even social organization of the Soviet Union.

The Soviet Union has itself acquired a commensurate capability and on both sides the capacity for offensive destruction vastly and decisively exceeds any ability of the opponent to conduct direct defense. Both countries are committed to preventing war by exchanging threats of punishment. However uncomfortable we may be about this result, and however strongly we might dissent from its implicit premises, we collectively have made it our highest security priority and the enduring centerpiece of international security arrangements.

There have been at least two important corollaries of this development and several as yet unmastered implications. First, the wide dispersion of weapons in an advanced state of operational readiness placed the United States in a technical sense at a perpetual brink of war and made the effective exercise of political authority a practical question of supreme importance. Weapons design safeguards and procedural arrangements have been worked out to assure that nuclear weapons will not be used without the legitimate authority of the President. Belief in the effectiveness of these measures against any meaningful prospect of accidental or induced failure has become an axiom of security. In addition, the deployed forces have been given protection against prior attack intended to assure that they could effectively operate in retaliation. Commitment to credible retaliation has also become a generally accepted axiom of American security.

The as-yet unmastered implications arise from a tension between these two principles that has been difficult to grasp and that at any rate we have been reluctant to acknowledge. The offensive power now embodied in the U.S. and Soviet nuclear weapons arsenals is so swift and so destructive that a practical conflict has been created between the assured exercise of political control and the guarantee of effective retaliation. If a reasonable number of protected and dispersed weapons can be expected to survive even the heaviest initial attack, the command arrangements that enable coherent political authority cannot be, not for very long at any rate. Recognizing that coherent political control might be lost in the initial stages of war, the United States has prepared for a very efficient and very rapid exercise of political authority in response to the first indications of attack. The responsiveness is in fact so rapid that it borders on preemptive initiation and has inherent potential to become preemptive initiation under the pressures of severe crisis, whatever prior intentions may have been. The Soviet Union appears to be acutely sensitive to that possibility, and is itself subject to the same pressures.

Under normal peacetime circumstances, these tensions between the major principles of security are dormant, and the possibility of a sudden, catastrophic breakdown in the established regime of deterrence appears to be acceptably remote. It is difficult and arbitrary, however, to extend such assurance to the conditions of severe crisis. We have no directly pertinent experience. Since the Cuban affair in 1962, which occurred well before either side had established a fully matured capability, there has been no crisis confrontation of nuclear forces. Moreover, it should be obvious even without direct experience that the highly dispersed actions of a large military establishment cannot be completely known or entirely controlled from the apex of its command network. Despite extraordinary procedures for central control, nuclear weapons operations are not immune from the realities that produce spontaneous, unpredictable behavior in any large and dispersed organization—the frequent cause of surprise, frustration, and tragedy throughout the history of warfare.

The highly developed preparation of offensive nuclear weapons that has been our main security commitment has certainly deterred war and rendered it intuitively unlikely. We have not absolutely prevented war, however, and in the means of deterrence we have chosen we have exposed ourselves to a risk of catastrophe whose dimensions cannot be completely determined and whose magnitude cannot be measured.

After thirty-five years of technically sophisticated weapons development, therefore, the security of the United States remains seriously incomplete and in seeking the necessary supplements we do well to remember Bohr's vision as a statesman.

The regime of deterrence that we have assiduously created and that we are destined either to live or to perish with for the indefinite future must be balanced by international reassurance of the sort that Bohr attempted to inspire. Our adversary must know not only that we are capable of terrible destruction but also that we are capable of restraint, and committed to preserving restraint under any circumstance short of actual attack. Though we have done a great deal to express this reassurance despite endemic political antagonism, the physical posture of our strategic forces is not securely reassuring and under conditions of crisis it could become dangerously provocative for spontaneous reasons that we could not reliably control. If the word attitude is used to refer not only to declared intentions but also to the physical and organizational state of our military establishment, then American attitudes toward nuclear weapons are aggressive to the limits of wisdom and moderation is our most serious unrealized security requirement. We threaten our adversary very effectively indeed; we must now learn how to reassure that adversary.

In addition and related to that, we must recognize the overriding interest we have in preventing serious crisis from arising and thereby carrying us beyond the peacetime conditions in which we have learned to manage nuclear forces tolerably well. That too requires an openness among nations, an exchange of information, a grasp of their politics, history and internal dynamics, a degree of understanding necessary for any stable relationship. Clearly we have a long way to go in that regard. We in the United States, enmeshed in our fears and absorbed in our assertion of conflicting political principles, do not understand the Soviet Union. Nor do we understand many of the other countries whose political turmoil brings us into conflict with the Soviet Union. Against the background of the security regime we have created, that is a very dangerous circumstance and it is important for us to correct it.

The greatest honor we could do to the memory of Niels Bohr and the greatest favor to ourselves is simply to learn these unmastered lessons.

Science, Technology, and the Arms Race: European Perspectives

Edoardo Amaldi

UNIVERSITÀ LA SAPIENZA

1. Introduction

At a number of conferences held during the course of 1985, and in particular in today's lecture by Professor Martin Sherwin, a deep analysis has been presented of Niels Bohr's idea that all the applications of nuclear energy had to be developed in a world without secrets and without barriers among different states. Unfortunately Bohr's views came to nothing when, in the summer of 1944, he presented them to Churchill and Roosevelt, as well as when he recast them, in a more complete and elaborate form, in the open letter which in 1950 he sent to the United Nations.[1]

The production of weapons for mass destruction has gone on during all the past years in a regime of secrecy; now, the arms race, in particular that of the Superpowers, has given rise to nuclear arsenals of incredible, crazy dimensions.

If Bohr were still alive today and could renew his proposal, modified to take into account the evolution of the situation, he would certainly have very little chance of being listened to.

However, in spite of this disillusionment, the idea of an 'open world,' as conceived of by Niels Bohr, was—and still is—of fundamental importance, and should remain a goal that mankind must try to reach in the future.

The economic development of the more advanced countries appears to be largely conditioned by—or at least correlated to—their scientific and technological potential, which, in turn, has almost always, sooner or later, some repercussion of military importance.

The policy of secrecy unavoidably favours the technologies that are of military interest. In support of such an approach, it is often affirmed that all military technologies give rise to technological spin-off that is of interest in civilian life.

In an 'open world' the arms race would automatically be avoided and the development of the technologies would naturally be oriented towards civilian goals and could easily be shaped toward the improvement of the living conditions of mankind.

In my contribution to this conference I would like to examine some historical aspects of the interrelation between science and technology on the one hand and the arms race on the other, within the frame of Western Europe, starting from the end of the Second World War.

In Section 2 of my speech, following this Introduction, I will give an outline of the most important organizations that were created, in Europe, to deal with various fields of science and technology immediately after the Second World War. In Section 3 I will try to present the problem of the arms race control as viewed by the Europeans. In order to carry out such a difficult task I will lean heavily on the conclusions which emerged during a recent international conference held on this specific subject in Italy. In this presentation I will also touch upon the Strategic Defense Initiative (SDI) proposed by President Reagan. Finally, in Section 4 I will mention a few, more recent developments, the last of which is the Eureka proposal.

As a preamble I shall recall a few facts of an economic nature but which are of considerable importance as a background to what I will say later, in particular in Section 2. At the end of the Second World War the political and economical situation of all European States appeared to be seriously compromised. The central problem of the economic reconstruction of these states led the United States of America to create a plan of economic assistance, called the 'European Recovery Programme' but better known as the 'Marshall Plan,' from the name of its promoter. This plan was subordinate to the condition that the realization of economic assistance from the United States be carried out jointly by the European States.

Sixteen countries adhered to the Marshall Plan,[2] and in April 1948 signed a Convention for European economic co-operation, the implementation of which required the creation of the *Organization for European Economic Co-operation* (OEEC). In September 1961 this body was transformed into the *Organization for Economic Co-operation and Development* (OECD), which includes, in addition to all states of Western Europe, many other countries such as Canada, Finland, Japan, New Zealand, Turkey, and the USA, for a total of 24 members.[3]

The OEEC was the right tool for the implementation of the Marshall Plan, but it could not fullfil the requirements for a closer economic collaboration between the European States, in particular the major ones. Thus at a press conference in May 1950, the French Minister of Foreign Affairs, Robert Schuman, proposed to put in common the French and German production of coal and steel. The proposal was primarily addressed to West Germany, but was open also to other European States. Its development took the form of the so-called 'Schuman Plan' and culminated in April 1951 in Paris, in the signing of a treaty creating the 'Communauté Européenne du Charbon et de l'Acier' (CECA). Six countries signed the treaty: Belgium, France, the Federal Republic of Germany, Italy, Luxemburg and the Netherlands. The United Kingdom abstained in consideration of the difficulties arising from its preferential Commonwealth ties.

In the wake of the success of CECA, in May 1952 another treaty was signed in Europe. This constituted the 'Communauté Européenne de Défense' (CED), which foresaw the integration of the military forces of the six member states under a unified command. This attempt to unify Europe on the military plane was a failure.

2. European Initiatives in Science and Technology Immediately after the Second World War

Already at the end of the 1940's in many European and overseas countries scientists were becoming aware of the continuously increasing gap between the means available in Europe to the various fields of science and those available in the United States. Passing from fundamental sciences to applied sciences and technologies, the gap between the United States and Western Europe was even greater. It was also evident to a great number of scientists and technologists that the unsatisfactory European situation could be changed only by a considerable common effort made by many European countries.[4]

These views expressed by scientists and technologists were also in accord with those of many politicians of a number of European countries, in particular France, Italy, the Federal Republic of Germany and Belgium, who had the idea of moving towards some form of economic—and, perhaps later, political—unification of at least a part of the Old Continent. The Schuman Plan and the Coal and Steel Community (CECA) mentioned above are the most important practical manifestations of this general tendency.

In this climate of unification, a number of mission-oriented organizations were set up for tackling, through a joint European effort, those problems appartaining to specific scientific or technological areas.

CERN was the first European organization of this type to be created.[5] The idea of constructing a large European laboratory for the experimental investigation of high energy physics was proposed in 1950. *CERN* (Conseil Européen pour la Recherche Nucleaire) came into being, in a provisional form, in 1952, and became a permanent organization at the beginning of October 1954. For a little more than 30 years *CERN* has been very successful, and has contributed enormously to bringing Europe to a very important and worldwide recognized position in the investigation of the ultimate constituents of matter.

After this first step, which concerned the fundamental sciences, the next one was focused on the modern technologies.

Euratom was created in 1957 by the same six countries which, three years before, had given life to CECA. The long and detailed list of Euratom's objectives included all technical items necessary for the peaceful applications of nuclear energy and the creation of the basis for the formation and rapid development of a European nuclear industry. It set up four Common Centres of Research: one multidisciplinary centre in Ispra (Italy) and three specialized centres, in Karlsruhe (Federal Republic of Germany), Petten (Belgium) and Geel (The Netherlands).[6] The last three fulfilled their functions, whilst the first one lacked the clear general line which would have enabled it to orient its activities and to bring to important results within the rather short time limits imposed by the rapid development of the US nuclear industry.

The influence of Euratom on the development of a European fission reactor industry was only indirect, through the formation of a number of people who were competent in many aspects of the nuclear technologies. On the other hand, Euratom's activity in the field of plasma physics and fusion research was very successful. At the beginning, Euratom very efficiently co-ordinated the research activities in the national laboratories of the member countries. Then, following the unanimous advice of the people involved in these research activities, it was decided to construct the *Joint European*

Torus (JET) as a common project. This was designed between 1972 and 1978 and built between 1978 and 1983 in an area, allocated by the British Authorities, in the Culham Plasma Laboratory near Oxford.

This machine has a medium magnetic field, and a large volume of plasma crossed by a current of 5 MA.[7] It is one of the most advanced machines for the study of fusion reactors.

In those years an industry for the design and construction of fission reactors was developed in many European countries, in particular in France, the Federal Republic of Germany, and the United Kingdom.

Support for these activities came from the *European Nuclear Energy Agency* (ENEA) of the OECD, which, for example, in 1958 took the boiling heavy water reactor constructed by the Norwegians at Halden, south of Oslo, and transformed it into a European facility for testing the fuel elements of boiling water reactors and pressurized water reactors; this facility is still in use today. Another plant whose installation was promoted by ENEA was EUROCHEMIC, a European company designed for the reprocessing of irradiated fuel elements, and which was set up in Mol (Belgium). This company operated from the end of July 1959 until very recently.

The French Commissariat à l'Energie Atomique (CEA) was very successful in the construction, between 1968 and 1974, at its Marcoule Centre, of a prototype fast breeder reactor of 200 MWe called *Phenix*. As the result of this enterprise, two national utilities, Electricité de France (EDF) and Ente Nazionale Energia Elettrica (ENEL, Italy) and the German private company Rheinisch-Westfalisches Elektrizitätswerk (RWE),[8] decided to construct two fast breeder reactors of 1200 MWe each, based on different technologies, one located in France (Superphenix), the other in West Germany (SNR-2). Superphenix became critical in September 1985.

In 1957 the *European Southern Observatory* (ESO) was set up by six countries, followed by others some years later.[9] Its goal was to study the southern sky by creating an observatory at La Silla in the Chilean Andes. For various reasons the life of ESO was rather uncertain until 1970–72, but later the design, construction, and installation of its main instrument (a 3.6 m reflector) was completed. The headquarters was established in Garching, near Munich, in 1981; since then it has been operating very successfully as a European centre for astrophysics.

In particular, ESO is charged with the supervision of the 'European Coordinating Facility,' founded in Garching by ESA (see below) for the analysis of the *Space Telescope data* by astrophysicists of the West European countries. Furthermore ESO is ready to mount another 3.5 m reflector, in the La Silla Observatory (Chile's Andes) for testing new technologies involved in the design of a *Very Large Telescope* consisting of four reflectors of 8 m diameter each, arranged along a 150 m base.

The Federal Republic of Germany, Spain, Sweden and the United Kingdom have recently opened, at Las Palmas (Gran Canaria), the *Canary Islands Observatory*, where the most important telescope is a 4.2 m reflector.

The Soviet Union successfully launched its first satellite, Sputnik I, on 4 October 1957. In 1958, the National Aeronautics and Space Administration (NASA) was created in the USA. And in Europe, two separate organizations were set up between 1960 and 1964, in a few successive steps: the European Space Research Organization (ESRO) and the European Launcher Development Organization (ELDO). ESRO pretty soon started to fullfil its research functions— if not perfectly, at least adequately. ELDO, on the contrary, came to nothing, for reasons that many people had foreseen and denounced before its constitution.[10]

Within a few years the need to reorganize the European space programme became clear to everybody. This reorganization took place in 1971, when ESRO and ELDO were replaced by the *European Space Agency* (ESA), which takes care of applied as well as of fundamental research, but not of the construction of launchers. In so doing the decision was taken that fundamental research, had to absorb not more than 17% of the total budget, a restriction which, during the late 70's started to be too stringent.[11,12] Notwithstanding, since 1968, first ESRO, and later its successor ESA, have been increasingly successful: five application satellites (two meteorological and three for telecommunications) and fourteen scientific satellites have been launched, all of them—except one—by means of launchers provided by NASA.

The Ministries responsible for space matters in ten European countries decided, at the European Space Conference held in Brussels at the end of July 1973, that it was essential for Europe to have a competitive launcher if it wished to ensure its independence and enter the international market for application satellites. In December of the same year an Agreement between the participating states and ESA on the development of the *Ariane launcher* came into force.[13]

ESA was entrusted with monitoring the execution of the programme. The technical experience on which the design of the first Ariane launcher was based, and the need to maintain technical consistency, led to the selection of the Centre National d'Études Spatiales (CNES) as the prime contractor.

A series of improved versions of Ariane were then constructed. Starting from Ariane 2 and 3, the manufacturing and the launching of all these space devices were entirely the responsibility of *Arianspace*. This is the first commercial space transport company, set up on 26 March 1980 by 36 main European manufacturers in the aerospace and electronic industries, 13 important European Banks, and CNES.

The last ESA mission to Halley's Comet (Giotto) was set in orbit, on 2 July 1985, by Ariane V 14 from the Guyana Space Centre.

Another important organization that I should like to mention is the *European Molecular Biology Organization* (EMBO), created, in successive steps, between 1962 and 1972. This organization is formed by 17 member states[14] and its research laboratory (EMBL) is located in Heidelberg (Federal Republic of Germany).

As an example of a purely industrial achievement of considerable technological relevance, I should cite the construction of the commercial airplane *Concorde*. Its design was started in 1962 through the combined effort of British and French aerospace industry, i.e. for the airframe, the British Aerospace Company and the French Aérospatiale; for the motors, Rolls Royce and SNECMA. This realization was based on the technologies used in the construction of the military fighter planes Farley and Mirage. The first test flight took place in 1969, and since 1976 the Concorde has been flying commercially.

A contribution to the development of European scientific and technical capabilities came also from the *NATO Science Committee*, established in 1958 in recognition of the crucial role of science and technology in maintaining the economic, political, and military strength of the North Atlantic Communities. The activities of this Committee form an important element of what is sometimes referred to as the 'third dimension' of the *North Atlantic Alliance*[15]: the non-military dimension, concerned with the enhancement of contacts between member nations in the areas of science and technology and the problems of modern society.

The broad policy and the direction of the Science Programme are established by the NATO Science Committee, whose membership is drawn from amongst the most eminent scientists of Each Member State and chaired by the Assistant Secretary General for Scientific and Environmental Affairs of NATO.

The total budget available in 1984, for example, was about eleven million dollars,[16] 52.9% of which was used for the Science Fellowships Programme, 22% for the Advanced Study Institutes and Advanced Research Workshop Programme, 13.7% for the Research Grants Programme, etc.

My presentation, although rather sketchy and incomplete, is sufficient to give an idea of the overall effort made in Europe in the fields of science and technology. The organizations I have mentioned above range from rather large to small, both from the budgetary point of view and from the number of people involved. Their efficiency in pursuing the alloted institutional tasks was, unfortunately, different from case to case, irrespective of the size of the organization. In a number of cases the efficiency was quite satisfactory, whilst in others the results were considerably inferior to the expectations. A detailed analysis of all these unfortunate instances shows that the lack of success was never due to the intrinsic difficulty of the problems to be solved or to incompetence of the staff but to a number of other factors such as: a) insufficient commitment by the governments of some of the Member States who at times, faced with specific, brilliant, and promising project, preferred to tackle it as part of their national programme rather than as part of the European common effort; b) the general presumption of some governments of their technological superiority, accompanied by a strong reluctance to share the corresponding benefits with other members of the organization; c) the desire of some governments to transfer to a European Organization a project already begun as a national endeavour but which was either becoming too expensive, or was technically not sufficiently well planned, or both; d) a too heavy and complex bureaucracy within the European Organization; and finally e) a weakness of the leadership of the European Organization, resulting in available funds being spread over too wide a programme of activities.

These criticisms should clearly be taken only as a reminder of what one should try to do (or not to do) in the future. A number of European organizations have been operated very efficiently from the beginning; others which started off incorrectly, were later completely set right. Furthermore, we should never forget that in Western Europe, less than fifty years ago, the majority of children were still being taught to believe that all other people outside the frontiers of their own country were bad and ugly.

3. Arms Race and Arms Control

The arms race and the control of armaments is such a wide and complex subject that any attempt to summarize it in a talk of reasonable duration would certainly be inadequate and only partial. It is, however, fortunate that between 21 and 25 October 1985, an International Conference was held, in Castiglioncello, near Livorno, on 'Nuclear Weapons and Arms Control in Europe'; this was organized by the Unione Scienziati per il Disarmo (USPID: Union of Scientists for Disarmament[17]). About one hundred people originating from the Western and Eastern European countries, the USA, the USSR, etc., were present, many of whom are well recognized experts in this kind of problem.

A round-table on 'European Defence' was to be presided over by the Italian Minister of Defence G. Spadolini, who at the last moment could not attend because of the Achille Lauro crisis. He was replaced by Field Marshal Lord Carver, who, a few years ago, had been Chief of Staff of the UK armed forces.

'Arms Control and Europe' was dealt with in speeches by F. Calogero of the University of Rome, J.P. Holdren from the University of California at Berkeley (USA) and J. Baldauf, from the Federal Republic of Germany; this topic was also discussed by A. Handler-Chayes, R. Garthoff and J. Mendelsohn, all from United States.

A full day was devoted to the 'Strategic Defense Initiative' with speeches by P.S. Brown of the Lawrence Livermore National Laboratory ('Why Research on Defensive Weapons is Important'), by P.F. Prilutski of the Space Research Institute in Moscow, by J. Ruina of MIT, by S. Drell of Stanford University, and others.

It emerged from the meeting that in Europe there is a strong preoccupation with the possible use of tactical nuclear weapons. Their number and the various scenarios considered for their possible use in a local war were examined by P. Cotta Ramusino of the University of Milan, and the fallacy of the strategic concepts about their use was presented by Lord Carver in his speech on the 'Military and Political Role of Nuclear Weapons in Europe.'

In the following I will lean heavily on the speech by Calogero, which presented not only his personal opinions but also a summary of widespread European points of view. He helped me in trying to condense them into a few items which appear to be important and non-controversial for a large part of the European public.

1. The security arrangement in Western Europe, based on NATO, has worked well so far, at least in the sense of having avoided any military conflict. Of course, much could and should be done to lower the risks of nuclear confrontation—a topic that was much debated during the Castiglioncello meeting. NATO enjoys ample popular support in Western Europe as an organization providing a viable security arrangement, and it should be viewed also by the Soviet Union in the same light, since the only other realistic alternative would be a less reliable equilibrium based on a much-enhanced military role of West Germany.

2. The prospect of arms control in Europe is usually considered in the context of the confrontation between the two Superpowers. A large part of the general public, in Western as well as in Eastern Europe, realizes that this is in some way unavoidable, but would like to see a much more active role being played by the European governments in bringing the two Superpowers at the table of discussion, to accept a serious commitment to reach an agreement. The nuclear arsenals are over-large by any reasonable standard. The scope for arms reduction is therefore vast, and there is ample opportunity for either side to make concessions, without risk. The fact that for the first time substantial reductions—by factors of two to begin with—are mentioned and seriously contemplated by political leaders, is of course a hopeful sign. Whether these advances will bear fruit or will degenerate into propaganda moves, remains to be seen—and will be seen pretty soon.

3. The viability and the achievement of arms control agreements have been largely impeded by focusing on the question of numbers. This has increased the difficulty of reaching an agreement by giving the impression that numerical parity can be achieved. Given the variety of weapons systems such an approach has a very limited and illusory meaning. The qualitative aspects of weapons systems is no less important than the number of warheads. Agreements about the qualitative aspects are difficult; but there are some measures relevant to this effect which are quite viable and could be easily agreed upon, if there were the political willingness to do so. An example of this is the *Comprehensive Test Ban*, which would prohibit all nuclear explosions (including those underground). Another measure would be an agreement to ban altogether the introduction

of new weapons systems such as, for instance, *cruise missiles*, or of complete technological sectors such as *chemical weapons*. This last goal might appear almost Utopian; yet complete success has been achieved in banning biological weapons, thereby ridding a whole technological and scientific sector of those activities that were aimed at the development of destructive capabilities. Still another limitation of this kind would be an agreement to avoid the 'weaponization' of space.

4. Also relevant to these considerations is the Strategic Defense Initiative (SDI) launched by President Reagan. As I said before, this also was amply discussed at the Castiglioncello meeting by European and American participants. A large part of the European Scientific Community shares the doubts expressed by many experts in the USA regarding the technological feasibility of SDI to 'render nuclear weapons impotent and obsolete,' and is worried that this development rather stimulates the nuclear arms race and impedes progress in arms control. Whether, and to what extent, these worries are justified will be seen in the next few days of discussion.

5. One aspect of arms control where the two Superpowers have a clearly recognized common interest—an interest shared by all peoples—is in preventing the world-wide proliferation of the capability to produce nuclear weapons. Indeed, many see in the prospect of widespread proliferation an omen of assured catastrophe—and the time scale of this development is measured in decades not centuries. Article VI of the Non Proliferation Treaty reads as follows: 'Each of the Parties to the Treaty undertakes to pursue negotiations in good faith on effective measures relating to cessation of the nuclear arms race at an early date and to nuclear disarmament, and on a treaty on general and complete disarmament under strict and effective international control.' Clearly during the 15 years, passed since the entry into force of this Treaty, the two Superpowers have forgotten their commitments towards all the *non-nuclear* states that were also signatories to it.

4. More Recent Developments in Europe

Some time after Euratom was instituted it became clear that a coordination of the European Communities, the Coal and Steel Communities, and Euratom, would be beneficial for several reasons. Their unification took place in Brussels, on 8 April 1965. Since then, the original communities have been merged to form the European Communities (ECC), the principal organs of which are:

a) the *European Parliament*, which since 1979 is elected by direct universal suffrage;

b) the *Council*, composed of ten members, which takes decisions and promulgates the Communities' laws. It is composed of Ministers of the Member States, which differ according to the nature of the problems appearing on the agenda of the meeting;

c) the *Commission*, composed of fourteen independent members, which proposes the Communities' laws, checks that the treaties are respected, and administers the common policies;

d) the *Court of Justice*, which ensures the interpretation of the Communities's law and solves any dispute connected with its application;

e) the *Economic and Social Committee*; and

f) the *Audit Office*.

Later, the Economic Communities were enlarged by the adhesion of the United Kingdom, the Republic of Ireland, and Denmark in 1973, of Greece in 1981, and of Spain and Portugal in the course of 1985.[18]

The overall scenario of the scientific and technological activity developed in common by the West European States may appear to be rather complex, but it is complete and promising. In addition, the legal frame work of the ECC is formally so impressive that a superficial observer would easily be led to conclude that great progress has been made towards some kind of unification of a large part of Western Europe through a series of non-aggressive and peaceful steps. In reality the structures are much weaker than they seem.

If we limit our considerations to science and technology, and if we do not look at the structures and the programmes but at the concrete results, we will see that the progress is really satisfactory only in certain restricted areas such as those involving fundamental science—and therefore far removed from applications of wide economic interest. A few years ago most people, even at a very high level, appeared to be unaware of the inadequacy of the European effort. They were too busy with national economic difficulties, with unemployment, with the national and international drug traffic and youth problems.

Between 1978 and 1980, almost suddenly, everybody became aware that Japan had conquered a large fraction of the markets all over the world with the products of its industries: automobiles, motorcycles, electronic gadgets, instruments, etc.

This came as a kind of shock, which, however, had also some beneficial consequences. Everywhere, and particularly in Europe, the experts in economics, industrial management, etc., started to discuss the 'Japanese case,' and to compare it with the 'American case,' which, until then, had been considered the only example that had led to—and perhaps the only one that could lead to—a satisfactory economic development.

I remember, for example, two meetings, both held in November 1981. The first, on 'Scientific Research in the European Community: Perspectives and Prospects,' was organized jointly by the Solvay Foundation and the ECC and took place in Brussels at the beginning of November; the second meeting, held in Paris about two weeks later, on 'Innovation and Society: Interaction between Scientific, Technological, Socio-Economic and Cultural Innovations,' was organized jointly by UNESCO and the European Academy of Sciences, Arts and Letters.[19]

In 1982 the Commission of the ECC presented a 'Framework Programme' for the period 1984–1987, the aim of which was to provide research funds for specific objectives which had clearly emerged from recent international analyses.

The main lines of this programme were approved in June 1983 by the Council of Ministers of Research of the Member Countries but without, however, taking a final decision about the allocation of the funds asked for by the Commission (3750 million ECU).[12] Up to now only 2860 million ECU have been appropriated for the four years 1984–1987 (with a reduction of 24%). This sum will be devoted to a list of specific objectives, the most interesting of which are listed below:

ESPRIT (1984–1987; 750 million ECU) is a European programme for research and development in information technologies.

FUSION (1985–1988; 690 million ECU), is in part devoted to the experimentation with JET, in part to the initial phase of NET (the Next European Torus), which has been started in Garching, near Munich. This machine will have a blanket containing lithium compounds for the production of tritium under neutron bombardment. A preliminary project for determining the principal parameters has been started; these will probably be frozen in 1989. Final decisions regarding the construction design of NET will be arrived at around

1992, taking into account the experimental results obtained in the meantime with JET and other USA, USSR, and Japanese machines.

BRITE (1985–1988; 125 million ECU) concerns a long range research and development programme in the fields of basic technological research and applications of new technologies.

NON-NUCLEAR ENERGIES (1985–1988; 175 million ECU); and so on.[20]

Another, unexpected, external stimulus to re-examine the problem of the scientific and technological development of Western Europe was provided by President Reagan's speech of 23 March 1983 about the SDI.

I have already mentioned that many nations, including European ones, expressed doubts about this venture. To these should be added the concern of Western Europe: that a large participation in the SDI programme would imply a renunciation of the development of its own initiatives which could avoid some of the political objections and would fall in those sectors where the Old Continent needs to be present.

Thus the European public opinion was, to some extent, ready to receive other proposals, when a new programme, called *Eureka*, was announced by the President of the French Republic, François Mitterrand, on 17 April 1985 in Paris.

The information provided on that occasion was very scanty, and the international press gave contradictory interpretations even of the nature and the goals of the proposed programme.

These latter were clarified in a Report entitled '*Eureka, the Technological Renaissance of Europe,*' prepared by the Minister of Foreign Affairs and the Minister of Research and Technology of the French Republic and dated June 1985. Two points were immediately clear and should be stressed here: first, all Eureka projects will serve civilian purposes; and, second, in setting down a preliminary list of possible projects, the proponents had used and, in some way, tried to condense, the conclusions reached during the last few years by various European organizations. In the June report these projects were divided into five technological areas, in each of which Europe was recognized to be still behind the USA and Japan. Later, this division into large areas was dropped for the sake of simplicity and flexibility, but the single projects have remained substantially the same, without unnecessary detail.

On 17 July 1985, Eureka was established by the Paris Conference of Ministers of 17 countries and Members of the Commission of the European Communities.

At the Paris Conference, each country nominated a High Representative, usually a qualified technologist, who, together with a few high-level collaborators, was given the responsibility of carrying on with the work in between the meetings of the Ministers.

At the national level these tasks are complemented by work carried out in each Member states, where the Minister of Science and Technology (or an equivalent authority) organizes meetings with technologists and leaders of industries, institutes and, whenever useful, university laboratories.

A second Meeting of Ministers of 18 countries and Members of the Commission of the European Communities was held in Hanover on 5 and 6 November 1985. The participants not only agreed on all important points of principle and on the implementation procedures but also approved the first ten Eureka projects.

Concerning the points of principle, I should like to quote a few lines of the 'Declaration of principle relating to Eureka':

i. Objective

The objective of Eureka is to raise, through closer co-operation among enterprises and research institutes in the field of advanced technologies, the productivity and competitiveness of Europe's industries and national economies on the world market, and hence strengthen the basis for lasting prosperity and employment: Eureka will enable Europe to master and exploit the technologies that are important for its future, and to build up its capability in crucial areas.

This will be achieved by encouraging and facilitating increased industrial, technological, and scientific co-operation on projects directed at developing products, systems, and services having a world-wide market potential and based on advanced technologies.

Eureka projects will serve civilian purposes and be directed both at private and public sector markets.

ii) Focus and criteria

1. Eureka projects will initially relate primarily to products, processes, and services in the following areas of advanced technologies: information and telecommunication, robotics, manufacturing, biotechnology, marine technology, lasers, environmental protection, and transport technologies. . . .

Before explaining the implementation procedures adopted at the Hanover meeting, it may be useful to recall that during the past years Europe has had a lot of experience with organizations which carry out their scientific and/or technological research and development activities in laboratories financed by public money provided by the Member States. CERN, ESO, EMBO—and even JET within the wider structure of the European Communities—are successful examples of such organizations.

In the case of the European Space Agency (ESA), a scheme has been adopted whose aim is to stimulate the acquisition by industry of the know-how required by many modern technologies. The structure of the organization is reduced to the minimum which is necessary for taking competent decisions about the scientific and technical policy, for making an efficient study of the corresponding space missions, and their implementation, and for establishing realistic cost estimates. The prototypes of the scientific instruments are developed by the universities and national laboratories. The manufacturing of the final instruments and the detailed planning and construction of the space vehicle is assigned, as the result of a normal tender, to one of the competing 'consortia' of industries. Each consortium is made up of a number of industries which belong to different countries, and which together have the capability to build a whole space vehicle and its scientific payload.[21] The scientific and technical services of ESA—in particular ESTEC, its technological laboratory in the Netherlands—guide and control the work, and test the results reached in the intermediate and final phases of the detailed planning and construction process.

In the case of Eureka, the Ministers involved have decided to go a step further in reducing the centralized organization, which will then consist only of a secretariat. The launching of a specific Eureka project starts with an agreement between a few industries, belonging to different Member States, that decide to work together towards a common goal proposed by them or by a university, a national laboratory, a single scientist, or a group of scientists or technologists.

These industries should inform their national Ministers of Research and Technology, who examine the proposal, and, if they consider it acceptable, submit it to the Council of Ministers for approval. The financing of the project should come from the countries involved in the project, and can be public, or private, or both, depending on the circumstances.

The British involved in the preparatory work have pressed the point that Eureka projects should mainly involve 'market led research,' a phrase expressing the confidence—or perhaps the hope—

that, at least a great part, if not all, of the necessary funds should be of private origin.

Such a procedure has actually been followed for the ten projects agreed upon in Hanover on 6 November 1985. These are concerned with computers, amorphous silicon, robotics, lasers, transportation, research networks, etc.

The 'market-led research' approach is quite a reasonable one for developments which will make financial benefits in a few years time and which, most probably, many European industries would have made in any case, even without a multigovernmental label.

Projects of this type, however, cannot compete with those proposed, for example, through the SDI scheme, which are backed by considerable sums of public money.

Furthermore, 'a market-led research' approach can be reasonable only as a first step. But if the declaration of the beginning of a new 'era' is to correspond to something concrete and valid, then we would also have to start thinking about technological projects that are so advanced that they will unavoidably require a number of years for their implementation, and therefore also an appreciable amount of financial support from public sources. I, and many other people, strongly hope that this point will be reconsidered at a not too far distant Ministers' Meeting.

As a conclusion to my contribution, I should like to insist on the 'civil goals' of all Eureka projects and of all the European organizations that I have mentioned in the second part of my speech.

The collaboration between the countries of Western and Eastern Europe has for many years been very active and fruitful at *CERN*, at ESA, and in many other scientific organizations. As everybody knows, the European Physical Society counts as collective members the national Physical Societies of all the Western and Eastern European countries.

All these examples represent only minor realizations with respect to the 'Open World' of Niels Bohr, but they are all oriented in the right direction, and we may hope that they will contribute to paving the way towards this still very far but always fundamental goal.

Acknowledgment

The author expresses his thanks to Mrs Marie-Suzy Vascotto and Kitty Wakley of *CERN* for the revision of the English.

Notes and References

1. See, for example, a) Margaret Gowing, Niels Bohr and nuclear weapons, in The Lesson of Quantum Theory, ed. by I. de Boer, E. Dal and O. Ulfbeck (North Holland for the Royal Danish Academy of Sciences and Letters, 1986) p. 343. b) Martin Sherwin, Niels Bohr and the origins of nuclear arms control, invited paper given at this symposium.

2. These countries were: Austria, Belgium, Denmark, France, Greece, Iceland, the Republic of Ireland, Italy, Luxemburg, the Netherlands, Norway, Portugal, Sweden, Switzerland, Turkey, and the United Kingdom. To them must be added the three western zones of occupied Germany.

3. The member countries of the OECD are: Australia, Austria, Belgium, Canada, Denmark, Finland, France, Federal Republic of Germany, Greece, Iceland, the Republic of Ireland, Italy, Japan, Luxemburg, the Netherlands, New Zealand, Norway, Portugal, Spain, Sweden, Switzerland, Turkey, the United Kingdom and the United States.

4. See, for example, E. Amaldi: First international collaborations between Western European countries after World War II in the field of high energy physics, in History of twentieth century physics, ed. by C. Weiner, 57th Course of the 'Enrico Fermi' International School of Physics, Varenna, 1972 (Academic Press, New York, 1977), p. 326. L. Kowarski, New forms of organization in physical research after 1945, ibid., p. 370.

5. Information on the history of CERN can be found in Ref. 4, and in much more detail in The history of CERN—from the begining to 1965—by Armin Hermann, Professor of History of Science and Technology at the University of Stuttgart and his junior collaborators: John Krige, Dominique Pestre, Ulrike Mersits and Lamberto Belloni (in preparation). The two volumes will be published by North Holland Publishing Co., the first one in 1987, the second about one year later. The Member States of CERN are: Austria, Belgium, Denmark, France, the Federal Republic of Germany, Greece, Italy, the Netherlands, Norway; Portugal, Spain, Sweden, Switzerland, and the United Kingdom. Originally, also Yugoslavia was a Member of CERN; a few years later, however, for financial reasons, Yugoslavia preferred to pass to the status of Observer, a position held also by Poland and Turkey. CERN's 1985 budget amounts to about 724 million Swiss francs, corresponding to about 340 million dollars. Its staff numbers about 3600 persons, amongst whom are 90 research scientists and around 500 engineers and applied physicists. In addition, there are about 300 Research Fellows of various types, and a user community of about 3000 scientists from external institutes.

6. The Transuranic Institute in Karlsruhe, the test of material in high neutron fluxes in Petten, and the Community Bureau of Reference in Geel.

7. Amongst the most important parameters of JET, are the following:

Minor radius of plasma (horizontal)   1.25 m
Minor radius of plasma (vertical)   2.10 m
Major radius of plasma   2.96 m
Toroidal magnetic field   3.5 T
Duration of flat top   20 s
Plasma current   5 MA
Measured parameters: $n_e = 3 \times 10^{19}$ m^{-3};
                        $T_i = 3$ keV,
                        $\tau_{conf} = 0.8$ s.

The research programme of JET is foreseen to be extended beyond 1990 and probably up to about 1992.

8. RWE was later replaced by SBK, a pool of Belgian, British, Dutch and German utilities.

9. Originally the member countries of ESO were: Belgium, Denmark, France, the Federal Republic of Germany, the Netherlands and Sweden. The UK did not participate because it was already involved in the construction of the Anglo-Australian Telescope. Italy and Switzerland entered ESO in 1982. The 1985 budget of ESO amounts to 56 million Deutch mark, (22 million dollars). The total staff consists of 260 persons, 60 of whom are scientists working at Garching and at La Silla (Chile). The total number of external scientists using the facilities located in these two places is close to 600.

10. At the time when initiatives were being taken to establish a European organization for the launching of scientific satellites (ESRO), the British Ministries of Foreign Affairs, Health, and of Aviation, Thorneycroft (of the MacMillan government), saw an opportunity to get rid of the Blue Streak, a military rocket which had been intended as a ballistic weapon with a range of 3000 km, built around two liquid-fuel rocket engines. This missile was proving too expensive and was being bypassed by the USSR and by the American ICBM (Intercontinental Ballistic Missiles). After suitable diplomatic preparation, the UK invited a number of European countries to participate in the membership of an organization for the launching of satellites (1960). A conference was held in Strasburg on 30 January 1961, and was attended by representatives of eleven European countries and observers from four more countries (including Canada). At the conference it was agreed that, if such a European organization were to be created, the first programme would consist in the development of a three-stage rocket: stage one would consist in a development of the Blue Streak; the second stage would be built by France as a development of its military missile Emeraude (also open to criticism by international experts); the third stage and any satellite to be launched would be built by other European countries. Tests of the complete vehicle would be carried out at Woomera in Australia. This became the

initial programme of ELDO; it was a complete failure. None of the three launchings made from Woomera brought the satellite into orbit. Furthermore, already at the beginning of 1965 the original cost estimate (70 million pounds sterling) turned out to be over-optimistic, and a further 22 million pounds extra could be necessary. At the same time, the technical working group proposed to replace the original programme, based on the Blue Streak and on upper stages with conventional fuel, by a new programme called ELDOB, which was much more promising but which would have brought the total cost to more than 140 million pounds. The dissatisfaction of the British Government with the rising cost of ELDO led to its decision to withdraw from the current ELDO programme, and also to the end of ELDO (see, for example, European Launcher Development Organization—Beginning and End, Nature (London) **210** (1966) 1091; ELDO under Scrutiny, Nature (London) **211** (1966) 787).

11. See, for example: Space Science Horizon 2000, ESA SP-1070, December 1984 (European Space Agency, 8-10 rue Mario Nikis, 75738 Paris Cedex 15, France). The member States of ESA are: Belgium, Denmark, France, the Federal Republic of Germany, the Republic of Ireland, Italy, the Netherlands, Norway, Spain, Sweden, Switzerland and the United Kingdom. Austria, Canada and Norway have the status of Observers; however, starting from 1 January 1987, Austria and Norway will become full members. The total 1985 budget of ESA amounts to 1124 million AU[12] (~910 million dollars), of which 142 million AU (115 million dollars; 13%) are devoted to the scientific programme and 32 million AU (26 million dollars; 3%) to microgravity research. The total staff consists of 1440 persons of whom 147 are involved in the scientific programme. the external scientific community which participates in ESA scientific activities consists of 40–50 European groups with hardware experience, and 60–70 groups with interest in the utilization of the data provided by space missions (e.g., IUE, EXOSAT, etc.); i.e. a total of 100–120 European groups for a total of about 1500 scientists + technicians.

12. The ECU (European Currency Unit) is a weighted average of the national currencies of the Member States of the ECC, used in a number of investments and quoted daily on the International Stock Exchanges. The AU (account unit) is a conventional monetary unit defined as the weighted mean value of the currencies of the ECU basket plus the Swedish kroner, averaged over the month of June of the preceding year. The values of these two units differ only on the third significant figure, so that here they can be assumed to be equal. For 1985, one can assume that 1 AU = 1 ECU = US dollar 0.81.

13. The Member States' participation in the Ariane 1 development was (in percent)

Belgium	5.00	Italy	1.74
Denmark	0.50	the Netherlands	2.00
France	63.87	Spain	2.00
Federal Republic		Sweden	1.10
of Germany	20.12	Switzerland	1.20
		the United Kingdom	2.47
			100.00%

14. The Member States of EMBO are: Austria, Belgium, Denmark, the Federal Republic of Germany, Finland, France, Greece, Iceland, the Republic of Ireland, Israel, Italy, the Netherlands, Norway, Spain, Sweden, Switzerland, the United Kingdom. For 1985 the budget of EMBO amounts to 4.2 million AU (~3.4 million dollars) and is almost completely devoted to fellowships, courses, workshops, and symposia. The budget of EMBL amounts to about 45 million Deutschmark (~18 million dollars); its staff is close to 300, to which about 100 Fellows and visiting scientists should be added.

15. The North Atlantic Treaty was signed on 4 April 1949. The countries members of the North Atlantic Treaty Organization are: Belgium, Canada, Denmark, (France), Federal Republic of Germany, Greece, Italy, Luxemburg, the Netherlands, Portugal, Spain, Turkey, the United Kingdom and the United States. France is a member of the Alliance but does not take part in the integrated military structure.

16. The official budget is given in Belgian francs (BF 612,263,000).

17. The USPID (Unione Scienziati per il Disarmo) is similar to the Federation of American Scientists and the Union of Concerned Scientists, also in the USA. A document 'On the American SDI and research in Italy' made public by USPID in May 1985, was distributed at the Castiglioncello Conference.

18. The adhesion of Spain and Portugal to the ECC and the corresponding legal and bureaucratic procedures have been completed in 1985, but will start to be effective from 1 January 1986.

19. The first conference was held on 5 and 6 November 1981, the second one from 19 to 21 November.

20. The part of the budget of the ECC devoted to research activities in 1985 amounts to 561 million ECU (450 million dollars) and involves a staff of 2640 persons employed on permanent programmes, plus 282 on temporary programmes.

21. As an example, I will give here the composition of the Consortium that is taking care of the construction of Hipparcos, a space-astrometry mission which will determine the trigonometric parallaxes, proper motions ($0''.002/y$) and positions ($0''.002$: present $0''.04$ at best) of about 100,000 selected stars, with the aim of providing a uniform high precision, whole-sky, stellar reference frame suitable for astrometric and astrophysical studies.

Project direction: ESA

First contractor: Matra (France)

Study of the system: AERITALIA (Italy) with CRA (Denmark)—BAC (United Kingdom)—ERNO,ANT (Federal Republic of Germany)—Fokker (Netherlands)—BTM (Belgium)—Matra (France)—INTA (Spain)—SAAB (Sweden)—SELENIA (Italy)—SEP,SAFT (France)—Galileo (Italy).

Pay-load: MATRA (France) with SRU, TNO, Fokker (Netherlands)—Zeiss, Donnier, ERNO (Federal Republic of Germany)—CEH, CON-TRAVES (Switzerland)—CRA (Denmark)—IAL (Belgium)—LAS, REOSC (France)—CASA (Spain)

Integration: AERITALIA (Italy) with INTA, SENER (Spain)—CIR (Switzerland)—LABEN (Italy) ERNO (Federal Republic of Germany).

Index